Student Solutions Manual

Prealgebra
An Applied Approach

SIXTH EDITION

Richard N. Aufmann
Palomar College

Joanne S. Lockwood
Nashua Community College

Prepared by

Jeremiah Gilbert
San Bernardino Valley College

Chris Schroeder
Morehead State University

BROOKS/COLE
CENGAGE Learning

Australia • Brazil • Japan • Korea • Mexico • Singapore • Spain • United Kingdom • United States

For product information and technology assistance, contact us at
**Cengage Learning Customer & Sales Support,
1-800-354-9706**

For permission to use material from this text or product, submit all requests online at **www.cengage.com/permissions**
Further permissions questions can be emailed to
permissionrequest@cengage.com

ISBN-13: 978-1-285-09233-1
ISBN-10: 1-285-09233-3

Brooks/Cole
20 Davis Drive
Belmont, CA 94002-3098
USA

Cengage Learning is a leading provider of customized learning solutions with office locations around the globe, including Singapore, the United Kingdom, Australia, Mexico, Brazil, and Japan. Locate your local office at: **www.cengage.com/global**

Cengage Learning products are represented in Canada by Nelson Education, Ltd.

To learn more about Brooks/Cole, visit
www.cengage.com/brookscole

Purchase any of our products at your local college store or at our preferred online store
www.cengagebrain.com

Printed in the United States of America
1 2 3 4 5 6 7 16 15 14 13 12

Table of Contents

Chapter 1: Whole Numbers

Prep Test

1. 8

2. 1 2 3 4 5 6 7 8 9 10

3. a and D; b and E; c and A; d and B; e and F; f and C

4. 0

5. fifty

Section 1.1

Concept Check

1. is greater than

3. thousand

5. less

Objective A Exercises

7.

9.

11.

13.

5 is 4 units to the left of 9.

15.

5 is 3 units to the right of 2.

17.

0 is 7 units to the left of 7.

19. 27 < 39

21. 0 < 52

23. 273 > 194

25. 2,761 < 3,857

27. 4,610 > 4,061

29. 8,005 < 8,050

31. Yes

33. 11, 14, 16, 21, 32

35. 13, 48, 72, 84, 93

37. 26, 49, 77, 90, 106

39. 204, 399, 662, 736, 981

41. 307, 370, 377, 3,077, 3,700

Objective B Exercises

43. five hundred eight

45. six hundred thirty-five

47. four thousand seven hundred ninety

49. fifty-three thousand six hundred fourteen

51. two hundred forty-six thousand fifty-three

53. three million eight hundred forty-two thousand nine hundred five

55. 496

57. 53,340

59. 502,140

61. 9,706

63. 5,012,907

65. 8,005,010

67. 7,000 + 200 + 40 + 5

69. 500,000 + 30,000 + 2,000 + 700 + 90 + 1

71. 5,000 + 60 + 4

73. 20,000 + 300 + 90 + 7

75. 400,000 + 2,000 + 700 + 8

77. 8,000,000 + 300 + 10 + 6

79. Tens

Objective C Exercises

81. 7,108 $\lceil$ Given place value
$\lfloor$ 8 > 5

7,108 rounded to the nearest ten is 7,110.

83. 4,962 $\lceil$ Given place value
$\lfloor$ 6 > 5

4,962 rounded to the nearest hundred is 5,000.

85. 28,551 ⌐Given place value
5 = 5

28,551 rounded to the nearest hundred is 28,600.

87. 6,808 ⌐Given place value
8 > 5

6,809 rounded to the nearest thousand is 7,000.

89. 93,825 ⌐Given place value
8 > 5

93,825 rounded to the nearest thousand is 94,000.

91. 629,513 ⌐Given place value
5 = 5

629,513 rounded to the nearest thousand is 630,000.

93. 352,876 ⌐Given place value
2 < 5

352,876 rounded to the nearest ten-thousand is 350,000.

95. Sometimes true

97. Never true

Objective D Exercises

99. Strategy   To find the person with the greater number of stolen bases, compare the numbers 892 and 937.

Solution   892 < 937
Billy Hamilton had the greater number of stolen bases.

101. Strategy   To find the greater number of performances, compare the numbers 2,844 and 3,242.

Solution   2,844 < 3,242
"Fiddler on the Roof" was performed the greater number of times.

103. Strategy   To find the food which contains more calories, compare the numbers 190 and 114.

Solution   190 > 114
Two tablespoons of peanut butter contain more calories.

105. Strategy

a. To determine during which school year enrollment was lowest, use the line graph to find the school year corresponding to the lowest point on the line.

b. To determine if enrollment increased or decreased between 1990 and 2000, read the graph at 1990 and at 2000.

Solution
a. Student enrollment was lowest during the 1980 school year.

b. Enrollment was greater in 2000 than in 1990. Enrollment increased between 1990 and 2000.

107. Strategy

a. To determine the most often mentioned complaint, find the complaint with the largest number of responses.

b. To determine the least often mentioned complaint, find the complaint with the smallest number of responses.

Solution
a. The complaint with the largest number of responses is people talking. The most often mentioned complaint is people talking.

b. The complaint with the smallest number of responses is uncomfortable seats. The least often mentioned complaint was uncomfortable seats.

109. Strategy   To find which chain had more restaurants, compare the two numbers 33,749 and 32,737.

Solution   33,749 > 32,737
There were more Subway restaurants at the end of 2010.

111. Strategy   To find the land area to the nearest ten-thousand acres, round 161,546 to the nearest ten-thousand.

Solution   161,546 rounded to the nearest ten-thousand is 160,000.
To the nearest ten-thousand acres, the land area of the Appalachian Trail is 160,000 acres.

113. Strategy   To find the U.S. population to the nearest million, round 308,745,538 to the nearest million.

Solution   308,745,538 rounded to the nearest million is 309,000,000.
To the nearest million, the population of the United States on April 10, 2010 was 309,000,000.

Critical Thinking 1.1

115. The largest three-digit number is 999. (The next largest whole number is 1,000, which has four digits.) The smallest five-digit number is 10,000. (The next smallest whole number if 9,999, which has 4 digits.)

Projects or Group Activities 1.1

117. Asia (17,266,000 mi^2)
Africa (11,667,000 mi^2)
North America (9,355,000 mi^2)
South America (6,878,000 mi^2)
Antarctica (5,500,000 mi^2)
Europe (4,056,000 mi^2)
Australia (2,966,144 mi^2)

119. Answers to the question of the total enrollment of your school will vary. Generally, the population of a small school (less than 5,000 students) is rounded to the nearest hundred; the population of a medium-size school (5,000-20,000 students) or a large school (more than 20,000 students) is rounded to the nearest thousand.
The population of your state can be found in most annual almanacs. The place value to which the population is rounded will vary. For example, the population of New York might be rounded to the nearest million (18,000,000) while the population of Wyoming might be rounded to the nearest hundred thousand (500,000).
Generally the population of the United States is rounded to the nearest ten million (280,000,000).

Section 1.2

Concept Check

1. 24 ; 15 ; 39

3. 523; 448; 971

5. addition

Objective A Exercises

7.
$$
\begin{array}{r}
732,453 \\
+\,651,206 \\
\hline
1,383,659
\end{array}
$$

9.
$$
\begin{array}{r}
\scriptstyle 1\ 1\ 1 \\
2,879 \\
+\,3,164 \\
\hline
6,043
\end{array}
$$

11.
$$
\begin{array}{r}
\scriptstyle 1\ 1 \\
4,037 \\
3,342 \\
+\,5,169 \\
\hline
12,548
\end{array}
$$

13.
$$
\begin{array}{r}
\scriptstyle 1\ 1\ 1 \\
67,390 \\
42,761 \\
+\,89,405 \\
\hline
199,556
\end{array}
$$

15.
$$
\begin{array}{r}
\scriptstyle 2\ 1\ 1\ 2 \\
54,097 \\
33,432 \\
97,126 \\
64,508 \\
78,310 \\
\hline
327,473
\end{array}
$$

17.
$$
\begin{array}{r}
88,123 \\
+\,80,451 \\
\hline
168,574
\end{array}
$$

19.
$$
\begin{array}{r}
\scriptstyle 1 \\
7,293 \\
+\,654 \\
\hline
7,947
\end{array}
$$

21.
$$
\begin{array}{r}
\scriptstyle 1\ \ \ 1 \\
216 \\
8,707 \\
+\,90,714 \\
\hline
99,637
\end{array}
$$

23.
$$
\begin{array}{r}
585 \\
497 \\
412 \\
+\,378 \\
\hline
1,872
\end{array}
$$

The total number of undergraduates enrolled at the college in 2001 was 1,872.

25.
$$
\begin{array}{rcr}
6,742 & \rightarrow & 7,000 \\
+\,8,298 & \rightarrow & +\,8,000 \\
\hline
15,040 & & 15,000
\end{array}
$$

27.
$$
\begin{array}{rcr}
972,085 & \rightarrow & 1,000,000 \\
+\,416,832 & \rightarrow & +\,400,000 \\
\hline
1,388,917 & & 1,400,000
\end{array}
$$

29.

$$\begin{array}{rcl} 387 & \to & 400 \\ 295 & \to & 300 \\ 614 & \to & 600 \\ +\,702 & \to & +\,700 \\ \hline 1{,}998 && 2{,}000 \end{array}$$

31.

$$\begin{array}{rcl} 224{,}196 & \to & 200{,}000 \\ 7{,}074 & \to & 7{,}000 \\ +\,98{,}531 & \to & +\,100{,}000 \\ \hline 329{,}801 && 307{,}000 \end{array}$$

33. $x + y$

574 + 698

$$\begin{array}{r} 1\ 1 \\ 574 \\ +\,698 \\ \hline 1{,}272 \end{array}$$

35. $x + y$

4,752 + 7,398

$$\begin{array}{r} 1\ 1\ 1 \\ 4{,}752 \\ +\,7{,}398 \\ \hline 12{,}150 \end{array}$$

37. $x + y$

38,229 + 51,671

$$\begin{array}{r} 1\ 1 \\ 38{,}229 \\ +\,51{,}671 \\ \hline 89{,}900 \end{array}$$

39. $a + b + c$

693 + 508 + 371

$$\begin{array}{r} 1\ 1 \\ 693 \\ 508 \\ +\,371 \\ \hline 1{,}572 \end{array}$$

41. $a + b + c$

4,938 + 2,615 + 7,038

$$\begin{array}{r} 1\quad 2 \\ 4{,}938 \\ 2{,}615 \\ +\,7{,}038 \\ \hline 14{,}591 \end{array}$$

43. $a + b + c$

12,897 + 36,075 + 7,038

$$\begin{array}{r} 1\,1\quad 2\,2 \\ 12{,}897 \\ 36{,}075 \\ +\,7{,}038 \\ \hline 56{,}010 \end{array}$$

45. The Commutative Property of Addition

47. The Associative Property of Addition

49. The Addition Property of Zero

51. $28 + 0 = 28$

53. $9 + (4 + 17) = (9 + 4) + 17$

55. $15 + 34 = 34 + 15$

57. Commutative Property of Addition

59. $m + 6 = 13$

$$\overline{17 + 6}\ \big|\ 13$$

$23 \ne 13$

No, 17 is not a solution of the equation $m + 6 = 13$.

61. $n = 17 + 24$

$$41\ \big|\ \overline{17 + 24}$$

$41 = 41$

Yes, 41 is a solution of the equation $n = 17 + 24$.

63. $38 = 11 + z$

$$38\ \big|\ \overline{11 + 29}$$

$38 \ne 40$

No, 29 is not a solution of the equation $38 = 11 + z$.

Objective B Exercises

65.

$$\begin{array}{r} 8\ 11 \\ 5\not{9}\not{1} \\ -\,238 \\ \hline 353 \end{array}$$

67.

$$\begin{array}{r} 8\ 14\ 10 \\ 9\not{5}\not{0} \\ -\,483 \\ \hline 467 \end{array}$$

69.

$$\begin{array}{r} 5\ 12 \\ 7\not{6}2 \\ -\,659 \\ \hline 103 \end{array}$$

71.

$$\begin{array}{r} 7\ 9\ 15 \\ 8\not{0}\not{5} \\ -\,147 \\ \hline 658 \end{array}$$

73.

$$\begin{array}{r} 6\ 12\ 15\ 11 \\ 7{,}\not{3}\not{6}\not{1} \\ -\,4{,}575 \\ \hline 2{,}786 \end{array}$$

75.

$$\begin{array}{r} 3\ 9\ 9\ 10 \\ 4{,}\not{0}\not{0}\not{0} \\ -\,1{,}873 \\ \hline 2{,}127 \end{array}$$

77.
$$\begin{array}{r} \overset{6\ 12\ 9\ 10}{7,\cancel{3}\cancel{0}\cancel{0}} \\ -\ 2,562 \\ \hline 4,738 \end{array}$$

79.
$$\begin{array}{r} \overset{6\ 9\ 9\ 9\ 13}{7\cancel{0},\cancel{0}\cancel{0}\cancel{3}} \\ -\ 8,246 \\ \hline 61,757 \end{array}$$

81.
$$\begin{array}{r} \overset{5\ 11\ 13}{1,\cancel{6}\cancel{2}\cancel{3}} \\ -\ 287 \\ \hline 1,336 \end{array}$$

83.
$$\begin{array}{r} \overset{7\ 9\ 11}{14,\cancel{8}\cancel{0}\cancel{1}} \\ -\ 3,522 \\ \hline 11,279 \end{array}$$

85. $x - y$

87.
$$\begin{array}{r} \overset{1\ 9\ 10}{\cancel{2}\cancel{0}\cancel{0}} \\ -\ 175 \\ \hline 25 \end{array}$$

The eruption of the Giant is 25 ft higher than the eruption of Old Faithful.

89.
$$\begin{array}{r} 8,953 \\ -\ 2,217 \\ \hline 6,736 \end{array} \rightarrow \begin{array}{r} 9,000 \\ 2,000 \\ \hline 7,000 \end{array}$$

91.
$$\begin{array}{r} 63,051 \\ 29,478 \\ \hline 33,573 \end{array} \rightarrow \begin{array}{r} 60,000 \\ -\ 30,000 \\ \hline 30,000 \end{array}$$

93.
$$\begin{array}{r} 58,316 \\ -19,072 \\ \hline 39,244 \end{array} \rightarrow \begin{array}{r} 60,000 \\ -\ 20,000 \\ \hline 40,000 \end{array}$$

95.
$$\begin{array}{r} 873,925 \\ -\ 28,744 \\ \hline 845,181 \end{array} \rightarrow \begin{array}{r} 900,000 \\ 30,000 \\ \hline 870,000 \end{array}$$

97. $x - y$
$$80 - 33$$
$$\begin{array}{r} \overset{7\ 10}{\cancel{8}\cancel{0}} \\ -\ 33 \\ \hline 47 \end{array}$$

99. $x - y$
$$623 - 197$$
$$\begin{array}{r} \overset{5\ 11\ 13}{\cancel{6}\cancel{2}\cancel{3}} \\ -\ 197 \\ \hline 426 \end{array}$$

101. $x - y$
$$870 - 243$$
$$\begin{array}{r} \overset{6\ 10}{8\cancel{7}\cancel{0}} \\ -243 \\ \hline 627 \end{array}$$

103. $x - y$
$$7,814 - 3,512$$
$$\begin{array}{r} 7,814 \\ -\ 3,512 \\ \hline 4,302 \end{array}$$

105. $x - y$
$$1,406 - 968$$
$$\begin{array}{r} \overset{0\ 13\ 9\ 16}{1,\cancel{4}\cancel{0}\cancel{6}} \\ -\ 968 \\ \hline 438 \end{array}$$

107. $x - y$
$$56,397 - 8,249$$
$$\begin{array}{r} \overset{4\ 16\quad 8\ 17}{5\cancel{6},3\cancel{9}\cancel{7}} \\ -\ 8,249 \\ \hline 48,148 \end{array}$$

109.
$$48 - p = 17$$
$$\overline{48 - 31\ |\ 17}$$
$$17 = 17$$
Yes, 31 is a solution of the equation $48 - p = 17$.

111.
$$34 = x - 9$$
$$\overline{34\ |\ 25 - 9}$$
$$34 \ne 16$$
No, 25 is not a solution of the equation $34 = x - 9$.

113.
$$72 = 100 - d$$
$$\overline{72\ |\ 100 - 28}$$
$$72 = 72$$
Yes, 28 is a solution of the equation $72 = 100 - d$.

Objective C Exercises

115. Strategy   To find the difference, subtract 99 (the largest two-digit number) from 1,000 (the smallest four-digit number).

$$\text{Solution} \quad \begin{array}{r} 1,000 \\ -\ 99 \\ \hline 901 \end{array}$$

The difference is 901.

117. Strategy   To find the number of calories, add the number of calories in one apple (80), one cup of cornflakes (95), one tablespoon of sugar (45), and one cup of milk (150).

Solution $80 + 95 + 45 + 150 = 370$
The breakfast contained 370 calories.

119. Strategy   To find the perimeter of a rectangle, replace L with 24 and W with 15 in the given formula and solve for P.

Solution $P = L + W + L + W$
$P = 24 + 15 + 24 + 15$
$P = 78$
The perimeter is 78 m.

121. Strategy   To find the perimeter of a triangle, replace a, b, and c with 16, 12, and 15 in the given formula and solve for P.

Solution $P = a + b + c$
$P = 16 + 12 + 15$
$P = 43$
The perimeter is 43 in.

123. Strategy   To find the amount, add the deposit (870) to the balance in the checking account (1,054).

Solution $\begin{array}{r} 870 \\ +1,054 \\ \hline 1,924 \end{array}$
You have $1,924 in your checking account.

125. Strategy   To find the total repair bill, add the cost of parts (358), labor (156) and tax (30).

Solution $\begin{array}{r} 358 \\ 156 \\ +30 \\ \hline 544 \end{array}$
The total repair bill is $544.

127. Strategy   To find the estimate of the number of miles driven:
→Round the two readings of the odometer.
→Subtract the rounded reading from the beginning of the year from the rounded reading at the end of the year.

Solution $\begin{array}{l} 77,912 \ \rightarrow \quad \ \ 80,000 \\ 58,376 \ \rightarrow \ \underline{-60,000} \\ \qquad\qquad\qquad 20,000 \end{array}$
The car was driven approximately 20,000 mi during the year.

129. Strategy   To determine between which two months car sales increased the most, find the difference, if there was an increase, between sales for January and February, February and March, and March and April for 2005.

Solution   Between January and February, 2005: $132 - 108 = 24$
Between February and March, 2005: $152 - 132 = 20$
Between March and April, 2005: not an increase
Car sales increased the most between January and February of 2005. The increase was 24 cars.

131. Strategy   To find the number of U.S. households, add the number of households with broadband Internet access to the number of households without broadband Internet access.

Solution $37 + 78 = 115$
There are 115 million U.S. households.

133. Strategy   To determine between which two years electric car sales are projected to increase the most, find the difference between projected sales for each pair of consecutive years.

Solution   Between 2015 and 2016:
230,000 – 188,000 = 42,000
Between 2016 and 2017:
312,000 – 230,000 = 82,000
Between 2017 and 2018:
359,000 – 312,000 = 47,000
Between 2018 and 2019:
406,000 – 359,000 = 47,000
Between 2019 and 2020:
414,000 – 406,000 = 8,000
Car sales are projected to increase the most between 2016 and 2017. The increase is projected to be 82,000 cars.

135. Strategy   To find the value of the investment, replace P by 12,500 and I by 775 in the given formula and solve for A.

Solution $A = P + I$
$A = 12,500 + 775$
$A = 13,275$
The value of the investment is $13,275.

137. Strategy   To find the mortgage loan, replace S by 290,000 and D by 29,000 in the given formula and solve for M.

Solution $M = S - D$
$M = 290,000 - 29,000$
$M = 261,000$
The mortgage loan on the home is $261,000.

139. Strategy   To find the ground speed, replace a by 375 and h by 25 in the given formula and solve for g.

Solution $g = a - h$
$g = 375 - 25$
$g = 350$
The ground speed of the airplane is 350 mph.

141. It is not possible to tell how many motorists were driving at 70 mph. The data shows that 3,717 drivers were traveling at 66–70 mph. The number 3,717 includes the motorists driving at 66–69 mph as well as those driving at 70 mph. We cannot separate those driving at 70 mph from the others.

143. $b - a$; $b - a$ represents how much longer the side of length b is than the side of length a.

Critical Thinking 1.2

145. The two-digit numbers are the numbers 10 through 99.
99 – 9 = 90.
There are 90 two-digit numbers.
The three-digit numbers are the numbers 100 through 999. 999 – 99 = 900.
There are 900 three-digit numbers.

Projects or Group Activities 1.2

147.

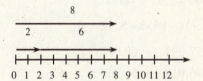

Check Your Progress: Chapter 1

1.

2. 199, 247, 462, 506, 831

3. 397>246; 898<1594

4.

3 is 2 units to the left of 5.

5. six thousand seven hundred two

6. 32,518

7. 900,000 + 3000 + 400 + 80 + 7

8. 16,000

9. 433,000

10. 172,621

11. 1683

12.
$$\begin{array}{ll} 473 \rightarrow & 500 \\ 879 \rightarrow & 900 \\ 215 \rightarrow & 200 \\ +306 \rightarrow & +300 \\ \hline 1,873 & 1,900 \end{array}$$

13. The Associative Property of Addition

14. $x + y + z$
$16 + 42 + 39$

$$\begin{array}{r} {}^{1}16 \\ 42 \\ +39 \\ \hline 97 \end{array}$$

15. $24 = 18 + n$

$$24 \mid 18 + 42$$

$24 \neq 60$
No, 42 is not a solution of the equation
$24 = 18 + n$.

16. $4600 - 2781 = 1819$

17. $13,904 - 8655 = 5249$

18.
$$\begin{array}{ll} 57,293 \rightarrow & 60,000 \\ -46,018 \rightarrow & 50,000 \\ \hline 11,275 & 10,000 \end{array}$$

19. $x - y$
$704 - 279$

$$\begin{array}{r} 704 \\ -279 \\ \hline 425 \end{array}$$

20. $82 = 143 - x$

$$82 \mid 143 - 61$$

$82 = 82$
Yes, 61 is a solution of the equation
$82 = 143 - 61$.

21. Strategy   To find the difference, subtract 6530 from 7460.

Solution
$$\begin{array}{r} 7460 \\ -6530 \\ \hline 930 \end{array}$$

The difference is $930.

22. Strategy   To find the total amount of charitable contributions, add each of her contributions for the six-month period.

Solution $25 + 30 + 13 + 15 + 20 + 27 = 130$
The total amount of charitable contributions is $130.

23. Strategy   To find which planet is smaller, compare the diameter of Uranus (32,200) with the diameter of Neptune (30,800).

Solution $32,200 > 30,800$
Neptune is smaller than Uranus.

24. Strategy   To find the perimeter of a triangle, replace a, b, and c with 6, 9, and 12 in the given formula and solve for P.

Solution $P = a + b + c$
$P = 6 + 9 + 12$
$P = 27$
The perimeter is 27 m.

25. Strategy   To find the price, replace C by 119 and M by 79 in the given formula and solve for P.

Solution $P = C + M$
$P = 119 + 79$
$P = 198$
The price is $198.

Section 1.3

Concept Check

1. Students should note that there are five 6's. Therefore, the addition $6 + 6 + 6 + 6 + 6$ can be written as 5 times 6: 5×6.

3. a. 3; 4

   b. 3; 3; 3; 3; 81

5. 495; 6

7. factors; 1; 3; 5; 15

9. division

Objective A Exercises

11.
$$\begin{array}{r} {}^{2\,6} \\ 127 \\ \times \;\; 9 \\ \hline 1,143 \end{array}$$

13.
$$
\begin{array}{r}
^{4\ \ 6} \\
6,709 \\
\times \quad 7 \\
\hline
46,963
\end{array}
$$

15.
$$
\begin{array}{r}
^{7\,6\ 5\ 7} \\
58,769 \\
\times \qquad 8 \\
\hline
470,152
\end{array}
$$

17.
$$
\begin{array}{r}
683 \\
\times\, 71 \\
\hline
683 \\
47\,81 \\
\hline
48,493
\end{array}
$$

19.
$$
\begin{array}{r}
7,053 \\
\times\, 46 \\
\hline
42\,3\,18 \\
282\ 12 \\
\hline
324,438
\end{array}
$$

21.
$$
\begin{array}{r}
3,285 \\
\times\, 976 \\
\hline
19\,710 \\
229\,95 \\
2\,956\,5 \\
\hline
3,206,160
\end{array}
$$

23. $500 \cdot 3 = 1,500$

25. $40 \cdot 50 = 2,000$

27. $400 \cdot 3 \cdot 20 \cdot 0 = 1,200 \cdot 20 \cdot 0$
                    $= 24,000 \cdot 0$
                    $= 0$

29. **a.**
$$
\begin{array}{r}
^{3\,7\,3} \\
373 \\
\times\, 6 \\
\hline
2,238
\end{array}
$$
You would burn 2,238 calories by working out vigorously on the stair climber for 6 h.

    b.
$$
\begin{array}{r}
^{3\,5\,3} \\
353 \\
\times\ 12 \\
\hline
706 \\
3\,53 \\
\hline
4,236
\end{array}
$$
You would burn 4,236 calories by working out moderately on a treadmill for 12 h.

31. $8,745 \rightarrow 9,000$
       $63 \rightarrow \quad\ \ 60$
      $8,745 \cdot 63 = 550,935$
      $9,000 \cdot 60 = 540,000$

33. $64,409 \rightarrow 60,000$
          $67 \rightarrow \qquad 70$
      $64,409 \cdot 67 = 4,315,403$
      $60,000 \cdot 70 = 4,200,000$

35. $432 \rightarrow 400$
       $91 \rightarrow \ \ 90$
      $432 \cdot 91 = 39,312$
      $400 \cdot 90 = 36,000$

37. $2,837 \rightarrow 3,000$
        $216 \rightarrow \quad 200$
      $2,837 \cdot 216 = 612,792$
      $3,000 \cdot 200 = 600,000$

39. cd
      $381 \cdot 25 = 9,525$

41. $6n$
      $6 \cdot 382 = 2,292$

43. abc
      $4 \cdot 20 \cdot 50 = 80 \cdot 50$
                  $= 4,000$

45. $4ab$
      $4 \cdot 95 \cdot 33 = 380 \cdot 33$
                  $= 12,540$

47. For example, 20 and 500

49. The Associative Property of Multiplication

51. The Multiplication Property of Zero

53. $(5 \cdot 6)100 = 5(6 \cdot 100)$

55. $1 \cdot 77 = 77$

57. $4 = 4n$
      $4 \,|\, 4(0)$
      $4 \neq 0$
      No, 0 is not a solution of the equation $4 = 4n$.

59. $56 = 4c$
      $56 \,|\, 4(14)$
      $56 = 56$
      Yes, 14 is a solution of the equation $56 = 4c$.

61. $44 = 3a$
      $44 \,|\, 3(11)$
      $44 \neq 33$
      No, 11 is not a solution of the equation $44 = 3a$.

Objective B Exercises

63. $3 \cdot 3 \cdot 3 \cdot 3 \cdot 3 \cdot 3 \cdot 5 \cdot 5 \cdot 5 = 3^6 \cdot 5^3$

65. $7 \cdot 7 \cdot 11 \cdot 11 \cdot 11 \cdot 19 \cdot 19 \cdot 19 \cdot 19 = 7^2 \cdot 11^3 \cdot 19^4$

67. $d \cdot d \cdot d = d^3$

69. $a \cdot a \cdot b \cdot b \cdot b \cdot b = a^2 b^4$

71. $2^6 = 2 \cdot 2 \cdot 2 \cdot 2 \cdot 2 \cdot 2 = 64$

73. $10^9 = 1,000,000,000$

75. $2^4 \cdot 3^2 = (2 \cdot 2 \cdot 2 \cdot 2) \cdot (3 \cdot 3) = 16 \cdot 9 = 144$

77. $2^4 \cdot 10^2 = (2 \cdot 2 \cdot 2 \cdot 2) \cdot (10 \cdot 10)$
$= 16 \cdot 100 = 1,600$

79. $4^3 \cdot 0^3 = (4 \cdot 4 \cdot 4) \cdot (0 \cdot 0 \cdot 0) = 64 \cdot 0 = 0$

81. $5^2 \cdot 2 \cdot 3^4 = (5 \cdot 5) \cdot 2 \cdot (3 \cdot 3 \cdot 3 \cdot 3)$
$= 25 \cdot 2 \cdot 81 = 50 \cdot 81 = 4,050$

83. $6^3 = 6 \cdot 6 \cdot 6 = 36 \cdot 6 = 216$

85. $11^2 = 11 \cdot 11 = 121$

87. t^5

89. $x^2 y$
$3^2 \cdot 4 = (3 \cdot 3) \cdot 4$
$= 9 \cdot 4$
$= 36$

91. ab^3
$7 \cdot 4^3 = 7 \cdot (4 \cdot 4 \cdot 4)$
$= 7 \cdot 64$
$= 448$

93. $m^3 n^3$
$5^3 \cdot 10^3 = (5 \cdot 5 \cdot 5)(10 \cdot 10 \cdot 10)$
$= 125 \cdot 1,000$
$= 125,000$

Objective C Exercises

95.
```
       540
   4) 2,160
     −20
       16
      −16
        0
       −0
        0
```

97.
```
      204 r 4
   8) 1,636
     −16
        3
       −0
       36
      −32
        4
```

99.
```
      8,700
   5) 43,500
     −40
       35
      −35
        0
       −0
        0
       −0
        0
```

101.
```
       21 r 18
   41) 879
      −82
       59
      −41
       18
```

103.
```
      180 r 21
   23) 4,161
      −23
      186
     −184
       21
       −0
       21
```

105.
```
      200 r 7
   26) 5,207
      −52
        0
       −0
        7
       −0
        7
```

107.
```
       609
   64) 38,976
      −384
        57
       −0
       576
      −576
         0
```

109.
$$
\begin{array}{r}
40 \text{ r } 27 \\
223{\overline{\smash{\big)}\,8{,}947}} \\
\underline{-8\ 92} \\
27 \\
\underline{-0} \\
27
\end{array}
$$

111.
$$
\begin{array}{r}
908 \\
9{\overline{\smash{\big)}\,8{,}172}} \\
\underline{-81} \\
7 \\
\underline{-0} \\
72 \\
\underline{-72} \\
0
\end{array}
$$

113.
$$
\begin{array}{r}
461 \text{ r } 4 \\
9{\overline{\smash{\big)}\,4{,}153}} \\
\underline{-36} \\
55 \\
\underline{-54} \\
13 \\
\underline{-9} \\
4
\end{array}
$$

115. **a.** $300{,}000 \div 12 = 25{,}000$
The average monthly claim for theft would be $25,000.

  b. $(560{,}000 + 300{,}000 + 80{,}000 + 50{,}000 + 20{,}000 + 20{,}000 + 110{,}000) \div 12 = 95{,}000$
The average claims per month for all sources combined would be $95,000.

117. $62{,}176 \rightarrow 60{,}000$
$\qquad 58 \rightarrow \qquad 60$
$62{,}176 \div 58 = 1{,}072$
$60{,}000 \div 60 = 1{,}000$

119. $637{,}072 \rightarrow 600{,}000$
$\qquad 29 \rightarrow \qquad 30$
$637{,}072 \div 29 = 21{,}968$
$600{,}000 \div 30 = 20{,}000$

121. $11{,}792 \rightarrow 10{,}000$
$\qquad 53 \rightarrow \qquad 50$
$11{,}792 \div 53 = 222 \text{ r}26$
$10{,}000 \div 50 = 200$

123. $632{,}124 \rightarrow 600{,}000$
$\qquad 324 \rightarrow \qquad 300$
$632{,}124 \div 324 = 1{,}951$
$600{,}000 \div 300 = 2{,}000$

125. $\dfrac{x}{y}$

$\dfrac{56}{56} = 1$

127. $\dfrac{x}{y}$

$\dfrac{0}{23} = 0$

129. $\dfrac{x}{y}$

$\dfrac{16{,}200}{3}$

$$
\begin{array}{r}
5{,}400 \\
3{\overline{\smash{\big)}\,16{,}200}} \\
\underline{-15} \\
12 \\
\underline{-12} \\
0 \\
\underline{-0} \\
0 \\
\underline{-0} \\
0
\end{array}
$$

131. $\dfrac{n}{12} = 5$

$$
\dfrac{60}{12} \Big| 5
$$
$$
5 = 5
$$
Yes, 60 is a solution of the equation.

133. $6 = \dfrac{48}{y}$

$$
4 \Big| \dfrac{48}{16}
$$
$$
6 \ne 3
$$
No, 16 is not a solution of the equation.

Objective D Exercises

135. $20 \div 1 = 20$
$20 \div 2 = 10$
$20 \div 4 = 5$
$20 \div 5 = 4$
The factors of 20 are 1, 2, 4, 5, 10, and 20.

137. $9 \div 1 = 9$
$9 \div 3 = 3$
The factors of 9 are 1, 3, and 9.

139. $16 \div 1 = 16$
$16 \div 2 = 8$
$16 \div 4 = 4$
The factors of 16 are 1, 2, 4, 8, and 16.

141. 17 is a prime number.
The factors of 17 are 1 and 17.

143. $24 \div 1 = 24$
$24 \div 2 = 12$
$24 \div 3 = 8$
$24 \div 4 = 6$
$24 \div 6 = 4$
The factors of 24 are 1, 2, 3, 4, 6, 8, 12, and 24.

145. $36 \div 1 = 36$
$36 \div 2 = 18$
$36 \div 3 = 12$
$36 \div 4 = 9$
$36 \div 6 = 6$
The factors of 36 are 1, 2, 3, 4, 6, 9, 12, 18, and 36.

147. $45 \div 1 = 45$
$45 \div 3 = 15$
$45 \div 5 = 9$
$45 \div 9 = 5$
The factors of 45 are 1, 3, 5, 9, 15, and 45.

149. $32 \div 1 = 32$
$32 \div 2 = 16$
$32 \div 4 = 8$
$32 \div 8 = 4$
The factors of 32 are 1, 2, 4, 8, 16, and 32.

151. $64 \div 1 = 64$
$64 \div 2 = 32$
$64 \div 4 = 16$
$64 \div 8 = 8$
The factors of 64 are 1, 2, 4, 8, 16, 32, and 64.

153. $75 \div 1 = 75$
$75 \div 3 = 25$
$75 \div 5 = 15$
$75 \div 15 = 5$
The factors of 75 are 1, 3, 5, 15, 25, and 75.

155.
$$2\overline{)6}$$
$$2\overline{)12}$$
$$2\overline{)24}$$
$$24 = 2 \cdot 2 \cdot 2 \cdot 3 = 2^3 \cdot 3$$

157.
$$3\overline{)9}$$
$$3\overline{)27}$$
$$27 = 3 \cdot 3 \cdot 3 = 3^3$$

159.
$$3\overline{)9}$$
$$2\overline{)18}$$
$$2\overline{)36}$$
$$36 = 2 \cdot 2 \cdot 3 \cdot 3 = 2^2 \cdot 3^2$$

161.
$$5\overline{)25}$$
$$2\overline{)50}$$
$$50 = 2 \cdot 5 \cdot 5 = 2 \cdot 5^2$$

163. 83 is a prime number.

165.
$$2\overline{)10}$$
$$2\overline{)20}$$
$$2\overline{)40}$$
$$2\overline{)80}$$
$$80 = 2 \cdot 2 \cdot 2 \cdot 2 \cdot 5 = 2^4 \cdot 5$$

167.
$$7\overline{)49}$$
$$49 = 7 \cdot 7 = 7^2$$

169.
$$3\overline{)9}$$
$$3\overline{)27}$$
$$3\overline{)81}$$
$$81 = 3 \cdot 3 \cdot 3 \cdot 3 = 3^4$$

171. 89 is a prime number.

173.
$$3\overline{)15}$$
$$2\overline{)30}$$
$$2\overline{)60}$$
$$2\overline{)120}$$
$$120 = 2 \cdot 2 \cdot 2 \cdot 3 \cdot 5 = 2^3 \cdot 3 \cdot 5$$

Objective E Exercises

175. Strategy To find the average number of funerals per day, divide the number of funerals per year (10,200) by the number of days in a year (365). Then round to the nearest whole number.

Solution $\frac{10,200}{365} \approx 27.945$

There are approximately 28 funerals per day.

177. Strategy

a. To determine the number of marriages per week, multiply the number of marriages per day (542) by the number of days in a week (7).

b. To determine the number of marriages per year, multiply the number of marriages per day (542) by the number of days in a year (365).

Solution

a. $542 \cdot 7 = 3,794$
There are 3,794 marriages per week.

b. $542 \cdot 365 = 197,830$
There are 197,830 marriages per year.

179. Strategy To find the length of fencing needed to surround a square corral, substitute 55 for s in the perimeter formula below.

Solution $P = 4 \cdot s = 4 \cdot 55 = 220$
The length of fencing needed to surround a square corral is 220 ft.

181. Strategy To find the area of the patio, substitute 9 for s in the area formula below.

Solution $A = s^2 = 9^2 = 81$
The area of the patio is 81 ft^2.

183. Strategy To find the estimate of the total cost:
→Round the number of suits to hundreds and the cost of each suit to tens.
→Multiply the rounded numbers.

Solution $215 \rightarrow 200$
$83 \rightarrow 80$
$200 \cdot 80 = 16,000$
The total cost of the suits is approximately $16,000.

185. Strategy To find Melissa's monthly starting salary, divide her yearly salary ($69,048) by the number of months in a year (12).

Solution $\frac{69,048}{12} = 5,754$

Melissa's monthly starting salary is $5,754.

187. Strategy To find the total wages paid, multiply the number of plumbers (4), the number of hours each will work (23), and the wage per hour for each ($30).

Solution $4 \cdot 23 \cdot 30 = 2,760$
The total wages paid are $2,760.

189. Strategy To find the total amount paid, replace M by 285 and N by 24 in the given formula and solve for A.

Solution $A = MN$
$A = 285 \cdot 24$
$A = 6,840$
The total amount paid on the loan is $6,840.

191. Strategy To find the time to drive the distance, replace d by 513 and r by 57 in the given formula and solve for t.

Solution $t = \frac{d}{r}$
$t = \frac{513}{57}$
$t = 9$
It takes 9 h to drive 513 mi.

193. Strategy To find the value per share, replace
C by 10,500,000 and S by 500,000
in the given formula and solve for
V.

Solution $V = \dfrac{C}{S}$

$V = \dfrac{10,500,000}{500,000}$

$V = 21$

The stock has a value of \$21 per
share.

Critical Thinking 1.3

195. The largest possible number that can be
written using the digits $1, 3, 8, 2,$ and 7 is
87,321.
87,321 is not divisible by 4.
The next largest possible number that can be
written using these digits is 87,312.
87,312 is divisible by 4.
87,312 is the largest possible number that
can be written using the digits $1, 3, 8, 2, 7$
and that is divisible by 4.

197. **a**. We need to consider each of the digits
from 1 through 9. We know that any number
ending with an even digit is not prime as it is
divisible by 2. Also, any number ending
with 5 is not prime as it is divisible by 5.
That leaves the four digits 1,3,7, and 9.

b. There are many possible answers. For
example, 21, 33, 27, and 29 are not prime
numbers.

Projects or Group Activities 1.3

199. To use the Sieve of Eratosthenes to find all
prime numbers less than 100, write out all
numbers from 1 to 100 in order. Cross out 1,
since 1 is not prime. Mark 2 as prime and
cross out all multiples of 2. Proceed in order
by marking the next crossed out number as
prime, and crossing out each of its multiples.
Using this method, we see that the prime
numbers less than 100 are $2, 3, 5, 7, 11, 13,$
$17, 19, 23, 29, 31, 37, 41, 43, 47, 53, 59, 61,$
$67, 71, 73, 79, 83, 89,$ and $97.$

Section 1.4

Concept Check

1. $5; 15$

3. $n + 8 = 13$

Objective A Exercises

5. $x + 9 = 23$
$x + 9 - 9 = 23 - 9$
$x + 0 = 14$
$x = 14$
The solution is 14.

7. $8 + b = 33$
$8 - 8 + b = 33 - 8$
$0 + b = 25$
$b = 25$
The solution is 25.

9. $3m = 15$
$\dfrac{3m}{3} = \dfrac{15}{3}$
$1m = 5$
$m = 5$
The solution is 5.

11. $52 = 4c$
$\dfrac{52}{4} = \dfrac{4c}{4}$
$13 = 1c$
$13 = c$
The solution is 13.

13. $16 = w + 9$
$16 - 9 = w + 9 - 9$
$7 = w + 0$
$7 = w$
The solution is 7.

15. $28 = 19 + p$
$28 - 19 = 19 - 19 + p$
$9 = 0 + p$
$9 = p$
The solution is 9.

17. $10y = 80$
$\dfrac{10y}{10} = \dfrac{80}{10}$
$1y = 8$
$y = 8$
The solution is 8.

19. $41 = 41d$
$\dfrac{41}{41} = \dfrac{41d}{41}$
$1 = 1d$
$1 = d$
The solution is 1.

21. $b + 7 = 7$
$b + 7 - 7 = 7 - 7$
$b + 0 = 0$
$b = 0$
The solution is 0.

23. $15 + t = 91$

$15 - 15 + t = 91 - 15$

$0 + t = 76$

$t = 76$

The solution is 76.

Objective B Exercises

25. The unknown number: n

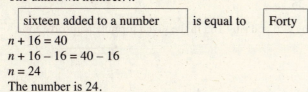

| sixteen added to a number | is equal to | Forty |

$n + 16 = 40$

$n + 16 - 16 = 40 - 16$

$n = 24$

The number is 24.

27. The unknown number: n

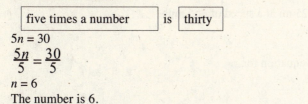

| five times a number | is | thirty |

$5n = 30$

$\dfrac{5n}{5} = \dfrac{30}{5}$

$n = 6$

The number is 6.

29. The unknown number: n

| fifteen | is | three more than a number |

$15 = n + 3$

$15 - 3 = n + 3 - 3$

$12 = n$

The number is 12.

31. Strategy   To find the width of the rectangle, write and solve an equation using w to represent the width.

Solution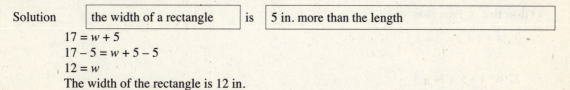

| the width of a rectangle | is | 5 in. more than the length |

$17 = w + 5$

$17 - 5 = w + 5 - 5$

$12 = w$

The width of the rectangle is 12 in.

33. Strategy   To find the number or iPods sold in January 2004, write and solve an equation using x to represent the number of iPods sold in January 2004.

Solution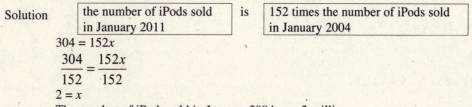

| the number of iPods sold in January 2011 | is | 152 times the number of iPods sold in January 2004 |

$304 = 152x$

$\dfrac{304}{152} = \dfrac{152x}{152}$

$2 = x$

The number of iPods sold in January 2004 was 2 million.

35. Strategy To find the number of payments, replace A by 17,460 and M by 485 in the given formula and solve for N.

Solution $A = MN$

$17,460 = 485N$

$$\frac{17,460}{485} = \frac{485N}{485}$$

$36 = N$

The number of payments is 36.

37. Strategy To find the time, replace d by 825 and r by 165 in the given formula and solve for t.

Solution $d = rt$

$825 = 165t$

$$\frac{825}{165} = \frac{165t}{165}$$

$5 = t$

It would take 5 h to travel 825 mi at a speed of 165 mph.

Critical Thinking 1.4

39. To find the value of $4x$, first solve the equation for x.

$x + 17 = 32$

$x + 17 - 17 = 32 - 17$

$x = 15$

Now multiply x by 4.

$4x = 4 \cdot 15$

$4x = 60$

Section 1.5

Concept Check

1. to multiply 7 times 2

Objective A Exercises

3. $8 \div 4 + 2 = 2 + 2$

$= 4$

5. $6 \cdot 4 + 5 = 24 + 5$

$= 29$

7. $4^2 - 3 = 16 - 3$

$= 13$

9. $5 \cdot (6 - 3) + 4 = 5 \cdot 3 + 4$

$= 15 + 4$

$= 19$

11. $9 + (7 + 5) \div 6 = 9 + 12 \div 6$

$= 9 + 2$

$= 11$

13. $13 \cdot (1 + 5) \div 13 = 13 \cdot 6 \div 13$

$= 78 \div 13$

$= 6$

15. $6 \cdot 3^2 + 7 = 6 \cdot 9 + 7$

$= 54 + 7$

$= 61$

17. $14 + 5 \cdot 2^3 = 14 + 5 \cdot 8$
$= 14 + 40$
$= 54$

19. $10 + (8 - 5) \cdot 3 = 10 + 3 \cdot 3$
$= 10 + 9$
$= 19$

21. $2^3 + 4(10 - 6) = 2^3 + 4 \cdot 4$
$= 8 + 4 \cdot 4$
$= 8 + 16$
$= 24$

23. $6(7) + 4^2 \cdot 3^2 = 6(7) + 16 \cdot 9$
$= 42 + 16 \cdot 9$
$= 42 + 144$
$= 186$

25. $18 + 3(7) = 18 + 21$
$= 39$

27. $6(8 - 3) - 12 = 6(5) - 12$
$= 30 - 12$
$= 18$

29. $16 - (13 - 5) \div 4 = 16 - 8 \div 4$
$= 16 - 2$
$= 14$

31. $17 + 1 - 8 \cdot 2 \div 4 = 17 + 1 - 16 \div 4$
$= 17 + 1 - 4$
$= 18 - 4$
$= 14$

33. $x - 2y$
$8 - 2 \cdot 3 = 8 - 6$
$= 2$

35. $x^2 + 3y$
$6^2 + 3 \cdot 7 = 36 + 3 \cdot 7$
$= 36 + 21$
$= 57$

37. $x^2 + y \div x$
$2^2 + 8 \div 2 = 4 + 8 \div 2$
$= 4 + 4$
$= 8$

39. $4x + (x - y)^2$
$4 \cdot 8 + (8 - 2)^2 = 4 \cdot 8 + 6^2$
$= 4 \cdot 8 + 36$
$= 32 + 36$
$= 68$

41. $x^2 + 3(x - y) + z^2$
$2^2 + 3(2 - 1) + 3^2 = 2^2 + 3 \cdot 1 + 3^2$
$= 4 + 3 \cdot 1 + 9$
$= 4 + 3 + 9$
$= 7 + 9$
$= 16$

43. $11 + (8 + 4) \div 6 = 11 + 12 \div 6$
$= 11 + 2$
$= 13$
$12 + (9 - 5) \cdot 3 = 12 + 4 \cdot 3$
$= 12 + 12$
$= 24$
Since 24 > 13,
$12 + (9 - 5) \cdot 3 > 11 + (8 + 4) \div 6$

45. $5 + 7 \cdot (3 - 1)$

47. $5 + (7 \cdot 3) - 1$

Critical Thinking 1.5

49. $15 + (8 - 3)(2^4) = 15 + (8 - 3)(16)$
$= 15 + (5)(16) = 15 + 80 = 95$
We are looking for the smallest prime
number greater than 95.
96 is not prime (for example, it is divisible
by 2).
97 is prime.
97 is the smallest prime number greater than
$15 + (8 - 3)(2^4)$

Projects and Group Activities 1.5

51. Using 3 as the digit:
$3 + 3^3 = 3 + 27 = 30$

Chapter Review

1.

2. $10^4 = 10,000$

3. $\begin{array}{r} 4,207 \\ -1,624 \\ \hline 2,583 \end{array}$

4. $3 \cdot 3 \cdot 5 \cdot 5 \cdot 5 \cdot 5 = 3^2 \cdot 5^4$

5. $\begin{array}{r} 319 \\ 358 \\ + 712 \\ \hline 1,389 \end{array}$

6. $\overbrace{38,729}^{\text{Given place value}}$
   $\underbrace{29}_{2<5}$

   38,729 rounded to the nearest hundred is 38,700.

7. $247 > 163$

8. 32,509

9. $2xy$
   $2 \cdot 50 \cdot 7 = 100 \cdot 7$
   $= 700$

10.
$$
\begin{array}{r}
2,607 \\
6\overline{)15,642} \\
-12 \\
\hline
36 \\
-36 \\
\hline
4 \\
-0 \\
\hline
42 \\
-42 \\
\hline
0
\end{array}
$$

11.
$$
\begin{array}{r}
6,407 \\
-\ 2,359 \\
\hline
4,048
\end{array}
$$

12.
$$
\begin{array}{rcl}
482 & \to & 500 \\
319 & \to & 300 \\
570 & \to & 600 \\
146 & \to & +100 \\
\hline
& & 1,500
\end{array}
$$

13. $50 \div 1 = 50$
    $50 \div 2 = 25$
    $50 \div 5 = 10$
    $50 \div 10 = 5$
    The factors of 50 are 1, 2, 5, 10, 25, and 50.

14. $24 - y = 17$
    $24 - 7 \mid 17$
    $17 = 17$
    Yes, 7 is a solution of the equation.

15. $16 + 4(7-5)^2 \div 8 = 16 + 4 \cdot 2^2 \div 8$
    $= 16 + 4 \cdot 4 \div 8$
    $= 16 + 16 \div 8$
    $= 16 + 2$
    $= 18$

16. The Commutative Property of Addition

17. four million nine hundred twenty-seven thousand thirty-six

18. $x^3 y^2$
    $3^3 \cdot 5^2 = (3 \cdot 3 \cdot 3) \cdot (5 \cdot 5)$
    $= 27 \cdot 25$
    $= 675$

19. Strategy
    a. To determine by many times more PG-13 films were released than NC-17 films:
       →Find the number of PG-13 films (112) and the number of NC-17 films (7) from the pie chart.
       →Divide the number of PG-13 films by the number of NC-17 films.

    b. To determine by many times more R rated films were released than NC-17 films:
       →Find the number of R rated films (427) and the number of NC-17 films (7) from the pie chart.
       →Divide the number of R rated films by the number of NC-17 films.

    Solution

    a. $\dfrac{112}{7} = 16$

       There were 16 times more PG-13 films released than NC-17 films.

    b. $\dfrac{427}{7} = 61$

       There were 61 times more R rated films released than NC-17 films.

20.
$$
\begin{array}{r}
67 \text{ r } 70 \\
92\overline{)6,234} \\
-552 \\
\hline
714 \\
-644 \\
\hline
70
\end{array}
$$

21.
$$
\begin{array}{r}
659 \\
\times\ \ 4 \\
\hline
2,636
\end{array}
$$

22. $x - y$
    $270 - 133 = 137$

23.
$$
\begin{array}{r}
5 \\
3\overline{)15} \\
3\overline{)45} \\
2\overline{)90}
\end{array}
$$

    $90 = 2 \cdot 3 \cdot 3 \cdot 5 = 2 \cdot 3^2 \cdot 5$

24. $\dfrac{x}{y}$

$\dfrac{480}{6} = 80$

25. $1 \cdot 82 = 82$

26. $36 = 4x$

$\dfrac{36}{4} = \dfrac{4x}{4}$

$9 = x$

The solution is 9.

27. $x + y$

$683 + 249 = 932$

28.
```
    18
  × 24
  ----
    72
   36
  ----
  432
```

29. $(a + b)^2 - 2c$

$(5 + 3)^2 - 2 \cdot 4 = 8^2 - 2 \cdot 4$

$= 64 - 2 \cdot 4$

$= 64 - 8$

$= 56$

30. Strategy   To find the person with the greater number of rebounds, compare the numbers 17,440 and 16,279.

Solution $17{,}440 > 16{,}279$
Kareem Abdul-Jabbar had more rebounds.

31. Strategy   To find the total cost, multiply the number of square feet of floor space (2,800) by the cost per square feet (65).

Solution $2{,}800 \cdot 85 = 238{,}000$
The total cost of the contractor's work will be $238,000.

32. Strategy
  a. To find the perimeter of the rectangle, substitute 25 for L and 12 for W in the formula below.

  b. To find the area of the rectangle, substitute 25 for L and 12 for W in the formula below.

Solution
  a. $P = 2L + 2W$
  $P = 2 \cdot 25 + 2 \cdot 12$
  $P = 50 + 24$
  $P = 74$
  The perimeter is 74 m.

  b. $A = LW$
  $A = 25 \cdot 12$
  $A = 300$
  The area is 300 m^2.

33. Strategy   To determine the projected growth from 2016 to 2019, subtract the number of adults projected to be enrolled in 2016 (3,621,000) from the number of adults projected to be enrolled in 2019 (3,904,000).

Solution $3{,}904{,}000 - 3{,}621{,}000 = 283{,}000$
The enrollment is projected to increase by 283,000 adults.

34. Strategy   To find the distance traveled, substitute 3 for t and 14 for r in the given formula and solve for d.

Solution $d = rt$
$d = 14 \cdot 3$
$d = 42$
The cyclist traveled 42 mi.

35. Strategy   To find the markup, substitute 2,224 for S and 1,775 for C in the given formula and solve for M.

Solution $M = S - C$
$M = 2{,}224 - 1{,}775$
$M = 449$
The markup on the copy machine is $449.

Chapter Test

1. $3{,}297 \cdot 100 = 329{,}700$

2. $2 \cdot 2 \cdot 2 \cdot 2 \cdot 10 \cdot 10 \cdot 10 = 16{,}000$

3. $\begin{array}{r} 4{,}902 \\ -\,873 \\ \hline 4{,}029 \end{array}$

4. $x \cdot x \cdot x \cdot x \cdot y \cdot y \cdot y = x^4 y^3$

5. $23 = p + 16$
$23 - 16 = p + 16 - 16$
$7 = p$
Yes

6. 2,961 ⌐ Given place value
└ $6 > 5$

2,961 rounded to the nearest hundred is 3,000.

7. $7{,}177 < 7{,}717$

8. 8,490

9. three hundred eighty-two thousand nine hundred four

10. $\begin{array}{r} 392 \to 400 \\ 477 \to 500 \\ 519 \to 500 \\ +\,648 \to 600 \\ \hline 2{,}036 \quad 2{,}000 \end{array}$

11. $\begin{array}{r} 1{,}376 \\ \times\quad 8 \\ \hline 11{,}008 \end{array}$

12. $36{,}479 \to 40{,}000$
$50 \to 60$
$40{,}000 \cdot 60 = 2{,}400{,}000$

13. $92 \div 1 = 92$
$92 \div 2 = 46$
$92 \div 4 = 23$
$92 \div 23 = 4$
The factors of 92 are 1, 2, 4, 23, 46, 92.

14. $3\overline{)15}$ → 5

$2\overline{)30}$

$2\overline{)60}$

$2\overline{)120}$

$2\overline{)240}$

$240 = 2 \cdot 2 \cdot 2 \cdot 2 \cdot 3 \cdot 5 = 2^4 \cdot 3 \cdot 5$

15. $x - y$
$39{,}241 - 8{,}375$
$= 30{,}866$

16. The Commutative Property of Addition

17. $\dfrac{x}{y}$

$\dfrac{3{,}588}{4}$
$= 897$

18. $27 - (12 - 3) \div 9$
$= 27 - 9 \div 9$
$= 27 - 1$
$= 26$

19. Strategy
To determine between which two years girls grow the most, find the difference in median height for each pair of consecutive years.

Solution
Birth to age 1: $74 - 49 = 25$
Age 1 to age 2: $84 - 74 = 10$
Age 2 to age 3: $95 - 84 = 11$
Age 3 to age 4: $100 - 95 = 5$
Age 4 to age 5: $108 - 100 = 8$
Girls grow the most between birth and age 1.

20. $60 = 17 + d$
$60 - 17 = 17 + d - 17$
$51 = d$
The solution is 51.

21. $176 = 4t$
$\dfrac{176}{4} = \dfrac{4t}{4}$
$t = 44$
The solution is 44.

22. $5x + (x - y)^2$
$5 \cdot 8 + (8 - 4)^2$
$= 40 + 4^2$
$= 40 + 16$
$= 56$
The solution is 56.

23. 7

24. $12 + x = 90$
$12 + x - 12 = 90 - 12$
$x = 78$
The number is 78.

25. $6 \cdot 5 \cdot 4 \cdot 3 \cdot 2 \cdot 1 = 720$

26. Strategy   First, to find the total cost of the computer system, add the prices of the components. Then, to find the balance of the checking account, subtract the total cost of the computer system from $2,276.

Solution $850 + 270 + 175 + 425 = 1720$
$2,276 - 720 = 556$
The balance is $556.

27. Strategy
  a. To find the perimeter of the square, substitute 24 for s in the formula below.

  b. To find the area of the square, substitute 24 for s in the formula below.

Solution
  a. $P = 4s$
  $P = 4 \cdot 24$
  $P = 96$
  The perimeter is 96 cm.

  b. $A = s^2$
  $A = (24)^2$
  $A = 576$
  The area is 576 cm^2.

28. Strategy   To find the data processor's take home pay, subtract the sum of his deductions, taxes (854), retirement (272), and insurance (108), from the total salary (5,690).

Solution $5,690 - (854 + 272 + 108)$
$= 5,690 - 1,234$
$= 4,456$
The data processor's take home pay is $4,456.

29. Strategy   To determine the how many more registered users are on Facebook than Twitter, subtract the number users on Twitter (175 million) from the number of users of Facebook (640 million).

Solution $640 - 175 = 465$
There are 465 million more registered users on Facebook than on Twitter.

30. Strategy   To find the commission earned, substitute 3 for R and 480 for U in the given formula and solve for C.

Solution $C = U \cdot R$
$= 480 \cdot 3$
$= 1,440$
The commission earned is $1,440.

31. Strategy   To find the value per share of the fund, substitute 5,500,000 for C and 500,000 for S in the given formula and solve for V.

Solution $V = \dfrac{C}{S} = \dfrac{5,5000,000}{500,000} = 11$

The value per share is $11.

Chapter 2: Integers

Prep Test

1. $54 > 45$

2. 4 units

3. $7,654 + 8,193 = 15,847$

4. $6,097 - 2,318 = 3,779$

5. $472 \times 56 = 26,432$

6. $144 \div 24 = 6$

7. $\begin{aligned} 22 &= y + 9 \\ 22 - 9 &= y + 9 - 9 \\ 13 &= y \end{aligned}$

8. $\begin{aligned} 12b &= 60 \\ \frac{12b}{12} &= \frac{60}{12} \\ b &= 5 \end{aligned}$

9. $P = C + M$
$P = 129 + 43 = 172$
The price is $172.

10. $(8-6)^2 + 12 \div 4 \cdot 3^2$
$= 2^2 + 12 \div 4 \cdot 3^2$
$= 4 + 12 \div 4 \cdot 9$
$= 4 + 3 \cdot 9$
$= 4 + 27$
$= 31$

Section 2.1

Concept Check

1. **a.** left

   b. right

3. negative; positive

5. absolute value

Objective A Exercises

7.

9.

11.

13.

15.

1 is 3 units to the right of –2.

17.

–1 is 4 units to the left of 3.

19.

3 is 6 units to the right of –3.

21.

A is –4, and C is –2.

23.

A is –7, and D is –4.

25. $-2 > -5$

27. $3 > -7$

29. $-42 < 27$

31. $53 > -46$

33. $-51 < -20$

35. $-131 < 101$

37. $-7, -2, 0, 3$

39. $-5, -3, 1, 4$

41. $-4, 0, 5, 9$

43. $-10, -7, -5, 4, 12$

45. **a.** never true

   b. sometimes true

   c. sometimes true

   d. always true

Objective B Exercises

47. -45

49. 88

51. $-n$

53. d

55. the opposite of negative thirteen

57. the opposite of negative p

59. five plus negative ten

61. negative fourteen minus negative three

63. negative thirteen minus eight

65. m plus negative n

67. $-(-7) = 7$

69. $-(46) = -46$

71. $-(-73) = 73$

73. $-(-z) = z$

75. $-(p) = -p$

77. negative

Objective C Exercises

79. $|-4| = 4$

81. $|9| = 9$

83. $|-11| = 11$

85. $|-12| = 12$

87. $|-23| = 23$

89. $-|27| = -27$

91. $|25| = 25$

93. $-|-41| = -41$

95. $-|-93| = -93$

97. $|x|$
$|-10| = 10$

99. $|-x|$
$|-8| = 8$

101. $|-y|$
$|-(-6)| = |6| = 6$

103. $|-12| = 12, |8| = 8$
$12 > 8$
$|-12| > |8|$

105. $|6| = 6, |13| = 13$
$6 < 13$
$|6| < |13|$

107. $|-1| = 1, |-17| = 17$
$1 < 17$
$|-1| < |-17|$

109. $x = x$
$|x| = |-x|$

111. $-|6| = -6, -(4) = -4, |-7| = 7, -(-9) = 9$
$-|6|, -(4), |-7|, -(-9)$

113. $-|-7| = -7, -9 = -9, -(5) = -5, |4| = 4$
$-9, -|-7|, -(5), |4|$

115. $-(-3) = 3, -|-8| = -8, |5| = 5,$
$-|10| = -10, -(-2) = 2$
$-|10|, -|-8|, -(-2), -(-3), |5|$

Objective D Exercises

117. $-329,000,000$

119. Strategy   To determine which quarter had the greater loss, compare the absolute values of the numbers $-12,575$ and $-11,350$. The larger number corresponds to the quarter with the greater loss.

Solution $|-12,575| = 12,575$
$|-11,350| = 11,350$
$12,575 > 11,350$
The loss was greater during the first quarter.

121. Strategy
 a. To find the earnings per share in 2005, read the number in the bar graph below the bar corresponding to 2005.

 b. To find the earnings per share in 2007, read the number in the bar graph below the bar corresponding to 2007.

 Solution
 a. The earnings per share for 2005 were –27¢.

 b. The earnings per share for 2007 were –40¢.

123. Strategy To find a year in which Mycopen had a profit, use the bar graph to find a year in which earnings per share were positive.

 Solution The only recorded positive earnings per share were 11¢. Earnings per share were 11¢ in 2008.
   Mycopen did earn a profit during the years shown. Mycopen earned a profit in 2008.

125. Strategy To determine which stock showed the least net change, compare the absolute values of the numbers –1 and –2. The smaller number represents the least net change.

 Solution $|-1| = 1, |-2| = 2$
   $1 < 2$
   Stock B showed the least net change.

127. Strategy To find the wind chill factor, use the given table.

 Solution Find the number where the column with 10 and the row with 20 cross. Read the number –9.
   The wind chill factor is –9°F.

129. Strategy To find the cooling power, use the given table.

 Solution Find the number where the column with –15 and the row with 10 cross. Read the number –35.
   The cooling power is –35°F.

131. Strategy To find the situation which feels colder:
  →From the given table, find the wind chill factor with a temperature of –30°F with a 5-mph wind and the wind chill factor with a temperature of –20°F with a 10-mph wind.
  →Compare the wind chill factors.

 Solution The wind chill factor with a temperature of –30°F and 5-mph wind is –46°F.
   The wind chill factor with a temperature of –20°F and 10-mph wind is –41°F.
   $-46 < -41$
   –30°F with a 5-mph wind feels colder.

Critical Thinking 2.1

133. The absolute value of a number is the distance from zero to the number on the number line. If $|y| = 11$, then y must be a number that is 11 units from 0 on the number line. Therefore, y is –11 or 11.

135. x must be less than 7 and greater than –7.
$-6, -5, -4, -3, -2, -1, 0, 1, 2, 3, 4, 5, 6$

137. **a.** Two numbers that are 4 units from 2 on the number line are –2 and 6.

 b. Two numbers that are 5 units from 3 on the number line are –2 and 8.

Projects or Group Activities 2.1

139. Answers will vary. One example is:
Point A is a point on the number line
halfway between –3 and 5. Point B is a
point halfway between A and the graph of 3
on the number line.
Solution: If A is a point on the number line
halfway between –3 and 5, then A is the
graph of 1.
B is a point halfway between A and the
graph of 3 on the number line.
Therefore, B is a point halfway between 1
and 3.
B is the graph of 2.

Section 2.2

Concept Check

1. the same; negative

3. negative; negative

5. negative; minus

7. (-5)

Objective A Exercises

9. $-3 + (-8) = -11$

11. $-8 + 3 = -5$

13. $-5 + 13 = 8$

15. $6 + (-10) = -4$

17. $3 + (-5) = -2$

19. $-4 + (-5) = -9$

21. $-6 + 7 = 1$

23. $(-5) + (-10) = -15$

25. $-7 + 7 = 0$

27. $(-15) + (-6) = -21$

29. $0 + (-14) = -14$

31. $73 + (-54) = 19$

33. $2 + (-3) + (-4) = -1 + (-4)$
$= -5$

35. $-3 + (-12) + (-15) = -15 + (-15)$
$= -30$

37. $-17 + (-3) + 29 = -20 + 29$
$= 9$

39. $11 + (-22) + 4 + (-5)$
$= -11 + 4 + (-5)$
$= -7 + (-5)$
$= -12$

41. $-22 + 10 + 2 + (-18)$
$= -12 + 2 + (-18)$
$= -10 + (-18)$
$= -28$

43. $-25 + (-31) + 24 + 19$
$= -56 + 24 + 19$
$= -32 + 19$
$= -13$

45. $3 + (-21) = -18$

47. $-5 + 16 = 11$

49. $(-3) + (-8) + 12 = -11 + 12$
$= 1$

51. $x + (-7)$

53. a. $-60,100,000,000 + (-50,100,000,000) =$
$-110,200,000,000$
The total of the U.S. balance of trade in
Japan and Mexico is
$-\$110,200,000,000$.

b. $-28,500,000,000 + (-50,100,000,000) =$
$-78,600,000,000$
The total of the U.S. balance of trade
with Canada and Mexico is
$-\$78,600,000,000$.

c. $-60,100,000,000 + (-273,100,000,000)$
$= -333,200,000,000$
The total of the U.S. balance of trade
with Japan and China is
$-\$333,200,000,000$.

55. $-a + b$
$-(-8) + (-3) = 8 + (-3)$
$= 5$

57. $-x + y$
$-(-5) + (-7) = 5 + (-7)$
$= -2$

59. $a + b + c$
$-10 + (-6) + 5 = -16 + 5$
$= -11$

61. $-x + (-y) + z$
$-(-2) + (-8) + (-11) = 2 + (-8) + (-11)$
$= -6 + (-11)$
$= -17$

63. The Addition Property of Zero

65. The Associative Property of Addition

67. $-13 + 0 = -13$

69. $18 + (-18) = 0$

71. $6 = -3 + z$
$$6 \mid -3 + (-8)$$
$6 \neq -11$
No, -8 is not a solution of the equation
$6 = -3 + z$.

73. $-7 + m = -15$
$$-7 + (-8) \mid -15$$
$-15 = -15$
Yes, -8 is a solution of the equation
$-7 + m = -15$.

75. $1 + z = z + 2$
$$1 + (-4) \mid -4 + 2$$
$-3 \neq -2$
No, -4 is not a solution of the equation
$1 + z = z + 2$.

77. sometimes true

79. always true

Objective B Exercises

81. $6 - 9 = 6 + (-9)$
$= -3$

83. $-9 - 4 = -9 + (-4)$
$= -13$

85. $3 - (-4) = 3 + 4$
$= 7$

87. $-4 - (-4) = -4 + 4$
$= 0$

89. $-10 - 7 = -10 + (-7)$
$= -17$

91. $(-7) - (-4) = -7 + 4$
$= -3$

93. $-4 - (-16) = -4 + 16$
$= 12$

95. $3 - (-24) = 3 + 24$
$= 27$

97. $(-41) - 65 = -41 + (-65)$
$= -106$

99. $-95 - (-28) = -95 + 28$
$= -67$

101. $-10 - (-4) = -10 + 4$
$= -6$

103. $-9 - 6 = -9 + (-6)$
$= -15$

105. $49 - (-33) = 49 + 33$
$= 82$
The difference between the highest and
lowest temperatures ever recorded in South
America is 82°C.

107. $-4 - 3 - 2 = -4 + (-3) + (-2)$
$= -7 + (-2)$
$= -9$

109. $12 - (-7) - 8 = 12 + 7 + (-8)$
$= 19 + (-8)$
$= 11$

111. $4 - 12 - (-8) = 4 + (-12) + 8$
$= -8 + 8$
$= 0$

113. $-16 - 47 - 63 - 12$
$= -16 + (-47) + (-63) + (-12)$
$= -63 + (-63) + (-12)$
$= -126 + (-12)$
$= -138$

115. $12 - (-6) + 8 = 12 + 6 + 8$
$= 18 + 8$
$= 26$

117. $-8 - (-14) + 7 = -8 + 14 + 7$
$= 6 + 7$
$= 13$

119. $9 - 12 + 0 - 5 = 9 + (-12) + 0 + (-5)$
$= -3 + 0 + (-5)$
$= -3 + (-5)$
$= -8$

121. $5 + 4 - (-3) - 7 = 5 + 4 + 3 + (-7)$
$= 9 + 3 + (-7)$
$= 12 + (-7)$
$= 5$

123. $-13 + 9 - (-10) - 4 = -13 + 9 + 10 + (-4)$
$= -4 + 10 + (-4)$
$= 6 + (-4)$
$= 2$

125. $-x - y$
$-(-3) - 9 = 3 + (-9)$
$= -6$

127. $-x - (-y)$
$-(-3) - (-9) = 3 + 9$
$= 12$

129. $a - b - c$
$4 - (-2) - 9 = 4 + 2 + (-9)$
$= 6 + (-9)$
$= -3$

131. $x - y - (-z)$
$-9 - 3 - (-30) = -9 + (-3) + 30$
$= -12 + 30$
$= 18$

133.
$$\begin{array}{c|c} x - 7 = -10 & \\ \hline -3 - 7 & -10 \\ -3 + (-7) & -10 \\ -10 = -10 \end{array}$$
Yes, -3 is a solution of the equation $x - 7 = -10$.

135.
$$\begin{array}{c|c} -5 - w = 7 & \\ \hline -5 - (-2) & 7 \\ -5 + 2 & 7 \\ -3 \neq 7 \end{array}$$
No, -2 is not a solution of the equation $-5 - w = 7$.

137.
$$\begin{array}{c|c} -t - 5 = 7 + t & \\ \hline -(-6) - 5 & 7 + (-6) \\ 6 - 5 & 7 + (-6) \\ 6 + (-5) & 7 + (-6) \\ 1 = 1 \end{array}$$
Yes, -6 is a solution of the equation $-t - 5 = 7 + t$.

139. sometimes true

Objective C Exercises

141. Strategy
   a. To find the difference, subtract the elevation of Death Valley (–86) from the elevation of Mt. Aconcagua (6,960).

   b. To find the difference, subtract the elevation of the Lake Assal (–156) from the elevation of Mt. Kilimanjaro (5,895).

Solution
   a. $6,960 - (-86) = 6,960 + 86$
   $= 7,046$
   The difference in elevation is 7,046 m.

   b. $5,895 - (-156) = 5,895 + 156$
   $= 6,051$
   The difference in elevation is 6,051 m.

143. Strategy   To determine for which continent the difference between the highest and lowest elevations is smallest:
   →Find the difference between the highest and lowest elevation for each continent.
   →Compare the differences.

Solution   Africa: $5,895 - (-156)$
   $= 5,895 + 156 = 6,051$
   Asia: $8,850 - (-411)$
   $= 8,850 + 411 = 9,261$
   Europe: $5,642 - (-28)$
   $= 5,642 + 28 = 5,670$
   America: $6,960 - (-86)$
   $= 6,960 + 86 = 7,046$
   $5,670 < 6,051 < 7,046 < 9,261$
   The difference between the highest and lowest elevations is smallest in Europe.

145. Strategy   To find the difference, subtract the average temperature at 40,000 ft (–70) from the average temperature at 12,000 ft (16).

Solution $16 - (-70) = 16 + 70 = 86$
The difference in temperature is 86°.

147. Strategy   To find the golfer's score, substitute 49 for N and 52 for P in the given formula and solve for S.

Solution $S = N - P$
$S = 49 - 52$
$S = 49 + (-52)$
$S = -3$
The golfer's score is –3.

149. Strategy
a. To find the difference, subtract the low temperature in the U.S. (–14) from the high temperature in the U.S. (93).

b. To find the difference, subtract the low temperature in the contiguous U.S. (–7) from the high temperature in the U.S. (93).

Solution
a. $93 - (-14) = 93 + 14 = 107$
The difference in temperature is 107°F.

b. $93 - (-7) = 93 + 7 = 100$
The difference in temperature is 100°F.

151. Strategy   To find d, replace a by 7 and b by –12 in the given formula and solve for d.

Solution $d = |a - b|$
$d = |7 - (-12)|$
$d = |7 + 12|$
$d = |19|$
$d = 19$
The distance between the two points is 19 units.

Critical Thinking 2.2

153. Answers will vary. Possible answers include –6 and –1, –5 and –2, –4 and –3.

Projects or Group Activities 2.2

155. Possible answers include:
$-11 - (-3) = -11 + 3 = -8$;
$-10 - (-2) = -10 + 2 = -8$;
$-9 - (-1) = -9 + 1 = 8$.
Choose a negative number which is less than –8 as the first number. Subtract 8 from the absolute value of that number, and then use the opposite of that result to get the second number.

Check Your Progress: Chapter 2

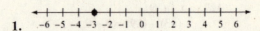

1. $-6\ -5\ -4\ -3\ -2\ -1\ \ 0\ \ 1\ \ 2\ \ 3\ \ 4\ \ 5\ \ 6$

2. –3 is 5 units to the left of 2

3. $-12 > -16$

4. $-19, -8, 4, 7$

5. a. 11

b. –13

c. m

6. negative five minus negative seven

7. a. $-(-42) = 42$

b. $-(t) = -t$

8. a. 18

b. 37

9. a. $|-51| = 51$

b. $-|67| = -67$

10. $-|x|$

$-|2| = -2$

11. $|-19| = 19, |7| = 7$
$19 > 7$
$|-19| > |7|$

12. $|-5| = 5, -(-6) = 6, |3| = 3,$
$-|-8| = -8, -|12| = -12$
$-|12|, -|-8|, |3|, |-5|, -(-6)$

13. $-8 + (-12) = -20$

14. $20 + (-3) + (-7) = 17 + (-7) = 10$

15. $5 - 40 = 5 + (-40)$
$= -35$

16. $-32 - (-16) = -32 + 16$
$= -16$

17. $4 - (-15) - 3 + 7$
$= 4 + 15 + (-3) + 7$
$= 19 + (-3) + 7$
$= 16 + 7$
$= 23$

18. $-11 - 16 = -11 + (-16) = -27$

19. $-6 + (-9) + 14 = -15 + 14 = -1$

20. $-x + y$
$-(-6) + (-2) = 6 + (-2) = 4$

21. $a - (-b)$
$-5 - (-7) = -5 + 7 = 2$

22. $-3 = y - 4$

$$\begin{array}{c|c} -3 & -7 - 4 \end{array}$$

$-3 \neq -11$
No, -8 is not a solution of the equation
$6 = -3 + z$.

23. Strategy   To determine which
temperature is colder, compare
the absolute values of the
numbers -16 and -4. The
larger number is the colder
temperature.

Solution $|-16| = 16, |-4| = 4$
$16 > 14$
-16 is the colder temperature.

24. Strategy   To find the temperature, add
the increase (8) to the previous
temperature (-3).

Solution $-3 + 8 = 5$
The temperature is 5°C.

25. Strategy   To find d, replace a by 9 and b
by -5 in the given formula and
solve for d.

Solution $d = |a - b|$
$d = |9 - (-5)|$
$d = |9 + 5|$
$d = |14|$
$d = 14$
The distance between the two
points is 21 units.

Section 2.3

Concept Check

1. a. different; negative

b. the same; positive

3. $\dfrac{-63}{9}$

Objective A Exercises

5. $-4 \cdot 6 = -24$

7. $-2(-3) = 6$

9. $(9)(2) = 18$

11. $5(-4) = -20$

13. $-8(2) = -16$

15. $(-5)(-5) = 25$

17. $(-7)(0) = 0$

19. $14(3) = 42$

21. $-32(4) = -128$

23. $(-8)(-26) = 208$

25. $9(-27) = -243$

27. $-5 \cdot (23) = -115$

29. $-7(-34) = 238$

31. $4 \cdot (-8) \cdot 3 = -32 \cdot 3$
$= -96$

33. $(-6)(5)(7) = -30(7)$
$= -210$

35. $-8(-7)(-4) = 56(-4)$
$= -224$

37. $2(-20) = -40$

39. $-30(-6) = 180$

41. $-q(r) = -qr$

43. **a.** $-170,000,000(4) = -680,000,000$
The annual net income for Sears
Holdings would be $-\$680,000,000$.

b. $-74,000,000(4) = -296,000,000$
The annual net income for
Rite Aid would be $-\$296,000,000$.

45. The Multiplication Property of One

47. The Associative Property of Multiplication

49. $-6 \cdot (5 \cdot 10) = (-6 \cdot 5) \cdot 10$

51. $1(-14) = -14$

53. $-xy$
$-(-3)(-8) = 3(-8)$
$= -24$

55. $-xyz$
$-(-6)(2)(-5) = 6(2)(-5)$
$= 12(-5)$
$= -60$

57. $-7n$
$-7(-51) = 357$

59. $8ab$
$8(7)(-1) = 56(-1)$
$= -56$

61. $-5st$
$-5(-40)(-8) = 200(-8)$
$= -1,600$

63. $-5x = -15$
$$\overline{-5(-3)\ |\ -15}$$
$15 \neq -15$
No, -3 is not a solution of the equation
$-5x = 15$.

65. $-8 = -8a$
$$\overline{-8\ |\ -8(0)}$$
$-8 \neq 0$
No, 0 is not a solution of the equation
$-8 = -8a$.

67. $-27 = -3c$
$$\overline{-27\ |\ -3(9)}$$
$-27 = -27$
Yes, 9 is a solution of the equation
$-27 = -3c$.

69. positive

Objective B Exercises

71. $18 \div (-3) = -6$

73. $(-64) \div (-8) = 8$

75. $-49 \div 1 = -49$

77. $-40 \div (-5) = 8$

79. $\dfrac{44}{-4} = -11$

81. $\dfrac{-98}{-7} = 14$

83. $-91 \div (-7) = 13$

85. $(-162) \div (-162) = 1$

87. $-130 \div (-5) = 26$

89. $(-92) \div (-4) = 23$

91. $\dfrac{550}{-5} = -110$

93. $\dfrac{-333}{-3} = 111$

95. $\dfrac{-9}{x}$

97. $-318,000,000 \div 3 = -106,000,000$
The average monthly net income
for Delta Airlines was $-\$106,000,000$.

99. $-a \div b$
$-(-36) \div (-4) = 36 \div (-4)$
$= -9$

101. $(-a) \div (-b)$
$-(-36) \div (-(-4)) = 36 \div 4$
$= 9$

103. $\dfrac{-x}{y}$
$\dfrac{-(-42)}{-7} = \dfrac{42}{-7}$
$= -6$

105. $\dfrac{-x}{-y}$

$\dfrac{-(-42)}{-(-7)} = \dfrac{42}{7}$

$= 6$

107. $6 = \dfrac{-c}{-3}$

$6 \,\big|\, \dfrac{-18}{-3}$

$6 = 6$

Yes, 18 is a solution of the equation.

109. $\dfrac{21}{n} = 7$

$\dfrac{21}{-3} \,\big|\, 7$

$-7 \neq 7$

No, -3 is not a solution of the equation.

111. $\dfrac{m}{-4} = \dfrac{-16}{m}$

$\dfrac{8}{-4} \,\big|\, \dfrac{-16}{8}$

$-2 = -2$

Yes, 8 is a solution of the equation.

113. $\dfrac{a}{b}$

115. $-\dfrac{a}{b}$

Objective C Exercises

117. Strategy To find the average score, divide the combined scores (-12) by the number of golfers (4).

Solution $-12 \div 4 = 3$
The average golf score was -3.

119. Strategy To find the average record low temperature for the first three months of the year:
→Add the average temperatures for January (-70), February (-66), and March (-50).
→Divide the sum by the number of months (3).

Solution $-70 + (-69) + (-50)$
$= -139 + (-50) = -189$
$-189 \div 3 = -63$
The average record low temperature for the first three months of the year is $-63°F$.

121. Strategy To find the average U.S. trade deficit for March, April, and May of 2011:
→Add the three trade deficits for March $(-47$ billion$)$, April $(-44$ billion$)$, and May $(-50$ billion$)$.
→Divide the sum by the number of months (3).

Solution $-47 + (-44) + (-50)$
$= -91 + (-50) = -141$
$-141 \div 3 = -47$
The average U.S. trade deficit for March, April, and May of 2011 is $-\$47$ billion.

123. Strategy To find the average daily low temperature for the week:
→Add the seven temperature readings.
→Divide by 7.

Solution $4 + (-5) + 8 + (-1) + (-12) + (-14) + (-8) = -28$
$-28 \div 7 = -4$
The average daily low temperature for the week was $-4°$.

Critical Thinking 2.3

125. Answers will vary. For example:
5 and –2 have different signs.
The product of 5 and –2 is $5(-2) = (-2) + (-2) + (-2) + (-2) + (-2) = -10$, which is a negative number.

Projects or Group Activities 2.3

127. Strategy   To find the next four numbers in the sequence:
→Find the multiplier by dividing the second number in the sequence (–4) by the first number (2).
→Use the multiplier to find the successive numbers in the sequence.

   Solution $\dfrac{-4}{2} = -2$
$8 \cdot (-2) = -16$
$-16 \cdot (-2) = 32$
$32 \cdot (-2) = -64$
$-64 \cdot (-2) = 128$
The next four number-s in the sequence are –16, 32, –64, and 128.

129. Strategy   To find the next four numbers in the sequence:
→Find the multiplier by dividing the second number in the sequence (–5) by the first number (–1).
→Use the multiplier to find the successive numbers in the sequence.

   Solution $\dfrac{-5}{-1} = 5$
$-25 \cdot 5 = -125$
$-125 \cdot 5 = -625$
$-625 \cdot 5 = -3,125$
$-3,125 \cdot 5 = -15,625$
The next four numbers in the sequence are –125, –625, –3125, and –15,625.

Section 2.4

Concept Check

1. 3

3. 7

5. $-6 = n + 12$

Objective A Exercises

7. $x - 6 = 9$
$x - 6 + 6 = 9 + 6$
$x = 15$
The solution is 15.

9. $8 = y - 3$
$8 + 3 = y - 3 + 3$
$11 = y$
The solution is 11.

11. $x - 5 = -12$
$x - 5 + 5 = -12 + 5$
$x = -7$
The solution is –7.

13. $-10 = z + 6$
$-10 - 6 = z + 6 - 6$
$-16 = z$
The solution is –16.

15. $x + 12 = 4$
$x + 12 - 12 = 4 - 12$
$x = -8$
The solution is -8.

17. $-12 = c - 12$
$-12 + 12 = c - 12 + 12$
$0 = c$
The solution is 0.

19. $6 + x = 4$
$6 + x - 6 = 4 - 6$
$x = -2$
The solution is -2.

21. $12 = n - 8$
$12 + 8 = n - 8 + 8$
$20 = n$
The solution is 20.

23. $3m = -15$
$\dfrac{3m}{3} = \dfrac{-15}{3}$
$m = -5$
The solution is -5.

25. $-10 = 5v$
$\dfrac{-10}{5} = \dfrac{5v}{5}$
$-2 = v$
The solution is -2.

27. $-8x = -40$
$\dfrac{-8x}{-8} = \dfrac{-40}{-8}$
$x = 5$
The solution is 5.

29. $-60 = -6v$
$\dfrac{-60}{-6} = \dfrac{-6v}{-6}$
$10 = v$
The solution is 10.

31. $5x = -100$
$\dfrac{5x}{5} = \dfrac{-100}{5}$
$x = -20$
The solution is -20.

33. $4x = 0$
$\dfrac{4x}{4} = \dfrac{0}{4}$
$x = 0$
The solution is 0.

Objective B Exercises

35. The unknown number: n

ten less than a number	is	fifteen

$n - 10 = 15$
$n - 10 + 10 = 15 + 10$
$n = 25$
The number is 25.

37. The unknown number: n

zero	is equal to	fifteen more than some number

$0 = n + 15$
$0 - 15 = n + 15 - 15$
$-15 = n$
The number is -15.

39. The unknown number: n

sixteen	equals	negative two times a number

$16 = -2n$
$\dfrac{16}{-2} = \dfrac{-2n}{-2}$
$-8 = n$
The number is -8.

41. Strategy To find the opening-weekend box-office earnings for *The Dark Knight*, write and solve an equation using *x* to represent the opening-weekend box-office earnings for *The Dark Night*.

Solution

The Dark Night earnings	were	$11 million less than the *Harry Potter* earnings.

$x = 169{,}000{,}000 - 11{,}000{,}000$
$x = 158{,}000{,}000$
The opening-weekend box-office earnings for *The Dark Night* were $158 million.

43. Strategy To find the selling price of the car, replace *P* by 925 and *C* by 12,600 in the given formula and solve for *S*.

Solution
$P = S - C$
$925 = S - 12{,}600$
$925 + 12{,}600 = S - 12{,}600 + 12{,}600$
$13{,}525 = S$
The selling price of the car should be $13,525.

45. Strategy To find the assets, replace *N* by 11 and *L* by 4 in the given formula and solve for *A*.

Solution
$N = A - L$
$11 = A - 4$
$11 + 4 = A - 4 + 4$
$15 = A$
The assets are $15 million.

Critical Thinking 2.4

47. a. False. For example, the solution of the equation $5x = 0$ is 0.

b. False. For example, the solution of the equation $-5x = -5$ is 1, a positive number.

Explanations will vary.

Projects or Group Activities 2.4

Answers will vary. Possible answers are given.

49. $x + 8 = 5$

51. $x - 2 = -5$

Section 2.5

Concept Check

1. division

Objective A Exercises

3. $3 - 12 \div 2 = 3 - 6$
   $= 3 + (-6)$
   $= -3$

5. $2(3 - 5) - 2 = 2(-2) - 2$
   $= -4 - 2$
   $= -4 + (-2)$
   $= -6$

7. $4 - (-3)^2 = 4 - 9$
   $= 4 + (-9)$
   $= -5$

9. $4 \cdot (2 - 4) - 4 = 4 \cdot (-2) - 4$
   $= -8 - 4$
   $= -8 + (-4)$
   $= -12$

11. $4 - (-2)^2 + (-3) = 4 - 4 + (-3)$
    $= 4 + (-4) + (-3)$
    $= 0 + (-3)$
    $= -3$

13. $3^3 - 4(2) = 27 - 4(2)$
    $= 27 - 8$
    $= 27 + (-8)$
    $= 19$

15. $3 \cdot (6 - 2) \div 6 = 3 \cdot 4 \div 6$
    $= 12 \div 6$
    $= 2$

17. $2^3 - (-3)^2 + 2 = 8 - 9 + 2$
    $= 8 + (-9) + 2$
    $= -1 + 2$
    $= 1$

19. $6 - 2(1 - 5) = 6 - 2(-4)$
    $= 6 - (-8)$
    $= 6 + 8$
    $= 14$

21. $6 - (-4)(-3)^2 = 6 - (-4)(9)$
    $= 6 - (-36)$
    $= 6 + 36$
    $= 42$

23. $4 \cdot 2 - 3 \cdot 7 = 8 - 3 \cdot 7$
    $= 8 - 21$
    $= 8 + (-21)$
    $= -13$

25. $-2^2 - 5(3) - 1 = 4 - 5(3) - 1$
    $= -4 - 15 - 1$
    $= -4 + (-15) + (-1)$
    $= -19 + (-1)$
    $= -20$

27. $3 \cdot 2^3 + 5 \cdot (3 + 2) - 17 = 3 \cdot 2^3 + 5 \cdot 5 - 17$
    $= 3 \cdot 8 + 5 \cdot 5 - 17$
    $= 24 + 5 \cdot 5 - 17$
    $= 24 + 25 - 17$
    $= 24 + 25 + (-17)$
    $= 49 + (-17)$
    $= 32$

29. $-12(6 - 8) + 1^3 \cdot 3^2 \cdot 2 - 6(2)$
    $= -12(-2) + 1^3 \cdot 3^2 \cdot 2 - 6(2)$
    $= -12(-2) + 1 \cdot 9 \cdot 2 - 6(2)$
    $= 24 + 1 \cdot 9 \cdot 2 - 6(2)$
    $= 24 + 18 - 6(2)$
    $= 24 + 18 - 12$
    $= 24 + 18 + (-12)$
    $= 42 + (-12)$
    $= 30$

31. $-27 - (-3)^2 - 2 - 7 + 6 \cdot 3$
    $= -27 - 9 - 2 - 7 + 6 \cdot 3$
    $= -27 - 9 - 2 - 7 + 18$
    $= -27 + (-9) + (-2) + (-7) + 18$
    $= -36 + (-2) + (-7) + 18$
    $= -38 + (-7) + 18$
    $= -45 + 18$
    $= -27$

33. $16 - 4 \cdot 8 + 4^2 - (-18) - (-9)$
    $= 16 - 4 \cdot 8 + 16 - (-18) - (-9)$
    $= 16 - 32 + 16 - (-18) - (-9)$
    $= 16 + (-32) + 16 + 18 + 9$
    $= -16 + 16 + 18 + 9$
    $= 0 + 18 + 9$
    $= 18 + 9$
    $= 27$

35. $3a + 2b$
    $3(-2) + 2(4) = -6 + 2(4)$
    $= -6 + 8$
    $= 2$

37. $16 \div (ac)$
    $16 \div ((-2)(-1)) = 16 \div 2$
    $= 8$

39. $bc \div (2a)$

$4(-1) \div ((2)(-2)) = 4(-1) \div (-4)$

$= (-4) \div (-4)$

$= 1$

41. $b^2 - c^2$

$4^2 - (-1)^2 = 16 - 1$

$= 16 + (-1)$

$= 15$

43. $(b-a)^2 + 4c$

$(4-(-2))^2 + 4(-1) = (4+2)^2 + 4(-1)$

$= 6^2 + 4(-1)$

$= 36 + 4(-1)$

$= 36 + (-4)$

$= 32$

45. $\dfrac{d-b}{c}$

$\dfrac{3-4}{-1} = \dfrac{3+(-4)}{-1}$

$= \dfrac{-1}{-1}$

$= 1$

47. $\dfrac{b-d}{c-a}$

$\dfrac{4-3}{-1-(-2)} = \dfrac{4+(-3)}{-1+2}$

$= \dfrac{1}{1}$

$= 1$

49. $(d-a)^2 - 3c$

$(3-(-2))^2 - 3(-1)$

$= (3+2)^2 - 3(-1)$

$= 5^2 - 3(-1)$

$= 25 - 3(-1)$

$= 25 - (-3)$

$= 25 + 3$

$= 28$

51. Any odd power of a negative number is negative.

Critical Thinking 2.5

53. $-2^2 - (-3)^2 + 5(4) \div 10 - (-6)$

$= -4 - 9 + 5(4) \div 10 - (-6)$

$= -4 - 9 + 20 \div 10 - (-6)$

$= -4 - 9 + 2 - (-6)$

$= -4 + (-9) + 2 + 6$

$= -13 + 2 + 6$

$= -11 + 6 = -5$

We are looking for the smallest integer greater than –5.

The smallest integer greater than –5 is –4.

The smallest integer greater than

$-2^2 - (-3)^2 + 5(4) \div 10 - (-6)$ is –4.

55. a.

$$
\begin{array}{c|c}
x^2 - 2x - 8 = 0 & \\
\hline
(-4)^2 - 2(-4) - 8 & 0 \\
16 - 2(-4) - 8 & 0 \\
16 + 8 - 8 & 0 \\
24 - 8 & 0 \\
16 \neq 0 &
\end{array}
$$

No, –4 is not a solution of the equation.

b.

$$
\begin{array}{c|c}
x^3 + 3x^2 - 5x - 15 = 0 & \\
\hline
(-3)^3 + 3(-3)^2 - 5(-3) - 15 & 0 \\
-27 + 3(9) - 5(-3) - 15 & 0 \\
-27 + 27 + 15 - 15 & 0 \\
0 + 15 - 15 & 0 \\
15 - 15 & 0 \\
0 = 0 &
\end{array}
$$

Yes, –3 is not a solution of the equation.

Projects or Group Activities 2.5

57. $6 - 12 \div 2 \cdot (3-5)^2$

59. $6 - (12 \div 2) \cdot 3 - 5^2$

Chapter Review Exercises

1. eight minus negative one

2. $-|-36| = -36$

3. $(-40)(-5) = 200$

4. $-a \div b$

$-(-27) \div (-3) = 27 \div (-3)$

$= -9$

5. $-28 + 14 = -14$

6. $-(-13) = 13$

7.

8. $-24 = -6y$

$$\frac{-24}{-6} = \frac{-6y}{-6}$$

$4 = y$

The solution is 4.

9. $-51 \div (-3) = 17$

10. $\frac{840}{-4} = -210$

11. $-6 - (-7) - 15 - (-12)$

$= -6 + 7 + (-15) + 12$

$= 1 + (-15) + 12$

$= -14 + 12$

$= -2$

12. $-ab$

$-(-2)(-9) = 2(-9)$

$= -18$

13. $18 + (-13) + (-6) = 5 + (-6)$

$= -1$

14. $-18(4) = -72$

15. $(-2)^2 - (-3)^2 \div (1-4)^2 \cdot 2 - 6$

$= (-2)^2 - (-3)^2 \div (-3)^2 \cdot 2 - 6$

$= 4 - 9 \div 9 \cdot 2 - 6$

$= 4 - 1 \cdot 2 - 6$

$= 4 - 2 - 6$

$= 4 + (-2) + (-6)$

$= 2 + (-6)$

$= -4$

16. $-x - y$

$-(-1) - 3 = 1 - 3$

$= 1 + (-3)$

$= -2$

17. $4 - (-14) = 4 + 14$

$= 18$

The difference between Tom Lehman's score and Mark Calcavecchia's score is 18 strokes.

18. $-15 - (-28) = -15 + 28$

$= 13$

19. The Commutative Property of Multiplication

20.

$$-6 - t = 3$$

$$\begin{array}{c|c} -6 - (-9) & 3 \\ -6 + 9 & 3 \end{array}$$

$$3 = 3$$

Yes, -9 is a solution of the equation $-6 - t = 3$.

21. $-9 + 16 - (-7) = -9 + 16 + 7$

$= 7 + 7$

$= 14$

22. $\frac{0}{-17} = 0$

23. $-5(2)(-6)(-1) = -10(-6)(-1)$

$= 60(-1)$

$= -60$

24. $3 + (-9) + 4 + (-10) = -6 + 4 + (-10)$

$= -2 + (-10)$

$= -12$

25. $(a-b)^2 - 2a$

$(-2 - (-3))^2 - 2(-2)$

$= (-2 + 3)^2 - 2(-2)$

$= 1^2 - 2(-2)$

$= 1 - 2(-2)$

$= 1 - (-4)$

$= 1 + 4$

$= 5$

26. $-8 > -10$

27. $-21 + 21 = 0$

28. $|-27| = 27$

29. The unknown number: n

forty-eight	is	the product of negative six and some number

$48 = -6n$

$$\frac{48}{-6} = \frac{-6n}{-6}$$

$-8 = n$

The number is -8.

30. Strategy   To find the colder temperature, compare the numbers −4 and −12. The smaller number represents the colder temperature.

Solution   −4 > −12
The colder temperature is −12°C.

31. Strategy   To find the boiling point of neon:
→Find the highest boiling point shown in the table.
→Multiply the highest boiling point by 7.

Solution   The highest boiling point shown in the table is −34°.
−34(7) = −238
The boiling point of neon is −238°C.

32. Strategy   To find the temperature, add the increase (5) to the previous temperature (−8).

Solution   −8 + 5 = −3
The temperature is −3°C.

33. Strategy   To find d, replace a by 7 and b by −5 in the given formula and solve for d.

Solution $d = |a - b|$
$d = |7 - (-5)|$
$d = |7 + 5|$
$d = |12|$
$d = 12$
The distance between the two points is 12 units.

Chapter Test

1. negative three plus negative five

2. $-|-34| = -34$

3. $3 - (-15) = 3 + 15$
$= 18$

4. $a + b$
$(-11) + (-9) = -20$

5. $(-x)(-y)$
$(-(-4)) \cdot (-(-6)) = 4 \cdot 6$
$= 24$

6. The Commutative Property of Addition

7. $-360 \div -30 = 12$

8. $-3 + -6 + 11 = -9 + 11$
$= 2$

9. $16 > -19$

10. $7 - (-3) - 12 = 7 + 3 - 12$
$= 10 - 12$
$= -2$

11. $a - b - c$
$= 6 - (-2) - 11 = 6 + 2 - 11$
$= 8 - 11$
$= -3$

12. $-(-49) = 49$

13. $50 \cdot (-5) = -250$

14. $-|5|, -(3), -|-9|, -(-11)$

15. $17 - x = 8$
$$\frac{17 - (-9) \mid 8}{17 + 9 \mid 8}$$
$26 \neq 8$
No, −9 is not a solution of the equation $17 - x = 8$.

16.

−3 is 2 units to the right of −5.

17. Strategy   To find the difference in scores, subtract Tseng's score (−19) from Leon's score (2).

Solution $2 - (-19) = 2 + 19$
$= 21$
The difference in scores was 21 strokes.

18. $\dfrac{0}{-16} = 0$

19. $2bc - (c - a)^3$
$(2 \cdot 4 \cdot (-1)) - (-1 + (-2))^3 = -8 - (-3)^3$
$= -8 - (-27)$
$= -8 + 27$
$= 19$

20. -25

21. $c - 11 = 5$
$c - 11 + 11 = 5 + 11$
$c = 16$
The solution is 16.

22. $0 - 11 = -11$

23. $-96 \div (-4) = 24$

24. $16 \div 4 - 12 \div (-2) = 4 - (-6)$
$= 4 + 6$
$= 10$

25. $\dfrac{-x}{y}$

$\dfrac{-(-56)}{-8} = \dfrac{56}{-8}$
$= -7$

26. $3xy$
$3 \cdot (-2) \cdot (-10) = 3 \cdot 20$
$= 60$

27. $-11w = 121$
$\dfrac{-11w}{-11} = \dfrac{121}{-11}$
$w = -11$
The solution is -11.

28. $4 - 14 = -10$

29. Strategy   To find the temperature, add the increase (11) to the previous temperature (-6).

Solution $-6 + 11 = 5$
The temperature is 5°C.

30. The unknown number: n

the wind chill at -25°F with a 40 mph wind	is	Four times the wind chill factor at -25°F with a 40 mph wind

$n = 4 \cdot (-16)$
$n = -64$
The wind chill factor is -64°F.

31. Strategy   To find the temperature from yesterday, add the increase (8) to today's temperature (-13).

Solution $-13 + 8 = -5$
The temperature is -5°C.

32. $d = |a - b|$
$d = |4 - (-12)| = |4 + 12|$
$= |16|$
$= 16$
The solution is 16 units.

33. Strategy   To find the assets, substitute 18 for N and 6 for L in the given formula and solve for A.

Solution $N = A - L$
$18 = A - 6$
$18 + 6 = A - 6 + 6$
$24 = A$
The assets are worth $24 million.

Cumulative Review Exercises

1. $-27 - (-32) = -27 + 32$
$= 5$

2. $439 \rightarrow 400$
$28 \rightarrow 30$
$400 \cdot 30 = 12{,}000$

3.
$$
\begin{array}{r}
3{,}209 \\
6\overline{)\,19{,}254} \\
\underline{-18} \phantom{{,}254} \\
12 \phantom{{,}54} \\
\underline{-12} \phantom{{,}54} \\
5 \\
\underline{-0} \\
54 \\
\underline{-54} \\
0
\end{array}
$$

4. $16 \div (3 + 5) \cdot 9 - 2^4 = 16 \div 8 \cdot 9 - 2^4$
$= 16 \div 8 \cdot 9 - 16$
$= 2 \cdot 9 - 16$
$= 18 - 16$
$= 18 + (-16)$
$= 2$

5. $-|-82| = -82$

6. $309{,}480$

7. $5xy$
$5 \cdot 80 \cdot 6 = 400 \cdot 6$
$= 2{,}400$

8. $-294 \div (-14) = 21$

9. $-28 - (-17) = -28 + 17$
$= -11$

10. $-24 + 16 + (-32) = -8 + (-32)$
$= -40$

11. $44 \div 1 = 44$
$44 \div 2 = 22$
$44 \div 4 = 11$
$44 \div 11 = 4$
The factors of 44 are 1, 2, 4, 11, 22, and 44.

12. $x^4 y^2$
$2^4 \cdot 11^2 = (2 \cdot 2 \cdot 2 \cdot 2) \cdot (11 \cdot 11)$
$= 16 \cdot 121$
$= 1,936$

13. $629{,}874$ *Given place value*
$8 > 5$

629,874 rounded to the nearest thousand is
630,000.

14. $356 \rightarrow \quad 400$
$481 \rightarrow \quad 500$
$294 \rightarrow \quad 300$
$117 \rightarrow \underline{+100}$
$\qquad\qquad 1{,}300$

15. $-a - b$
$-(-4) - (-5) = 4 + 5$
$= 9$

16. $-100 \cdot 25 = -2,500$

17. $\begin{array}{r} 23 \\ 3\overline{)69} \end{array}$

$69 = 3 \cdot 23$

18. $3x = -48$
$\dfrac{3x}{3} = \dfrac{-48}{3}$
$x = -16$
The solution is -16.

19. $(1 - 5)^2 \div (-6 + 4) + 8(-3)$

$= (-4)^2 \div (-2) + 8(-3) = 16 \div (-2) + 8(-3)$
$= -8 + 8(-3)$
$= -8 + (-24)$
$= -32$

20. $-c \div d$
$-(-32) \div (-8) = 32 \div (-8)$
$= -4$

21. $\dfrac{a}{b}$

$\dfrac{39}{-13} = -3$

22. $-62 < 26$

23. $-18(-7) = 126$

24. $12 + p = 3$
$12 - 12 + p = 3 - 12$
$p = -9$
The solution is -9.

25. $2 \cdot 2 \cdot 2 \cdot 2 \cdot 2 \cdot 7 \cdot 7 = 2^5 \cdot 7^2$

26. $4a + (a - b)^3$
$4(5) + (5 - 2)^3 = 4(5) + 3^3$
$= 4(5) + 27$
$= 20 + 27$
$= 47$

27. $\begin{array}{r} 5{,}971 \\ 482 \\ +\,3{,}609 \\ \hline 10{,}062 \end{array}$

28. $-21 - 5 = -21 + (-5)$
$= -26$

29. $\begin{array}{r} 7{,}352 \rightarrow \quad 7{,}000 \\ 1{,}986 \rightarrow \underline{-\,2{,}000} \\ 5{,}000 \end{array}$

30. $3^4 \cdot 5^2 = (3 \cdot 3 \cdot 3 \cdot 3) \cdot (5 \cdot 5)$
$= 81 \cdot 25$
$= 2,025$

31. Strategy   To find the land area, add the land area prior to the purchase (891,364) to the amount of land purchased (831,321).

Solution $\begin{array}{r} 891{,}364 \\ +\,831{,}321 \\ \hline 1{,}722{,}685 \end{array}$

The land area of the United States after the Louisiana purchase was 1,722,685 mi^2.

32. Strategy   To find the age, subtract the year of the birth (1879) from the year of his death (1955).

Solution $\begin{array}{r} 1955 \\ -\,1879 \\ \hline 76 \end{array}$

Albert Einstein was 76 years old when he died.

33. Strategy   To find the amount, subtract the down payment (7,850) from the cost (35,500).

Solution
$$
\begin{array}{r}
35,500 \\
-\ 7,850 \\
\hline
27,650
\end{array}
$$
The amount to be paid is $27,650.

34. Strategy   To find the cost of the land, multiply the number of acres (25) times the cost per acre (11,270).

Solution
$$
\begin{array}{r}
11,270 \\
\times\ 25 \\
\hline
56\ 350 \\
225\ 40\ \\
\hline
281,750
\end{array}
$$
The cost of the land is $281,750.

35. Strategy   To find the temperature, add the increase (7) to the original temperature (−12).

Solution $-12 + 7 = -5$
The temperature is −5°C.

36. Strategy
a. To find the difference in Arizona, subtract the record low (−40) from the record high (128).

b. To find the greatest difference, subtract the record lows from the record highs for each state.

Solution
a. $128 - (-40) = 128 + 40$
$= 168$
The difference in temperatures is 168°F.

b. $112 - (-27) = 112 + 27 = 139$
$100 - (-80) = 100 + 80 = 180$
$128 - (-40) = 128 + 40 = 168$
$120 - (-29) = 120 + 29 = 149$
$180 > 168 > 149 > 139$
The greatest difference in temperatures is in Alaska.

37. Strategy   To find the amount:
→Add the sales figures for the first three quarters $(28,550 + 34,850 + 31,700)$.
→Subtract the sum from the goal for the year (120,000).

Solution
$$
\begin{array}{r}
28,550 \\
34,850 \\
+\ 31,700 \\
\hline
95,100
\end{array}
$$
$$
\begin{array}{r}
120,000 \\
-\ 95,100 \\
\hline
24,900
\end{array}
$$
You must sell $24,900 in the last quarter to meet the goal.

38. Strategy   To find the score, substitute 198 for N and 206 for P in the given formula and solve for S.

Solution $S = N - P$
$S = 198 - 206$
$S = -8$
The golfer's score is −8.

Chapter 3: Fractions

Prep Test

1. $4 \times 5 = 20$

2. $2 \cdot 2 \cdot 2 \cdot 3 \cdot 5 = 120$

3. $9 \times 1 = 9$

4. $-6 + 4 = -2$

5. $-10 - 3 = -13$

6. $63 \div 30 = 2 \text{ r } 3$

7. $24 \div 8 = 3$
   $24 \div 12 = 2$
   Both 8 and 12 divide evenly into 24.

8. $16 \div 4 = 4$
   $20 \div 4 = 5$
   4 divides into both 16 and 20.

9. $3 + 8 \times 7 = 3 + 56 = 59$

10. $8 = ? + 1$
    $8 = 7 + 1$

11. $44 < 48$

Section 3.1

Concept Check

1. **a.** 6, 12, 18, 24, 30, 36

   b. 9, 18, 27, 36, 45

   c. 18

3. **a.** 2, 4, 8

   b. 2, 4, 7, 14, 28

   c. 4

Objective A Exercises

5. $4 = 2^2$
   $8 = 2^3$
   The LCM $= 2^3 = 8$.

7. $2 = 2$
   $7 = 7$
   The LCM $= 2 \cdot 7 = 14$.

9. $6 = 2 \cdot 3$
   $10 = 2 \cdot 5$
   The LCM $= 2 \cdot 3 \cdot 5 = 30$.

11. $9 = 3^2$
    $15 = 3 \cdot 5$
    The LCM $= 3^2 \cdot 5 = 45$.

13. $12 = 2^2 \cdot 3$
    $16 = 2^4$
    The LCM $= 2^4 \cdot 3 = 48$.

15. $4 = 4^2$
    $10 = 2 \cdot 5$
    The LCM $= 2^2 \cdot 5 = 20$.

17. $14 = 2 \cdot 7$
    $42 = 2 \cdot 3 \cdot 7$
    The LCM $= 2 \cdot 3 \cdot 7 = 42$.

19. $24 = 2^3 \cdot 3$
    $36 = 2^2 \cdot 3^2$
    The LCM $= 2^3 \cdot 3^2 = 72$.

21. $30 = 2 \cdot 3 \cdot 5$
    $40 = 2^3 \cdot 5$
    The LCM $= 2^3 \cdot 3 \cdot 5 = 120$.

23. $3 = 3$
    $5 = 5$
    $10 = 2 \cdot 5$
    The LCM $= 2 \cdot 3 \cdot 5 = 30$.

25. $4 = 2^2$
    $8 = 2^3$
    $12 = 2^2 \cdot 3$
    The LCM $= 2^3 \cdot 3 = 24$.

27. $9 = 3^2$
    $36 = 2^2 \cdot 3^2$
    $45 = 5 \cdot 3^2$
    The LCM $= 2^2 \cdot 3^2 \cdot 5 = 180$.

29. $6 = 2 \cdot 3$
    $9 = 3^2$
    $15 = 3 \cdot 5$
    The LCM $= 2 \cdot 3^2 \cdot 5 = 90$.

31. $13 = 13$
    $26 = 2 \cdot 13$
    $39 = 3 \cdot 13$
    The LCM $= 2 \cdot 3 \cdot 13 = 78$.

33. True

Objective B Exercises

35. $9 = 3^2$
$12 = 2^2 \cdot 3$
The GCF = 3.

37. $18 = 2 \cdot 3^2$
$30 = 2 \cdot 3 \cdot 5$
The GCF = $2 \cdot 3 = 6$.

39. $14 = 2 \cdot 7$
$42 = 2 \cdot 3 \cdot 7$
The GCF = $2 \cdot 7 = 14$.

41. $16 = 2^4$
$80 = 2^4 \cdot 5$
The GCF = $2^4 = 16$.

43. $21 = 3 \cdot 7$
$55 = 5 \cdot 11$
The GCF = 1.

45. $8 = 2^3$
$36 = 2^2 \cdot 3^2$
The GCF = $2^2 = 4$.

47. $12 = 2^2 \cdot 3$
$76 = 2^2 \cdot 19$
The GCF = $2^2 = 4$.

49. $24 = 2^3 \cdot 3$
$30 = 2 \cdot 3 \cdot 5$
The GCF = $2 \cdot 3 = 6$.

51. $24 = 2^3 \cdot 3$
$36 = 2^2 \cdot 3^2$
The GCF = $2^2 \cdot 3 = 12$.

53. $45 = 3^2 \cdot 5$
$75 = 3 \cdot 5^2$
The GCF = $3 \cdot 5 = 15$.

55. $6 = 2 \cdot 3$
$10 = 2 \cdot 5$
$12 = 2^2 \cdot 3$
The GCF = 2.

57. $6 = 2 \cdot 3$
$15 = 3 \cdot 5$
$36 = 2^2 \cdot 3^2$
The GCF = 3.

59. $21 = 3 \cdot 7$
$63 = 3^2 \cdot 7$
$84 = 2^2 \cdot 3 \cdot 7$
The GCF = $3 \cdot 7 = 21$.

61. $24 = 2^3 \cdot 3$
$36 = 2^2 \cdot 3^2$
$60 = 2^2 \cdot 3 \cdot 5$
The GCF = $2^2 \cdot 3 = 12$.

63. a. For example, 6 and 8. The GCF is 2 and the LCM is 24.

    b. i

Objective C Exercises

65. LCM

67. GCF

69. Strategy   To find how often the machines are starting to fill a box at the same time, find the LCM of 2 and 3.

       Solution $2 = 2$
$3 = 3$
The LCM = $2 \cdot 3 = 6$.
Every 6 min, the machines will start to fill a box at the same time.

71. Strategy   To find the number of copies to be packaged together, find the GCF of 75, 100, and 150.

       Solution $75 = 3 \cdot 5^2$
$100 = 2^2 \cdot 5^2$
$150 = 2 \cdot 3 \cdot 5^2$
The GCF = $5^2 = 25$.
Each package should contain 25 copies of the magazine.

73. Strategy   To find the time when all
sessions begin again at the
same time.
→Add the time of the break
(10 min) to the 30-minute
and the 40-minute session
to find the time for each
session including the
break.
→Find the LCM of the two
sessions found in step 1.
→Add the time found in
step 2 to 9:00 A.M.

Solution $30 + 10 = 40$
$40 + 10 = 50$
$40 = 2^3 \cdot 5$
$50 = 2 \cdot 5^2$
The LCM $= 2^3 \cdot 5^2 = 200$.
200 min = 3 h 20 min
9 A.M. + 3 h 20 min
= 12:20 P.M.
All sessions will start at the
same time at 12:20 P.M.
Lunch should be scheduled at
12:20 P.M. if all participants
are to eat at the same time.

Critical Thinking 3.1

75. Students might rephrase the definition of
least common multiple so as to involve
division. For example, the least common
multiple of two numbers is the smallest
number that is evenly divisible by both of
the numbers; or the greatest common factor
of the two numbers is the largest number by
which both numbers are divisible.

Section 3.2

Concept Check

1. a. 9; 4

   b. improper

   c. divided by; 9; 4

3. 4; $\dfrac{5 \cdot 4}{6 \cdot 4} = \dfrac{20}{24}$

5. a. denominator; 10; 6; 30

   b. 9; 5; >

Objective A Exercises

7. $\dfrac{4}{5}$

9. $\dfrac{1}{4}$

11. $\dfrac{4}{3}; 1\dfrac{1}{3}$

13. $\dfrac{13}{5}; 2\dfrac{3}{5}$

15. $\begin{array}{r} 3 \\ 4\overline{)13} \\ \underline{-12} \\ 1 \end{array}$ $\dfrac{13}{4} = 3\dfrac{1}{4}$

17. $\begin{array}{r} 4 \\ 5\overline{)20} \\ \underline{-20} \\ 0 \end{array}$ $\dfrac{20}{5} = 4$

19. $\begin{array}{r} 2 \\ 10\overline{)27} \\ \underline{-20} \\ 7 \end{array}$ $\dfrac{27}{10} = 2\dfrac{7}{10}$

21. $\begin{array}{r} 7 \\ 8\overline{)56} \\ \underline{-56} \\ 0 \end{array}$ $\dfrac{56}{8} = 7$

23. $\begin{array}{r} 1 \\ 9\overline{)17} \\ \underline{-9} \\ 8 \end{array}$ $\dfrac{17}{9} = 1\dfrac{8}{9}$

25. $\begin{array}{r} 2 \\ 5\overline{)12} \\ \underline{-10} \\ 2 \end{array}$ $\dfrac{12}{5} = 2\dfrac{2}{5}$

27. $\begin{array}{r} 18 \\ 1\overline{)18} \\ \underline{-1} \\ 8 \\ \underline{-8} \\ 0 \end{array}$ $\dfrac{18}{1} = 18$

29. $\begin{array}{r} 2 \\ 15\overline{)32} \\ \underline{-30} \\ 2 \end{array}$ $\dfrac{32}{15} = 2\dfrac{2}{15}$

31. $\dfrac{8}{8} = 1$

33. $3\overline{)\,28\,}$ 9
$\underline{-27}$
1 $\dfrac{28}{3} = 9\dfrac{1}{3}$

35. $2\dfrac{1}{4} = \dfrac{(4\cdot 2)+1}{4} = \dfrac{8+1}{4} = \dfrac{9}{4}$

37. $5\dfrac{1}{2} = \dfrac{(2\cdot 5)+1}{2} = \dfrac{10+1}{2} = \dfrac{11}{2}$

39. $2\dfrac{4}{5} = \dfrac{(5\cdot 2)+4}{5} = \dfrac{10+4}{5} = \dfrac{14}{5}$

41. $7\dfrac{5}{6} = \dfrac{(6\cdot 7)+5}{6} = \dfrac{42+5}{6} = \dfrac{47}{6}$

43. $7 = \dfrac{7}{1}$

45. $8\dfrac{1}{4} = \dfrac{(4\cdot 8)+1}{4} = \dfrac{32+1}{4} = \dfrac{33}{4}$

47. $10\dfrac{1}{3} = \dfrac{(3\cdot 10)+1}{3} = \dfrac{30+1}{3} = \dfrac{31}{3}$

49. $4\dfrac{7}{12} = \dfrac{(12\cdot 4)+7}{12} = \dfrac{48+7}{12} = \dfrac{55}{12}$

51. $8 = \dfrac{8}{1}$

53. $12\dfrac{4}{5} = \dfrac{(5\cdot 12)+4}{5} = \dfrac{60+4}{5} = \dfrac{64}{5}$

55. $a > b$

Objective B Exercises

57. $12 \div 2 = 6$
$\dfrac{1}{2} = \dfrac{1\cdot 6}{2\cdot 6} = \dfrac{6}{12}$
$\dfrac{6}{12}$ is equivalent to $\dfrac{1}{2}$.

59. $24 \div 8 = 3$
$\dfrac{3}{8} = \dfrac{3\cdot 3}{8\cdot 3} = \dfrac{9}{24}$
$\dfrac{9}{24}$ is equivalent to $\dfrac{3}{8}$.

61. $51 \div 17 = 3$
$\dfrac{2}{17} = \dfrac{2\cdot 3}{17\cdot 3} = \dfrac{6}{51}$
$\dfrac{6}{51}$ is equivalent to $\dfrac{2}{17}$.

63. $32 \div 4 = 8$
$\dfrac{3}{4} = \dfrac{3\cdot 8}{4\cdot 8} = \dfrac{24}{32}$
$\dfrac{24}{32}$ is equivalent to $\dfrac{3}{4}$.

65. $18 \div 1 = 18$
$6 = \dfrac{6}{1} = \dfrac{6\cdot 18}{1\cdot 18} = \dfrac{108}{18}$
$\dfrac{108}{18}$ is equivalent to 6.

67. $90 \div 3 = 30$
$\dfrac{1}{3} = \dfrac{1\cdot 30}{3\cdot 30} = \dfrac{30}{90}$
$\dfrac{30}{90}$ is equivalent to $\dfrac{1}{3}$.

69. $21 \div 3 = 7$
$\dfrac{2}{3} = \dfrac{2\cdot 7}{3\cdot 7} = \dfrac{14}{21}$
$\dfrac{14}{21}$ is equivalent to $\dfrac{2}{3}$.

71. $49 \div 7 = 7$
$\dfrac{6}{7} = \dfrac{6\cdot 7}{7\cdot 7} = \dfrac{42}{49}$
$\dfrac{42}{49}$ is equivalent to $\dfrac{6}{7}$.

73. $18 \div 9 = 2$
$\dfrac{4}{9} = \dfrac{4\cdot 2}{9\cdot 2} = \dfrac{8}{18}$
$\dfrac{8}{18}$ is equivalent to $\dfrac{4}{9}$.

75. $4 \div 1 = 4$
$7 = \dfrac{7}{1} = \dfrac{7\cdot 4}{1\cdot 4} = \dfrac{28}{4}$
$\dfrac{28}{4}$ is equivalent to 7.

77. $\dfrac{3}{12} = \dfrac{3}{2\cdot 2\cdot 3} = \dfrac{1}{4}$

79. $\dfrac{33}{44} = \dfrac{3\cdot 11}{2\cdot 2\cdot 11} = \dfrac{3}{4}$

81. $\dfrac{4}{24} = \dfrac{2\cdot 2}{2\cdot 2\cdot 2\cdot 3} = \dfrac{1}{6}$

83. $\dfrac{8}{33} = \dfrac{2\cdot 2\cdot 2}{3\cdot 11} = \dfrac{8}{33}$

85. $\dfrac{0}{8} = 0$

87. $\dfrac{42}{36} = \dfrac{2 \cdot 3 \cdot 7}{2 \cdot 2 \cdot 3 \cdot 3} = \dfrac{7}{6}$

89. $\dfrac{16}{16} = 1$

91. $\dfrac{21}{35} = \dfrac{3 \cdot 7}{5 \cdot 7} = \dfrac{3}{5}$

93. $\dfrac{16}{60} = \dfrac{2 \cdot 2 \cdot 2 \cdot 2}{2 \cdot 2 \cdot 3 \cdot 5} = \dfrac{4}{15}$

95. $\dfrac{12}{20} = \dfrac{2 \cdot 2 \cdot 3}{2 \cdot 2 \cdot 5} = \dfrac{3}{5}$

97. $\dfrac{12m}{18} = \dfrac{\overset{1}{2} \cdot 2 \cdot 3 \cdot m}{\underset{1}{2 \cdot 3 \cdot 3}} = \dfrac{2m}{3}$

99. $\dfrac{4y}{8} = \dfrac{\overset{1}{2} \cdot \overset{1}{2} \cdot y}{\underset{1}{2 \cdot 2} \cdot 2} = \dfrac{y}{2}$

101. $\dfrac{24a}{36} = \dfrac{\overset{1}{2} \cdot \overset{1}{2} \cdot 2 \overset{1}{3} \cdot a}{\underset{1}{2 \cdot 2} \cdot \underset{1}{3} \cdot 3} = \dfrac{2a}{3}$

103. $\dfrac{8c}{8} = \dfrac{\overset{1}{2} \cdot \overset{1}{2} \cdot \overset{1}{2} \cdot c}{\underset{1}{2 \cdot 2} \cdot \underset{1}{2}} = c$

105. $\dfrac{18k}{3} = \dfrac{2 \cdot 3 \cdot \overset{1}{3} \cdot k}{\underset{1}{3}} = 6k$

107. ii

109. iii; One example is $\dfrac{5}{6}$ and $\dfrac{15}{18}$.

Objective C Exercises

111. $\dfrac{5}{7} = \dfrac{15}{21}$ $\dfrac{2}{3} = \dfrac{14}{21}$

$\dfrac{15}{21} > \dfrac{14}{21}$

$\dfrac{5}{7} > \dfrac{2}{3}$

113. $\dfrac{7}{12} = \dfrac{14}{24}$ $\dfrac{5}{8} = \dfrac{15}{24}$

$\dfrac{14}{24} < \dfrac{15}{24}$

$\dfrac{7}{12} < \dfrac{5}{8}$

115. $\dfrac{11}{14} = \dfrac{22}{28}$ $\dfrac{3}{4} = \dfrac{21}{28}$

$\dfrac{22}{28} > \dfrac{21}{28}$

$\dfrac{11}{14} > \dfrac{3}{4}$

117. $\dfrac{11}{12} = \dfrac{33}{36}$ $\dfrac{7}{9} = \dfrac{28}{36}$

$\dfrac{33}{36} > \dfrac{28}{36}$

$\dfrac{11}{12} > \dfrac{7}{9}$

119. $\dfrac{5}{8} = \dfrac{35}{56}$ $\dfrac{4}{7} = \dfrac{32}{56}$

$\dfrac{35}{56} > \dfrac{32}{56}$

$\dfrac{5}{8} > \dfrac{4}{7}$

121. $\dfrac{11}{30} = \dfrac{44}{120}$ $\dfrac{7}{24} = \dfrac{35}{120}$

$\dfrac{44}{120} > \dfrac{35}{120}$

$\dfrac{11}{30} > \dfrac{7}{24}$

123. $\dfrac{9}{11} = \dfrac{72}{88}$ $\dfrac{7}{8} = \dfrac{77}{88}$

$\dfrac{72}{88} < \dfrac{77}{88}$

$\dfrac{9}{11} < \dfrac{7}{8}$

125. $\dfrac{3}{4} = \dfrac{39}{52}$ $\dfrac{11}{13} = \dfrac{44}{52}$

$\dfrac{39}{52} < \dfrac{44}{52}$

$\dfrac{3}{4} < \dfrac{11}{13}$

127. $\dfrac{2}{3} = \dfrac{20}{30}$ $\dfrac{7}{10} = \dfrac{21}{30}$

$\dfrac{20}{30} < \dfrac{21}{30}$

$\dfrac{2}{3} < \dfrac{7}{10}$

129. $\dfrac{3}{10} = \dfrac{15}{50}$ $\dfrac{7}{25} = \dfrac{14}{50}$

$\dfrac{15}{50} > \dfrac{14}{50}$

$\dfrac{3}{10} > \dfrac{7}{25}$

131. One example is $\dfrac{4}{7}$ and $\dfrac{3}{5}$.

Objective D Exercises

133. Strategy   To find the fractions, write a fraction with 6 in the numerator and 16 in the denominator. Write the fraction in simplest form.

Solution $\dfrac{6}{16} = \dfrac{2 \cdot 3}{2 \cdot 2 \cdot 2 \cdot 2} = \dfrac{3}{8}$

6 oz is $\dfrac{3}{8}$ of a pound.

135. Strategy   To find the fraction, write a fraction with 8 in the numerator and the number of hours in one day (24) in the denominator. Write the fraction in simplest form.

Solution $\dfrac{8}{24} = \dfrac{2 \cdot 2 \cdot 2}{2 \cdot 2 \cdot 2 \cdot 3} = \dfrac{1}{3}$

8 h is $\dfrac{1}{3}$ of a day.

137. Strategy   To find which service is more important,
→Rewrite the food quality fraction $\dfrac{1}{4}$ as a fraction with a denominator of 100.
→Rewrite location fraction $\dfrac{13}{50}$ as a fraction with a denominator of 100.
→Compare the two fractions.

Solution $\dfrac{1}{4} = \dfrac{1 \cdot 25}{4 \cdot 25} = \dfrac{25}{100}$

$\dfrac{13}{50} = \dfrac{13 \cdot 2}{50 \cdot 2} = \dfrac{26}{100}$

$\dfrac{26}{100} > \dfrac{25}{100}$

The location was more important than the food quality.

139. Strategy   To find the fraction, write a fraction with the number of respondents who said that avoiding credit card debt was the best financial advice for graduates (600) in the numerator and the total number of respondents (1200) in the denominator. Write the fraction in simplest form.

Solution $\dfrac{600}{1200} = \dfrac{2^3 \cdot 3 \cdot 5^2}{2^4 \cdot 3 \cdot 5^2} = \dfrac{1}{2}$

The fraction of respondents who said that avoiding credit card debt was the best financial advice for graduates is $\dfrac{1}{2}$.

141. Strategy   To determine if you answered more or less than $\dfrac{8}{10}$ of the questions correctly:
→Write a fraction with the number of questions answered correctly (42) in the numerator and the total number of questions (50) in the denominator.
→Rewrite $\dfrac{8}{10}$ as a fraction with a denominator of 50.
→Compare the two fractions.

Solution $\dfrac{42}{50}$

$\dfrac{8}{10} = \dfrac{8 \cdot 5}{10 \cdot 5} = \dfrac{40}{50}$

$\dfrac{42}{50} > \dfrac{40}{50}$

You answered more than $\dfrac{8}{10}$ of the questions correctly.

143. Strategy   To determine the fraction:

→Find the number of attempts he did not have a field goal by subtracting the number of field goals (36) from the number of attempts (63).

→Write a fraction with the number of attempts he did not have a field goal in the numerator and the number of attempts in the denominator. Write the fraction in simplest form.

Solution $63 - 36 = 27$

$$\frac{27}{63} = \frac{3 \cdot 3 \cdot 3}{3 \cdot 3 \cdot 7} = \frac{3}{7}$$

$\frac{3}{7}$ of his attempts did not result in a field goal.

Critical Thinking 3.2

145. a. The midpoint or halfway point between $\frac{2}{a}$ and $\frac{4}{a}$ is one-half the sum of $\frac{2}{a} + \frac{4}{a}$.

$$\frac{1}{2}\left(\frac{2}{a} + \frac{4}{a}\right) = \frac{1}{2}\left(\frac{6}{a}\right) = \frac{3}{a}$$

The diagram below also indicates that the midpoint is $\frac{3}{a}$.

b. The diagram below shows p and q evenly spaced between A and B. The point p is $\frac{1}{3}$ of the distance between A and B.

$$\frac{1}{3}\left(\frac{8}{b} - \frac{5}{b}\right) = \frac{1}{3}\left(\frac{3}{b}\right) = \frac{1}{b}$$

Add $\frac{1}{b}$ to $\frac{5}{b}$ to find p. $\frac{1}{b} + \frac{5}{b} = \frac{6}{b}$

Add $\frac{1}{b}$ to $\frac{6}{b}$ to find q. $\frac{1}{b} + \frac{6}{b} = \frac{7}{b}$

Section 3.3

Concept Check

1. a. can

    b. product; product

3. $-\dfrac{6}{5}$

5. $246;\ 10\dfrac{1}{4}$

Objective A Exercises

7. $\dfrac{2}{3} \cdot \dfrac{9}{10} = \dfrac{2 \cdot 9}{3 \cdot 10}$

$\quad = \dfrac{2 \cdot 3 \cdot 3}{3 \cdot 2 \cdot 5}$

$\quad = \dfrac{3}{5}$

9. $-\dfrac{6}{7} \cdot \dfrac{11}{12} = -\dfrac{6 \cdot 11}{7 \cdot 12}$

$\quad = -\dfrac{2 \cdot 3 \cdot 11}{7 \cdot 2 \cdot 2 \cdot 3}$

$\quad = -\dfrac{11}{14}$

11. $\dfrac{14}{15} \cdot \dfrac{6}{7} = \dfrac{14 \cdot 6}{15 \cdot 7}$

$\quad = \dfrac{2 \cdot 7 \cdot 2 \cdot 3}{3 \cdot 5 \cdot 7}$

$\quad = \dfrac{4}{5}$

13. $-\dfrac{6}{7} \cdot \dfrac{0}{10} = -\left(\dfrac{6}{7} \cdot \dfrac{0}{10}\right)$

$\quad = -\dfrac{6 \cdot 0}{7 \cdot 10}$

$\quad = -\dfrac{0}{70} = 0$

15. $\left(-\dfrac{4}{15}\right) \cdot \left(-\dfrac{3}{8}\right) = \dfrac{4}{15} \cdot \dfrac{3}{8}$

$\quad = \dfrac{4 \cdot 3}{15 \cdot 8}$

$\quad = \dfrac{2 \cdot 2 \cdot 3}{3 \cdot 5 \cdot 2 \cdot 2 \cdot 2}$

$\quad = \dfrac{1}{10}$

17. $-\dfrac{3}{4} \cdot \dfrac{1}{2} = -\left(\dfrac{3}{4} \cdot \dfrac{1}{2}\right)$

$\quad = -\dfrac{3 \cdot 1}{4 \cdot 2}$

$\quad = -\dfrac{3 \cdot 1}{2 \cdot 2 \cdot 2}$

$\quad = -\dfrac{3}{8}$

19. $\dfrac{9}{x} \cdot \dfrac{7}{y} = \dfrac{9 \cdot 7}{x \cdot y}$

$\quad = \dfrac{63}{xy}$

21. $-\dfrac{y}{5} \cdot \dfrac{z}{6} = -\left(\dfrac{y}{5} \cdot \dfrac{z}{6}\right)$

$\quad = -\dfrac{y \cdot z}{5 \cdot 6}$

$\quad = -\dfrac{yz}{30}$

23. $\dfrac{2}{3} \cdot \dfrac{3}{8} \cdot \dfrac{4}{9} = \dfrac{2 \cdot 3 \cdot 4}{3 \cdot 8 \cdot 9}$

$\quad = \dfrac{2 \cdot 3 \cdot 2 \cdot 2}{3 \cdot 2 \cdot 2 \cdot 2 \cdot 3 \cdot 3}$

$\quad = \dfrac{1}{9}$

25. $-\dfrac{7}{12} \cdot \dfrac{5}{8} \cdot \dfrac{16}{25} = -\left(\dfrac{7}{12} \cdot \dfrac{5}{8} \cdot \dfrac{16}{25}\right)$

$\quad = -\dfrac{7 \cdot 5 \cdot 16}{12 \cdot 8 \cdot 25}$

$\quad = -\dfrac{7 \cdot 5 \cdot 2 \cdot 2 \cdot 2 \cdot 2}{2 \cdot 2 \cdot 3 \cdot 2 \cdot 2 \cdot 2 \cdot 5 \cdot 5}$

$\quad = -\dfrac{7}{30}$

27. $\left(-\dfrac{3}{5}\right) \cdot \dfrac{1}{2} \cdot \left(-\dfrac{5}{8}\right) = \dfrac{3}{5} \cdot \dfrac{1}{2} \cdot \dfrac{5}{8}$

$\quad = \dfrac{3 \cdot 1 \cdot 5}{5 \cdot 2 \cdot 8}$

$\quad = \dfrac{3 \cdot 1 \cdot 5}{5 \cdot 2 \cdot 2 \cdot 2 \cdot 2}$

$\quad = \dfrac{3}{16}$

29. $6 \cdot \dfrac{1}{6} = \dfrac{6}{1} \cdot \dfrac{1}{6}$

$\quad = \dfrac{6 \cdot 1}{1 \cdot 6}$

$\quad = \dfrac{6}{6} = 1$

31. $\dfrac{3}{4} \cdot 8 = \dfrac{3}{4} \cdot \dfrac{8}{1}$

$= \dfrac{3 \cdot 8}{4 \cdot 1}$

$= \dfrac{3 \cdot 2 \cdot 2 \cdot 2}{2 \cdot 2 \cdot 1} = \dfrac{6}{1} = 6$

33. $12 \cdot \left(-\dfrac{5}{8}\right) = -\left(12 \cdot \dfrac{5}{8}\right)$

$= -\left(\dfrac{12}{1} \cdot \dfrac{5}{8}\right) = -\dfrac{12 \cdot 5}{1 \cdot 8}$

$= -\dfrac{2 \cdot 2 \cdot 3 \cdot 5}{1 \cdot 2 \cdot 2 \cdot 2}$

$= -\dfrac{15}{2} = -7\dfrac{1}{2}$

35. $-16 \cdot \dfrac{7}{30} = -\left(\dfrac{16}{1} \cdot \dfrac{7}{30}\right)$

$= -\dfrac{16 \cdot 7}{1 \cdot 30}$

$= -\dfrac{2 \cdot 2 \cdot 2 \cdot 2 \cdot 7}{1 \cdot 2 \cdot 3 \cdot 5}$

$= -\dfrac{56}{15} = -3\dfrac{11}{15}$

37. $\dfrac{6}{7} \cdot 0 = 0$

39. $\dfrac{5}{22} \cdot 2\dfrac{1}{5} = \dfrac{5}{22} \cdot \dfrac{11}{5}$

$= \dfrac{5 \cdot 11}{22 \cdot 5}$

$= \dfrac{5 \cdot 11}{2 \cdot 11 \cdot 5}$

$= \dfrac{1}{2}$

41. $3\dfrac{1}{2} \cdot 5\dfrac{3}{7} = \dfrac{7}{2} \cdot \dfrac{38}{7}$

$= \dfrac{7 \cdot 38}{2 \cdot 7}$

$= \dfrac{7 \cdot 2 \cdot 19}{2 \cdot 7}$

$= \dfrac{19}{1} = 19$

43. $3\dfrac{1}{3} \cdot \left(-\dfrac{7}{10}\right) = -\left(\dfrac{10}{3} \cdot \dfrac{7}{10}\right)$

$= -\dfrac{10 \cdot 7}{3 \cdot 10}$

$= -\dfrac{2 \cdot 5 \cdot 7}{3 \cdot 2 \cdot 5}$

$= -\dfrac{7}{3} = -2\dfrac{1}{3}$

45. $-1\dfrac{2}{3} \cdot \left(-\dfrac{3}{5}\right) = \dfrac{5}{3} \cdot \dfrac{3}{5}$

$= \dfrac{5 \cdot 3}{3 \cdot 5}$

$= 1$

47. $3\dfrac{1}{3} \cdot 2\dfrac{1}{3} = \dfrac{10}{3} \cdot \dfrac{7}{3}$

$= \dfrac{10 \cdot 7}{3 \cdot 3}$

$= \dfrac{2 \cdot 5 \cdot 7}{3 \cdot 3}$

$= \dfrac{70}{9} = 7\dfrac{7}{9}$

49. $3\dfrac{1}{3} \cdot (-9) = -\left(\dfrac{10}{3} \cdot \dfrac{9}{1}\right)$

$= -\dfrac{10 \cdot 9}{3 \cdot 1}$

$= -\dfrac{2 \cdot 5 \cdot 3 \cdot 3}{3 \cdot 1}$

$= -\dfrac{30}{1} = -30$

51. $8 \cdot 5\dfrac{1}{4} = \dfrac{8}{1} \cdot \dfrac{21}{4}$

$= \dfrac{8 \cdot 21}{1 \cdot 4}$

$= \dfrac{2 \cdot 2 \cdot 2 \cdot 3 \cdot 7}{1 \cdot 2 \cdot 2}$

$= \dfrac{42}{1} = 42$

53. $3\dfrac{1}{2} \cdot 1\dfrac{5}{7} \cdot \dfrac{11}{12} = \dfrac{7}{2} \cdot \dfrac{12}{7} \cdot \dfrac{11}{12}$

$= \dfrac{7 \cdot 12 \cdot 11}{2 \cdot 7 \cdot 12}$

$= \dfrac{7 \cdot 2 \cdot 2 \cdot 3 \cdot 11}{2 \cdot 7 \cdot 2 \cdot 2 \cdot 3}$

$= \dfrac{11}{2} = 5\dfrac{1}{2}$

55. $\dfrac{3}{4} \cdot \dfrac{14}{15} = \dfrac{3 \cdot 14}{4 \cdot 15}$

$= \dfrac{3 \cdot 2 \cdot 7}{2 \cdot 2 \cdot 3 \cdot 5}$

$= \dfrac{7}{10}$

57. $-\dfrac{9}{16} \cdot \dfrac{4}{27} = -\left(\dfrac{9 \cdot 4}{16 \cdot 27}\right)$

$= -\dfrac{3 \cdot 3 \cdot 2 \cdot 2}{2 \cdot 2 \cdot 2 \cdot 2 \cdot 3 \cdot 3 \cdot 3}$

$= -\dfrac{1}{12}$

59. $-\dfrac{7}{24}\cdot\dfrac{8}{21}\cdot\dfrac{3}{7}=-\left(\dfrac{7}{24}\cdot\dfrac{8}{21}\cdot\dfrac{3}{7}\right)$

$=-\dfrac{7\cdot8\cdot3}{24\cdot21\cdot7}$

$=-\dfrac{7\cdot2\cdot2\cdot2\cdot3}{2\cdot2\cdot2\cdot3\cdot3\cdot7\cdot7}$

$=-\dfrac{1}{21}$

61. $4\dfrac{4}{5}\cdot\dfrac{3}{8}=\dfrac{24}{5}\cdot\dfrac{3}{8}$

$=\dfrac{24\cdot3}{5\cdot8}$

$\dfrac{2\cdot2\cdot2\cdot3\cdot3}{5\cdot2\cdot2\cdot2}$

$=\dfrac{9}{5}=1\dfrac{4}{5}$

63. $-2\dfrac{2}{3}\cdot\left(-1\dfrac{11}{16}\right)=\dfrac{8}{3}\cdot\dfrac{27}{16}$

$=\dfrac{8\cdot27}{3\cdot16}$

$=\dfrac{2\cdot2\cdot2\cdot3\cdot3\cdot3}{3\cdot2\cdot2\cdot2\cdot2}$

$=\dfrac{9}{2}=4\dfrac{1}{2}$

65. Strategy   To find the amount spent on housing, multiply the housing fraction $\left(\dfrac{13}{45}\right)$ by the income (45,000).

Solution $45,000\cdot\dfrac{13}{45}=13,000$

The typical household spends $13,000 per year on housing.

67. xy

$-\dfrac{5}{16}\cdot\dfrac{7}{15}=-\left(\dfrac{5\cdot7}{16\cdot15}\right)$

$=-\dfrac{5\cdot7}{2\cdot2\cdot2\cdot2\cdot3\cdot5}$

$=-\dfrac{7}{48}$

69. xy

$\dfrac{4}{7}\cdot6\dfrac{1}{8}=\dfrac{4}{7}\cdot\dfrac{49}{8}$

$=\dfrac{4\cdot49}{7\cdot8}$

$=\dfrac{2\cdot2\cdot7\cdot7}{7\cdot2\cdot2\cdot2}$

$=\dfrac{7}{2}=3\dfrac{1}{2}$

71. xy

$-49\cdot\dfrac{5}{14}=-\left(\dfrac{49}{1}\cdot\dfrac{5}{14}\right)$

$=-\dfrac{49\cdot5}{1\cdot14}$

$=-\dfrac{7\cdot7\cdot5}{1\cdot2\cdot7}$

$=-\dfrac{35}{2}=-17\dfrac{1}{2}$

73. xy

$1\dfrac{3}{13}\cdot\left(-6\dfrac{1}{2}\right)=-\left(\dfrac{16}{13}\cdot\dfrac{13}{2}\right)$

$=-\dfrac{16\cdot13}{13\cdot2}$

$=-\dfrac{2\cdot2\cdot2\cdot2\cdot13}{13\cdot2}$

$=-\dfrac{8}{1}=-8$

75. xyz

$\dfrac{3}{8}\cdot\dfrac{2}{3}\cdot\dfrac{4}{5}=\dfrac{3\cdot2\cdot4}{8\cdot3\cdot5}$

$=\dfrac{3\cdot2\cdot2\cdot2}{2\cdot2\cdot2\cdot3\cdot5}$

$=\dfrac{1}{5}$

77. xyz

$2\dfrac{3}{8}\cdot\left(-\dfrac{3}{19}\right)\cdot\left(-\dfrac{4}{9}\right)=\dfrac{19}{8}\cdot\dfrac{3}{19}\cdot\dfrac{4}{9}$

$=\dfrac{19\cdot3\cdot4}{8\cdot19\cdot9}$

$=\dfrac{19\cdot3\cdot2\cdot2}{2\cdot2\cdot2\cdot19\cdot3\cdot3}$

$=\dfrac{1}{6}$

79. xyz

$\dfrac{5}{6}\cdot(-3)\cdot1\dfrac{7}{15}=-\left(\dfrac{5}{6}\cdot\dfrac{3}{1}\cdot\dfrac{22}{15}\right)$

$=-\dfrac{5\cdot3\cdot22}{6\cdot1\cdot15}$

$=-\dfrac{5\cdot3\cdot2\cdot11}{2\cdot3\cdot1\cdot3\cdot5}$

$=-\dfrac{11}{3}=-3\dfrac{2}{3}$

81. $\dfrac{3}{4}y = -\dfrac{1}{4}$

$$\dfrac{\dfrac{3}{4}\left(-\dfrac{1}{3}\right)}{} \,\bigg|\, -\dfrac{1}{4}$$

$$-\dfrac{3\cdot 1}{4\cdot 3} \,\bigg|\, -\dfrac{1}{4}$$

$$-\dfrac{3\cdot 1}{2\cdot 2\cdot 3} \,\bigg|\, -\dfrac{1}{4}$$

$$-\dfrac{1}{4} = -\dfrac{1}{4}$$

Yes, $-\dfrac{1}{3}$ is a solution of the equation.

83. $\dfrac{4}{5}x = \dfrac{5}{3}$

$$\dfrac{\dfrac{4}{5}\cdot\dfrac{3}{4}}{} \,\bigg|\, \dfrac{5}{3}$$

$$\dfrac{4\cdot 3}{5\cdot 4} \,\bigg|\, \dfrac{5}{3}$$

$$\dfrac{2\cdot 2\cdot 3}{5\cdot 2\cdot 2} \,\bigg|\, \dfrac{5}{3}$$

$$\dfrac{3}{5} \neq \dfrac{5}{3}$$

No, $\dfrac{3}{4}$ is not a solution of the equation.

85. $6x = 1$

$$\dfrac{6\left(-\dfrac{1}{6}\right)}{} \,\bigg|\, 1$$

$$\dfrac{6}{1}\left(-\dfrac{1}{6}\right) \,\bigg|\, 1$$

$$-\dfrac{6\cdot 1}{1\cdot 6} \,\bigg|\, 1$$

$$-\dfrac{2\cdot 3\cdot 1}{1\cdot 2\cdot 3} \,\bigg|\, 1$$

$$-1 \neq 1$$

No, $-\dfrac{1}{6}$ is not a solution of the equation.

87. Less than

Objective B Exercises

89. $\dfrac{5}{7} \div \dfrac{2}{5} = \dfrac{5}{7}\cdot\dfrac{5}{2}$

$$= \dfrac{5\cdot 5}{7\cdot 2}$$

$$= \dfrac{25}{14} = 1\dfrac{11}{14}$$

91. $\dfrac{4}{7} \div \left(-\dfrac{4}{7}\right) = -\left(\dfrac{4}{7} \div \dfrac{4}{7}\right)$

$$= -\left(\dfrac{4}{7}\cdot\dfrac{7}{4}\right)$$

$$= -\dfrac{4\cdot 7}{7\cdot 4}$$

$$= -\dfrac{2\cdot 2\cdot 7}{7\cdot 2\cdot 2} = -\dfrac{1}{1} = -1$$

93. $0 \div \dfrac{7}{9} = 0\cdot\dfrac{9}{7} = 0$

Zero divided by a non-zero number is 0.

95. $\left(-\dfrac{1}{3}\right) \div \dfrac{1}{2} = -\left(\dfrac{1}{3} \div \dfrac{1}{2}\right)$

$$= -\left(\dfrac{1}{3}\cdot\dfrac{2}{1}\right)$$

$$= -\dfrac{1\cdot 2}{3\cdot 1} = -\dfrac{2}{3}$$

97. $-\dfrac{5}{16} \div \left(-\dfrac{3}{8}\right) = \dfrac{5}{16} \div \dfrac{3}{8}$

$$= \dfrac{5}{16}\cdot\dfrac{8}{3}$$

$$= \dfrac{5\cdot 8}{16\cdot 3}$$

$$= \dfrac{5\cdot 2\cdot 2\cdot 2}{2\cdot 2\cdot 2\cdot 2\cdot 3}$$

$$= \dfrac{5}{6}$$

99. $\dfrac{0}{1} \div \dfrac{1}{9} = \dfrac{0}{1}\cdot\dfrac{9}{1}$

$$= \dfrac{0\cdot 9}{1\cdot 1}$$

$$= \dfrac{0}{1} = 0$$

101. $6 \div \dfrac{3}{4} = \dfrac{6}{1}\cdot\dfrac{4}{3}$

$$= \dfrac{6\cdot 4}{1\cdot 3}$$

$$= \dfrac{2\cdot 3\cdot 2\cdot 2}{1\cdot 3}$$

$$= \dfrac{8}{1} = 8$$

103. $\dfrac{3}{4} \div (-6) = -\left(\dfrac{3}{4} \div \dfrac{6}{1}\right)$

$= -\left(\dfrac{3}{4} \cdot \dfrac{1}{6}\right)$

$= -\dfrac{3 \cdot 1}{4 \cdot 6}$

$= -\dfrac{3 \cdot 1}{2 \cdot 2 \cdot 2 \cdot 3}$

$= -\dfrac{1}{8}$

105. $\dfrac{9}{10} \div 0$

Division by zero is undefined.

107. $\dfrac{5}{12} \div \left(-\dfrac{15}{32}\right) = -\left(\dfrac{5}{12} \div \dfrac{15}{32}\right)$

$= -\left(\dfrac{5}{12} \cdot \dfrac{32}{15}\right)$

$= -\dfrac{5 \cdot 32}{12 \cdot 15}$

$= -\dfrac{5 \cdot 2 \cdot 2 \cdot 2 \cdot 2 \cdot 2}{2 \cdot 2 \cdot 3 \cdot 3 \cdot 5}$

$= -\dfrac{8}{9}$

109. $\left(-\dfrac{2}{3}\right) \div (-4) = \dfrac{2}{3} \div \dfrac{4}{1}$

$= \dfrac{2}{3} \cdot \dfrac{1}{4}$

$= \dfrac{2 \cdot 1}{3 \cdot 4}$

$= \dfrac{2 \cdot 1}{3 \cdot 2 \cdot 2}$

$= \dfrac{1}{6}$

111. $\dfrac{8}{x} \div \left(-\dfrac{y}{4}\right) = -\left(\dfrac{8}{x} \div \dfrac{y}{4}\right)$

$= -\left(\dfrac{8}{x} \cdot \dfrac{4}{y}\right)$

$= -\dfrac{8 \cdot 4}{x \cdot y}$

$= -\dfrac{32}{xy}$

113. $\dfrac{b}{6} \div \dfrac{5}{d} = \dfrac{b}{6} \cdot \dfrac{d}{5}$

$= \dfrac{b \cdot d}{6 \cdot 5}$

$= \dfrac{bd}{30}$

115. $3\dfrac{1}{3} \div \dfrac{5}{8} = \dfrac{10}{3} \cdot \dfrac{8}{5} = \dfrac{10 \cdot 8}{3 \cdot 5}$

$= \dfrac{2 \cdot 5 \cdot 2 \cdot 2 \cdot 2}{3 \cdot 5}$

$= \dfrac{16}{3} = 5\dfrac{1}{3}$

117. $5\dfrac{3}{5} \div \left(-\dfrac{7}{10}\right) = -\left(\dfrac{28}{5} \div \dfrac{7}{10}\right)$

$= -\left(\dfrac{28}{5} \cdot \dfrac{10}{7}\right)$

$= -\dfrac{28 \cdot 10}{5 \cdot 7}$

$= -\dfrac{2 \cdot 2 \cdot 7 \cdot 2 \cdot 5}{5 \cdot 7}$

$= -\dfrac{8}{1} = -8$

119. $-1\dfrac{1}{2} \div 1\dfrac{3}{4} = -\left(\dfrac{3}{2} \div \dfrac{7}{4}\right)$

$= -\left(\dfrac{3}{2} \cdot \dfrac{4}{7}\right)$

$= -\dfrac{3 \cdot 4}{2 \cdot 7}$

$= -\dfrac{3 \cdot 2 \cdot 2}{2 \cdot 7}$

$= -\dfrac{6}{7}$

121. $5\dfrac{1}{2} \div 11 = \dfrac{11}{2} \div \dfrac{11}{1}$

$= \dfrac{11}{2} \cdot \dfrac{1}{11}$

$= \dfrac{11 \cdot 1}{2 \cdot 11}$

$= \dfrac{1}{2}$

123. $5\dfrac{2}{7} \div 1 = \dfrac{37}{7} \div \dfrac{1}{1}$

$= \dfrac{37}{7} \cdot \dfrac{1}{1}$

$= \dfrac{37 \cdot 1}{7 \cdot 1}$

$= \dfrac{37}{7} = 5\dfrac{2}{7}$

125. $-16 \div 1\dfrac{1}{3} = -\left(\dfrac{16}{1} \div \dfrac{4}{3}\right)$

$= -\left(\dfrac{16}{1} \cdot \dfrac{3}{4}\right)$

$= -\dfrac{16 \cdot 3}{1 \cdot 4}$

$= -\dfrac{2 \cdot 2 \cdot 2 \cdot 2 \cdot 3}{1 \cdot 2 \cdot 2}$

$= -\dfrac{12}{1} = -12$

127. $2\dfrac{4}{13} \div 1\dfrac{5}{26} = \dfrac{30}{13} \div \dfrac{31}{26}$

$= \dfrac{30}{13} \cdot \dfrac{26}{31}$

$= \dfrac{30 \cdot 26}{13 \cdot 31}$

$= \dfrac{2 \cdot 3 \cdot 5 \cdot 2 \cdot 13}{13 \cdot 31}$

$= \dfrac{60}{31} = 1\dfrac{29}{31}$

129. Strategy To find the number of servings, divide the net weight of Kellogg Honey Crunch Corn Flakes (24) by $\dfrac{3}{4}$.

Solution $24 \div \dfrac{3}{4} = \dfrac{24}{1} \cdot \dfrac{4}{3} = \dfrac{24 \cdot 4}{1 \cdot 3} = 32$

There are 32 servings in a box of Kellogg Honey Crunch Corn Flakes.

131. $\dfrac{9}{10} \div \dfrac{3}{4} = \dfrac{9}{10} \cdot \dfrac{4}{3}$

$= \dfrac{9 \cdot 4}{10 \cdot 3}$

$= \dfrac{3 \cdot 3 \cdot 2 \cdot 2}{2 \cdot 5 \cdot 3}$

$= \dfrac{6}{5} = 1\dfrac{1}{5}$

133. $\left(-\dfrac{15}{24}\right) \div \dfrac{3}{5} = -\left(\dfrac{15}{24} \cdot \dfrac{5}{3}\right)$

$= -\dfrac{15 \cdot 5}{24 \cdot 3}$

$= -\dfrac{3 \cdot 5 \cdot 5}{2 \cdot 2 \cdot 2 \cdot 3 \cdot 3}$

$= -\dfrac{25}{24} = -1\dfrac{1}{24}$

135. $\dfrac{7}{8} \div 3\dfrac{1}{4} = \dfrac{7}{8} \div \dfrac{13}{4}$

$= \dfrac{7}{8} \cdot \dfrac{4}{13}$

$= \dfrac{7 \cdot 4}{8 \cdot 13}$

$= \dfrac{7 \cdot 2 \cdot 2}{2 \cdot 2 \cdot 2 \cdot 13}$

$= \dfrac{7}{26}$

137. $-3\dfrac{5}{11} \div 3\dfrac{4}{5} = -\left(\dfrac{38}{11} \div \dfrac{19}{5}\right)$

$= -\left(\dfrac{38}{11} \cdot \dfrac{5}{19}\right)$

$= -\dfrac{38 \cdot 5}{11 \cdot 19}$

$= -\dfrac{2 \cdot 19 \cdot 5}{11 \cdot 19}$

$= -\dfrac{10}{11}$

139. $x \div y$

$-\dfrac{5}{8} \div \left(-\dfrac{15}{2}\right) = \dfrac{5}{8} \div \dfrac{15}{2}$

$= \dfrac{5}{8} \cdot \dfrac{2}{15}$

$= \dfrac{5 \cdot 2}{8 \cdot 15}$

$= \dfrac{5 \cdot 2}{2 \cdot 2 \cdot 2 \cdot 3 \cdot 5}$

$= \dfrac{1}{12}$

141. $x \div y$

$\dfrac{1}{7} \div 0$

Division by zero is undefined.

143. $x \div y$

$-18 \div \dfrac{3}{8} = -\left(\dfrac{18}{1} \cdot \dfrac{8}{3}\right)$

$= -\dfrac{18 \cdot 8}{1 \cdot 3}$

$= -\dfrac{2 \cdot 3 \cdot 3 \cdot 2 \cdot 2 \cdot 2}{1 \cdot 3}$

$= -\dfrac{48}{1} = -48$

145. $x \div y$

$-\dfrac{1}{2} \div \left(-3\dfrac{5}{8}\right) = \dfrac{1}{2} \div \dfrac{29}{8}$

$= \dfrac{1}{2} \cdot \dfrac{8}{29}$

$= \dfrac{1 \cdot 8}{2 \cdot 29}$

$= \dfrac{1 \cdot 2 \cdot 2 \cdot 2}{2 \cdot 29}$

$= \dfrac{4}{29}$

147. $x \div y$

$$6\frac{2}{5} \div (-4) = -\left(\frac{32}{5} \div \frac{4}{1}\right)$$

$$= -\left(\frac{32}{5} \cdot \frac{1}{4}\right)$$

$$= -\frac{32 \cdot 1}{5 \cdot 4}$$

$$= -\frac{2 \cdot 2 \cdot 2 \cdot 2 \cdot 2 \cdot 1}{5 \cdot 2 \cdot 2}$$

$$= -\frac{8}{5} = -1\frac{3}{5}$$

149. $x \div y$

$$-3\frac{2}{5} \div \left(-1\frac{7}{10}\right) = \frac{17}{5} \div \frac{17}{10}$$

$$= \frac{17}{5} \cdot \frac{10}{17}$$

$$= \frac{17 \cdot 10}{5 \cdot 17}$$

$$= \frac{17 \cdot 2 \cdot 5}{5 \cdot 17}$$

$$= \frac{2}{1} = 2$$

151. greater than

Objective C Exercises

153. Strategy To find the length of time, multiply the length of time for one chukker $\left(7\frac{1}{2}\right)$ by 4.

Solution $7\frac{1}{2} \cdot 4 = \frac{15}{2} \cdot \frac{4}{1} = \frac{15 \cdot 4}{2 \cdot 1} = 30$

Four periods of play takes 30 min.

155. Strategy →To find the number of feet in one rod, multiply the number of yards in one rod $\left(5\frac{1}{2}\right)$ by 3.

→To find the number of inches in one rod, multiply the number of yards in one rod $\left(5\frac{1}{2}\right)$ by 36.

Solution $5\frac{1}{2} \cdot 3 = \frac{11}{2} \cdot \frac{3}{1} = \frac{11 \cdot 3}{2 \cdot 1} = 16\frac{1}{2}$

$5\frac{1}{2} \cdot 36 = \frac{11}{2} \cdot \frac{36}{1} = \frac{11 \cdot 36}{2 \cdot 1}$

$= 198$

One rod is equivalent to $16\frac{1}{2}$ ft.

One rod is equivalent to 198 in.

157. Strategy To find the amount of time, multiply the number of hours cleaning per week $\left(4\frac{1}{2}\right)$ by the number of weeks (52).

Solution $4\frac{1}{2} \cdot 52 = \frac{9}{2} \cdot \frac{52}{1} = \frac{9 \cdot 52}{2 \cdot 1} = 234$

The average couple spends 234 h a year cleaning house.

159. Strategy To find the number of house plots:

→Subtract the number of acres set aside (3) from the total number of acres $\left(25\frac{1}{2}\right)$ to find the number of acres to be sold in parcels.

→Divide the number of acres available by the number of acres in one parcel $\left(\frac{3}{4}\right)$.

Solution $25\frac{1}{2} - 3 = 22\frac{1}{2}$

$22\frac{1}{2} \div \frac{3}{4} = \frac{45}{2} \cdot \frac{4}{3} = \frac{45 \cdot 4}{2 \cdot 3} = 30$

The developer plans to build 30 houses.

161. **Strategy** To approximate the asteroid's distance from Earth at its closest point, multiply the distance the earth is from the moon in miles (250,000) by the fraction of that distance which the asteroid is from the earth $\left(\dfrac{9}{10}\right)$.

Solution

$$250,000 \cdot \frac{9}{10} = \frac{250,000}{1} \cdot \frac{9}{10}$$

$$= \frac{2,250,000}{10} = 225,000$$

At its closest point, the asteroid is approximately 225,000 miles from Earth.

163. **Strategy** To find the area, use the formula below, substitute 21 for b, 13 for h, and solve for A.

Solution

$$A = \frac{1}{2}bh$$

$$A = \frac{1}{2} \cdot 21 \cdot 13 = \frac{1}{2} \cdot \frac{21}{1} \cdot \frac{13}{1}$$

$$= \frac{651}{2} = 136\frac{1}{2}$$

The area of the vegetable garden is $136\dfrac{1}{2}$ ft².

165. **Strategy** To find the number of bags of seed:
→Find the area of the triangle, use the formula below, substitute 20 for b, 24 for h, and solve for A.
→Divide the area by 120.

Solution

$$A = \frac{1}{2}bh$$

$$A = \frac{1}{2} \cdot 20 \cdot 24 = \frac{1}{2} \cdot \frac{20}{1} \cdot \frac{24}{1} = 240$$

$240 \div 120 = 2$
Two bags of grass seed should be purchased.

167. **Strategy** To find the pressure, substitute $12\dfrac{1}{2}$ for D in the given formula and solve for P.

Solution

$$P = 15 + \frac{1}{2}D$$

$$P = 15 + \frac{1}{2}\left(12\frac{1}{2}\right)$$

$$P = 15 + \frac{1}{2}\left(\frac{25}{2}\right)$$

$$P = 15 + 6\frac{1}{4}$$

$$P = 21\frac{1}{4}$$

The pressure is $21\dfrac{1}{4}$ pounds per square inch.

Critical Thinking 3.3

169. Divide $3\dfrac{1}{8}$ by $\dfrac{1}{8}$ to find the number of 50-mile segments between the two cities.

$$3\frac{1}{8} \div \frac{1}{8} = \frac{25}{8} \div \frac{1}{8} = \frac{25}{8} \cdot \frac{8}{1} = 25$$

Multiply 25 times 50 to find the number of miles.
$25 \cdot 50 = 1250$
The distance between the two cities is 1250 mi.

Projects or Group Activities 3.3

171. Our calendar year is 365 years. A complete orbit around the sun takes $365\dfrac{1}{4}$ days, so one solar year is $365\dfrac{1}{4}$ days. Therefore, after 365 days, the Earth is $\dfrac{1}{4}$ day short of a complete orbit around the sun. After two years, the earth is $\dfrac{1}{4}$ day $+ \dfrac{1}{4}$ day $= \dfrac{1}{2}$ day short of a complete orbit around the sun. After 4 years, the earth is $\dfrac{1}{4}$ day $+ \dfrac{1}{4}$ day $+ \dfrac{1}{4}$ day $+ \dfrac{1}{4}$ day = 1 day short of a complete orbit around the sun. Therefore, every four years, our calendar incorporates a leap year, which is a 366-day year, in order to make up for the one extra day in the four solar years.

Check Your Progress: Chapter 3

1. $10 = 2 \cdot 5$

   $25 = 5^2$

   The LCM $= 2 \cdot 5^2 = 50$.

2. $5 = 5$

   $12 = 2^2 \cdot 3$

   $15 = 3 \cdot 5$

   The LCM $= 2^2 \cdot 3 \cdot 5 = 60$.

3. $26 = 2 \cdot 13$

   $52 = 2^2 \cdot 13$

   The GCF $= 2 \cdot 13 = 26$.

4. $41 = 41$

   $67 = 67$

   The GCF $= 1$.

5. $\begin{array}{r} 3 \\ 3\overline{)\,10} \\ -9 \\ \hline 1 \end{array}$ $\dfrac{10}{3} = 3\dfrac{1}{3}$

6. $\begin{array}{r} 9 \\ 9\overline{)\,81} \\ -81 \\ \hline 0 \end{array}$ $\dfrac{81}{9} = 9$

7. $3\dfrac{1}{4} = \dfrac{(4 \cdot 3) + 1}{4} = \dfrac{12 + 1}{4} = \dfrac{13}{4}$

8. $17 = \dfrac{17}{1}$

9. $12 \div 3 = 4$

   $\dfrac{2}{3} = \dfrac{2 \cdot 4}{3 \cdot 4} = \dfrac{8}{12}$

   $\dfrac{8}{12}$ is equivalent to $\dfrac{2}{3}$.

10. $\dfrac{24}{64} = \dfrac{2 \cdot 2 \cdot 2 \cdot 3}{2 \cdot 2 \cdot 2 \cdot 2 \cdot 2 \cdot 2} = \dfrac{3}{8}$

11. $\dfrac{2}{3} = \dfrac{16}{24}$ $\dfrac{5}{8} = \dfrac{15}{24}$

    $\dfrac{16}{24} > \dfrac{15}{24}$

    $\dfrac{2}{3} > \dfrac{5}{8}$

12. $-\dfrac{4}{15} \cdot \dfrac{5}{12} = -\dfrac{4 \cdot 5}{15 \cdot 12}$

    $= -\dfrac{2 \cdot 2 \cdot 5}{3 \cdot 5 \cdot 2 \cdot 2 \cdot 3}$

    $= -\dfrac{1}{9}$

13. $2\dfrac{1}{4} \cdot 1\dfrac{2}{9} = \dfrac{9}{4} \cdot \dfrac{11}{9}$

    $= \dfrac{9 \cdot 11}{4 \cdot 9}$

    $= \dfrac{3 \cdot 3 \cdot 11}{2 \cdot 2 \cdot 3 \cdot 3}$

    $= \dfrac{11}{4} = 2\dfrac{3}{4}$

14. $\dfrac{3}{4} \cdot \left(-\dfrac{2}{9}\right) \cdot \left(\dfrac{2}{5}\right) = -\dfrac{3}{4} \cdot \dfrac{2}{9} \cdot \dfrac{2}{5}$

    $= -\dfrac{3 \cdot 2 \cdot 2}{4 \cdot 9 \cdot 5}$

    $= -\dfrac{3 \cdot 2 \cdot 2}{2 \cdot 2 \cdot 3 \cdot 3 \cdot 5}$

    $= -\dfrac{1}{15}$

15. $-\dfrac{1}{4} \div \dfrac{1}{3} = -\left(\dfrac{1}{4} \div \dfrac{1}{3}\right)$

    $= -\left(\dfrac{1}{4} \cdot \dfrac{3}{1}\right)$

    $= -\dfrac{1 \cdot 3}{4 \cdot 1}$

    $= -\dfrac{3}{4}$

16. $\left(-\dfrac{6}{m}\right) \div \dfrac{n}{5} = -\left(\dfrac{6}{m} \div \dfrac{n}{5}\right)$

$= -\left(\dfrac{6}{m} \cdot \dfrac{5}{n}\right)$

$= -\dfrac{6 \cdot 5}{m \cdot n}$

$= -\dfrac{30}{mn}$

17. $8\dfrac{1}{3} \div 25 = \dfrac{25}{3} \cdot \dfrac{1}{25} = \dfrac{25 \cdot 1}{3 \cdot 25}$

$= \dfrac{1}{3}$

18. $-\dfrac{6}{7} \cdot \dfrac{14}{15} = -\dfrac{6 \cdot 14}{7 \cdot 15}$

$= -\dfrac{2 \cdot 3 \cdot 2 \cdot 7}{7 \cdot 3 \cdot 5}$

$= -\dfrac{4}{5}$

19. xy

$-\dfrac{2}{3} \cdot \left(-\dfrac{5}{6}\right) = \dfrac{2 \cdot 5}{3 \cdot 6}$

$= \dfrac{2 \cdot 5}{3 \cdot 2 \cdot 3}$

$= \dfrac{5}{9}$

20.

$$-\dfrac{3}{5}x = \dfrac{1}{5}$$

$$\begin{array}{c|c} -\dfrac{3}{5} \cdot \dfrac{1}{3} & \dfrac{1}{5} \\ \hline -\dfrac{3 \cdot 1}{5 \cdot 3} & \dfrac{1}{5} \\ \hline -\dfrac{1}{5} \ne & \dfrac{1}{5} \end{array}$$

No, $\dfrac{1}{3}$ is not a solution of the equation.

21. $-2\dfrac{5}{8} \div 5\dfrac{1}{4} = -\left(\dfrac{21}{8} \div \dfrac{21}{4}\right)$

$= -\left(\dfrac{21}{8} \cdot \dfrac{4}{21}\right)$

$= -\dfrac{21 \cdot 4}{8 \cdot 21}$

$= -\dfrac{3 \cdot 7 \cdot 2 \cdot 2}{2 \cdot 2 \cdot 2 \cdot 3 \cdot 7}$

$= -\dfrac{1}{2}$

22. $x \div y$

$\dfrac{2}{3} \div \left(-\dfrac{7}{9}\right) = -\dfrac{2}{3} \div \dfrac{7}{9}$

$= -\dfrac{2}{3} \cdot \dfrac{9}{7}$

$= -\dfrac{2 \cdot 9}{3 \cdot 7}$

$= -\dfrac{2 \cdot 3 \cdot 3}{3 \cdot 7}$

$= -\dfrac{6}{7}$

23. Strategy To find the number of pairs of booties to be packaged together, find the GCF of 200, 250, and 300.

Solution $200 = 2^3 \cdot 5^2$

$250 = 2 \cdot 5^3$

$300 = 2^2 \cdot 3 \cdot 5^2$

The GCF $= 2 \cdot 5 \cdot 5 = 50$.
Each package should contain 50 jerseys.

24. Strategy   To find the fraction, write a fraction with 12 in the numerator and 16 in the denominator. Write the fraction in simplest form.

       Solution $\dfrac{12}{16} = \dfrac{2 \cdot 2 \cdot 3}{2 \cdot 2 \cdot 2 \cdot 2} = \dfrac{3}{4}$

                  12 ounce is $\dfrac{3}{4}$ of a pound.

25. Strategy   To find the dimensions of the board when folded, divide the length (14) by 2 and multiply the height $\left(\dfrac{7}{8}\right)$ by 2.

       Solution $14 \div 2$

$$\frac{7}{8} \cdot 2 = \frac{7}{8} \cdot \frac{2}{1} = \frac{7 \cdot 2}{8 \cdot 1}$$

$$= \frac{7 \cdot 2}{2 \cdot 2 \cdot 2 \cdot 1} = \frac{7}{4} = 1\frac{3}{4}$$

             The dimensions of the board are 14 in. by 7 in. by $1\dfrac{3}{4}$ in.

Section 3.4

Concept Check

1. denominators

3. a. $\dfrac{1}{2}$; $-\dfrac{3}{7}$

    b. $\dfrac{1}{2}$; $\dfrac{3}{7}$

5. subtraction

Objective A Exercises

7. $\dfrac{4}{11} + \dfrac{5}{11} = \dfrac{4+5}{11} = \dfrac{9}{11}$

9. $\dfrac{2}{3} + \dfrac{1}{3} = \dfrac{2+1}{3}$

       $= \dfrac{3}{3} = 1$

11. $\dfrac{5}{6} + \dfrac{5}{6} = \dfrac{5+5}{6}$

        $= \dfrac{10}{6} = \dfrac{5}{3} = 1\dfrac{2}{3}$

13. $\dfrac{7}{18} + \dfrac{13}{18} + \dfrac{1}{18} = \dfrac{7+13+1}{18}$

        $= \dfrac{21}{18} = \dfrac{7}{6} = 1\dfrac{1}{6}$

15. $\dfrac{7}{b} + \dfrac{9}{b} = \dfrac{7+9}{b}$

        $= \dfrac{16}{b}$

17. $\dfrac{5}{c} + \dfrac{4}{c} = \dfrac{5+4}{c}$

        $= \dfrac{9}{c}$

19. $\dfrac{1}{x} + \dfrac{4}{x} + \dfrac{6}{x} = \dfrac{1+4+6}{x}$

        $= \dfrac{11}{x}$

21. $\dfrac{1}{4} + \dfrac{2}{3} = \dfrac{3}{12} + \dfrac{8}{12}$

        $= \dfrac{3+8}{12} = \dfrac{11}{12}$

23. $\dfrac{7}{15} + \dfrac{9}{20} = \dfrac{28}{60} + \dfrac{27}{60}$

        $= \dfrac{28+27}{60} = \dfrac{55}{60} = \dfrac{11}{12}$

25. $\dfrac{2}{3} + \dfrac{1}{12} + \dfrac{5}{6} = \dfrac{8}{12} + \dfrac{1}{12} + \dfrac{10}{12}$

        $= \dfrac{8+1+10}{12}$

        $= \dfrac{19}{12} = 1\dfrac{7}{12}$

27. $\dfrac{7}{12} + \dfrac{3}{4} + \dfrac{4}{5} = \dfrac{35}{60} + \dfrac{45}{60} + \dfrac{48}{60}$

        $= \dfrac{35+45+48}{60}$

        $= \dfrac{128}{60} = \dfrac{32}{15} = 2\dfrac{2}{15}$

29. $-\dfrac{3}{4} + \dfrac{2}{3} = \dfrac{-3}{4} + \dfrac{2}{3}$

        $= \dfrac{-9}{12} + \dfrac{8}{12}$

        $= \dfrac{-9+8}{12} = \dfrac{-1}{12} = -\dfrac{1}{12}$

31. $\dfrac{2}{5} + \left(-\dfrac{11}{15}\right) = \dfrac{2}{5} + \dfrac{-11}{15}$

$= \dfrac{6}{15} + \dfrac{-11}{15}$

$= \dfrac{6 + (-11)}{15} = \dfrac{-5}{15} = -\dfrac{1}{3}$

33. $\dfrac{3}{8} + \left(-\dfrac{1}{2}\right) + \dfrac{7}{12} = \dfrac{3}{8} + \dfrac{-1}{2} + \dfrac{7}{12}$

$= \dfrac{9}{24} + \dfrac{-12}{24} + \dfrac{14}{24}$

$= \dfrac{9 + (-12) + 14}{24}$

$= \dfrac{11}{24}$

35. $\dfrac{2}{3} + \left(-\dfrac{5}{6}\right) + \dfrac{1}{4} = \dfrac{2}{3} + \dfrac{-5}{6} + \dfrac{1}{4}$

$= \dfrac{16}{24} + \dfrac{-20}{24} + \dfrac{6}{24}$

$= \dfrac{16 + (-20) + 6}{24}$

$= \dfrac{2}{24} = \dfrac{1}{12}$

37. $8 + 7\dfrac{2}{3} = 15\dfrac{2}{3}$

39. $2\dfrac{1}{6} + 3\dfrac{1}{2} = 2\dfrac{1}{6} + 3\dfrac{3}{6}$

$= 5\dfrac{4}{6} = 5\dfrac{2}{3}$

41. $8\dfrac{3}{5} + 6\dfrac{9}{20} = 8\dfrac{12}{20} + 6\dfrac{9}{20}$

$= 14\dfrac{21}{20} = 15\dfrac{1}{20}$

43. $5\dfrac{5}{12} + 4\dfrac{7}{9} = 5\dfrac{15}{36} + 4\dfrac{28}{36}$

$= 9\dfrac{43}{36} = 10\dfrac{7}{36}$

45. $2\dfrac{1}{4} + 3\dfrac{1}{2} + 1\dfrac{2}{3}$

$= 2\dfrac{3}{12} + 3\dfrac{6}{12} + 1\dfrac{8}{12}$

$= 6\dfrac{17}{12} = 7\dfrac{5}{12}$

47. $-\dfrac{5}{6} + \dfrac{4}{9} = \dfrac{-5}{6} + \dfrac{4}{9}$

$= \dfrac{-15}{18} + \dfrac{8}{18}$

$= \dfrac{-15 + 8}{18} = \dfrac{-7}{18} = -\dfrac{7}{18}$

49. $\dfrac{2}{7} + \dfrac{3}{14} + \dfrac{1}{4} = \dfrac{8}{28} + \dfrac{6}{28} + \dfrac{7}{28}$

$= \dfrac{8 + 6 + 7}{28}$

$= \dfrac{21}{28} = \dfrac{3}{4}$

51. $-\dfrac{5}{6} + \left(-\dfrac{2}{3}\right) = \dfrac{-5}{6} + \dfrac{-2}{3}$

$= \dfrac{-5}{6} + \dfrac{-4}{6}$

$= \dfrac{-5 + (-4)}{6} = \dfrac{-9}{6}$

$= -\dfrac{3}{2} = -1\dfrac{1}{2}$

53. $3\dfrac{7}{12} + 2\dfrac{5}{8} = 3\dfrac{14}{24} + 2\dfrac{15}{24}$

$= 5\dfrac{29}{24} = 6\dfrac{5}{24}$

55. $x + y$

$\dfrac{3}{5} + \dfrac{4}{5} = \dfrac{3 + 4}{5}$

$= \dfrac{7}{5} = 1\dfrac{2}{5}$

57. $x + y$

$\dfrac{2}{3} + \left(-\dfrac{3}{4}\right) = \dfrac{2}{3} + \dfrac{-3}{4}$

$= \dfrac{8}{12} + \dfrac{-9}{12}$

$= \dfrac{8 + (-9)}{12} = \dfrac{-1}{12} = -\dfrac{1}{12}$

59. $x + y$

$\dfrac{5}{6} + \dfrac{8}{9} = \dfrac{15}{18} + \dfrac{16}{18}$

$= \dfrac{15 + 16}{18} = \dfrac{31}{18} = 1\dfrac{13}{18}$

61. $x + y$

$-\dfrac{5}{8} + \left(-\dfrac{1}{6}\right) = \dfrac{-5}{8} + \dfrac{-1}{6}$

$= \dfrac{-15}{24} + \dfrac{-4}{24}$

$= \dfrac{-15 + (-4)}{24}$

$= \dfrac{-19}{24} = -\dfrac{19}{24}$

63. $x + y + z$

$$\frac{3}{8} + \frac{1}{4} + \frac{7}{12} = \frac{9}{24} + \frac{6}{24} + \frac{14}{24}$$

$$= \frac{9 + 6 + 14}{24}$$

$$= \frac{29}{24} = 1\frac{5}{24}$$

65. $x + y + z$

$$1\frac{1}{2} + 3\frac{3}{4} + 6\frac{5}{12} = 1\frac{6}{12} + 3\frac{9}{12} + 6\frac{5}{12}$$

$$= 10\frac{20}{12} = 10\frac{5}{3} = 11\frac{2}{3}$$

67. $x + y + z$

$$4\frac{3}{5} + 8\frac{7}{10} + 1\frac{9}{20}$$

$$= 4\frac{12}{20} + 8\frac{14}{20} + 1\frac{9}{20}$$

$$= 13\frac{35}{20} = 13\frac{7}{4} = 14\frac{3}{4}$$

69.

$$z + \frac{1}{4} = -\frac{7}{20}$$

$$\begin{array}{c|c} -\frac{3}{5} + \frac{1}{4} & -\frac{7}{20} \\ -\frac{12}{20} + \frac{5}{20} & -\frac{7}{20} \\ \frac{-12 + 5}{20} & -\frac{7}{20} \\ -\frac{7}{20} & = -\frac{7}{20} \end{array}$$

Yes, $-\frac{3}{5}$ is a solution of the equation.

71.

$$\frac{1}{4} + x = -\frac{7}{12}$$

$$\begin{array}{c|c} \frac{1}{4} + \left(-\frac{5}{6}\right) & -\frac{7}{12} \\ \frac{3}{12} + \left(-\frac{10}{12}\right) & -\frac{7}{12} \\ \frac{3 + (-10)}{12} & -\frac{7}{12} \\ -\frac{7}{12} & = -\frac{7}{12} \end{array}$$

Yes, $-\frac{5}{6}$ is a solution of the equation.

73.

$$\frac{19}{50} + \frac{6}{25} = \frac{19}{50} + \frac{12}{50} = \frac{31}{50}$$

$\frac{31}{50}$ of money borrowed on home-equity loans are spent on debt consolidation and home improvement.

75. iii

Objective B Exercises

77. $\frac{7}{12} - \frac{5}{12} = \frac{7 - 5}{12} = \frac{2}{12} = \frac{1}{6}$

79. $\frac{11}{24} - \frac{7}{24} = \frac{11 - 7}{24} = \frac{4}{24} = \frac{1}{6}$

81. $\frac{8}{d} - \frac{3}{d} = \frac{8 - 3}{d} = \frac{5}{d}$

83. $\frac{5}{n} - \frac{10}{n} = \frac{5 - 10}{n} = \frac{-5}{n} = -\frac{5}{n}$

85. $\frac{3}{7} - \frac{5}{14} = \frac{6}{14} - \frac{5}{14}$

$$= \frac{6 - 5}{14} = \frac{1}{14}$$

87. $\frac{2}{3} - \frac{1}{6} = \frac{4}{6} - \frac{1}{6}$

$$= \frac{4 - 1}{6} = \frac{3}{6} = \frac{1}{2}$$

89. $\frac{11}{12} - \frac{2}{3} = \frac{11}{12} - \frac{8}{12}$

$$= \frac{11 - 8}{12} = \frac{3}{12} = \frac{1}{4}$$

91. $-\frac{1}{2} - \frac{3}{8} = \frac{-1}{2} - \frac{3}{8}$

$$= \frac{-4}{8} - \frac{3}{8}$$

$$= \frac{-4 - 3}{8} = \frac{-7}{8} = -\frac{7}{8}$$

93. $-\frac{3}{10} - \frac{4}{5} = \frac{-3}{10} - \frac{4}{5}$

$$= \frac{-3}{10} - \frac{8}{10}$$

$$= \frac{-3 - 8}{10} = \frac{-11}{10}$$

$$= -1\frac{1}{10}$$

95. $-\frac{5}{12} - \left(-\frac{2}{3}\right) = \frac{-5}{12} - \frac{-2}{3}$

$$= \frac{-5}{12} - \frac{-8}{12}$$

$$= \frac{-5 - (-8)}{12}$$

$$= \frac{-5 + 8}{12} = \frac{3}{12} = \frac{1}{4}$$

97. $-\dfrac{5}{9}-\left(-\dfrac{11}{12}\right)=\dfrac{-5}{9}-\dfrac{-11}{12}$

$=\dfrac{-20}{36}-\dfrac{-33}{36}$

$=\dfrac{-20-(-33)}{36}$

$=\dfrac{-20+33}{36}=\dfrac{13}{36}$

99. $4\dfrac{11}{18}-2\dfrac{5}{18}=2\dfrac{6}{18}=2\dfrac{1}{3}$

101. $8\dfrac{3}{4}-2=6\dfrac{3}{4}$

103. $8\dfrac{5}{6}-7\dfrac{3}{4}=8\dfrac{10}{12}-7\dfrac{9}{12}$

$=1\dfrac{1}{12}$

105. $7-3\dfrac{5}{8}=6\dfrac{8}{8}-3\dfrac{5}{8}$

$=3\dfrac{3}{8}$

107. $10-4\dfrac{8}{9}=9\dfrac{9}{9}-4\dfrac{8}{9}$

$=5\dfrac{1}{9}$

109. $7\dfrac{3}{8}-4\dfrac{5}{8}=6\dfrac{11}{8}-4\dfrac{5}{8}$

$=2\dfrac{6}{8}=2\dfrac{3}{4}$

111. $12\dfrac{5}{12}-10\dfrac{17}{24}=12\dfrac{10}{24}-10\dfrac{17}{24}$

$=11\dfrac{34}{24}-10\dfrac{17}{24}$

$=1\dfrac{17}{24}$

113. $6\dfrac{2}{3}-1\dfrac{7}{8}=6\dfrac{16}{24}-1\dfrac{21}{24}$

$=5\dfrac{40}{24}-1\dfrac{21}{24}$

$=4\dfrac{19}{24}$

115. $10\dfrac{2}{5}-8\dfrac{7}{10}=10\dfrac{4}{10}-8\dfrac{7}{10}$

$=9\dfrac{14}{10}-8\dfrac{7}{10}$

$=1\dfrac{7}{10}$

117. $-\dfrac{7}{12}-\dfrac{7}{9}=\dfrac{-7}{12}-\dfrac{7}{9}$

$=\dfrac{-21}{36}-\dfrac{28}{36}$

$=\dfrac{-21-28}{36}=\dfrac{-49}{36}=-1\dfrac{13}{36}$

119. $-\dfrac{7}{8}-\left(-\dfrac{2}{3}\right)=\dfrac{-7}{8}-\dfrac{-2}{3}$

$=\dfrac{-21}{24}-\dfrac{-16}{24}$

$=\dfrac{-21-(-16)}{24}$

$=\dfrac{-21+16}{24}=\dfrac{-5}{24}=-\dfrac{5}{24}$

121. $8-1\dfrac{7}{12}=7\dfrac{12}{12}-1\dfrac{7}{12}$

$=6\dfrac{5}{12}$

123. $x-y$

$\dfrac{8}{9}-\dfrac{5}{9}=\dfrac{8-5}{9}=\dfrac{3}{9}=\dfrac{1}{3}$

125. $x-y$

$-\dfrac{11}{12}-\dfrac{5}{12}=\dfrac{-11}{12}-\dfrac{5}{12}$

$=\dfrac{-11-5}{12}$

$=\dfrac{-16}{12}=\dfrac{-4}{3}=-1\dfrac{1}{3}$

127. $x-y$

$-\dfrac{2}{3}-\left(-\dfrac{3}{4}\right)=\dfrac{-2}{3}-\dfrac{-3}{4}$

$=\dfrac{-8}{12}-\dfrac{-9}{12}$

$=\dfrac{-8-(-9)}{12}$

$=\dfrac{-8+9}{12}=\dfrac{1}{12}$

129. $x-y$

$-\dfrac{3}{10}-\left(-\dfrac{7}{15}\right)=\dfrac{-3}{10}-\dfrac{-7}{15}$

$=\dfrac{-9}{30}-\dfrac{-14}{30}$

$=\dfrac{-9-(-14)}{30}$

$=\dfrac{-9+14}{30}=\dfrac{5}{30}=\dfrac{1}{6}$

131. $x - y$

$5\frac{7}{9} - 4\frac{2}{3} = 5\frac{7}{9} - 4\frac{6}{9}$

$= 1\frac{1}{9}$

133. $x - y$

$7\frac{9}{10} - 3\frac{1}{2} = 7\frac{9}{10} - 3\frac{5}{10}$

$= 4\frac{4}{10} = 4\frac{2}{5}$

135. $x - y$

$5 - 2\frac{7}{9} = 4\frac{9}{9} - 2\frac{7}{9}$

$= 2\frac{2}{9}$

137. $x - y$

$10\frac{1}{2} - 5\frac{7}{12} = 10\frac{6}{12} - 5\frac{7}{12}$

$= 9\frac{18}{12} - 5\frac{7}{12} = 4\frac{11}{12}$

139. $\frac{4}{5} = \frac{31}{20} - y$

$\frac{4}{5}$	$\frac{31}{20} - \left(-\frac{3}{4}\right)$
$\frac{4}{5}$	$\frac{31}{20} - \left(-\frac{15}{20}\right)$
$\frac{4}{5}$	$\frac{31 - (-15)}{20}$
$\frac{4}{5}$	$\frac{31 + 15}{20}$
$\frac{4}{5}$	$\frac{46}{20}$
$\frac{4}{5}$	$\frac{23}{10}$

$\frac{8}{10} \neq \frac{23}{10}$

No, $-\frac{3}{4}$ is not a solution of the equation.

141. $x - \frac{1}{4} = -\frac{17}{20}$

$-\frac{3}{5} - \frac{1}{4}$	$-\frac{17}{20}$
$-\frac{12}{20} - \frac{5}{20}$	$-\frac{17}{20}$
$\frac{-12 - 5}{20}$	$-\frac{17}{20}$
$-\frac{17}{20}$	$-\frac{17}{20}$

Yes, $-\frac{3}{5}$ is a solution of the equation.

143. i

Objective C Exercises

145. Strategy   To find the amount of property owned, subtract the amount sold $\left(1\frac{1}{2}\right)$ from the amount originally owned $\left(3\frac{1}{4}\right)$.

Solution $3\frac{1}{4} - 1\frac{1}{2} = 3\frac{1}{4} - 1\frac{2}{4}$

$= 2\frac{5}{4} - 1\frac{2}{4}$

$= 1\frac{3}{4}$

You now own $1\frac{3}{4}$ acres.

147. Strategy   To find the number of hours still required, subtract the number of hours already contributed $\left(12\frac{1}{4}\right)$ from the original amount required (20).

Solution $20 - 12\frac{1}{4} = 19\frac{4}{4} = 12\frac{1}{4}$

$= 7\frac{3}{4}$

$7\frac{3}{4}$ h of community service are still required.

149. Strategy   To find the amount of roofing that remains:
→Add the amounts already done $\left(\frac{1}{3} + \frac{1}{4}\right)$.
→Subtract the amount already done from the total job (1).
To determine if the roofers can finish in another day, compare the amount already done with $\frac{1}{2}$.

Solution $\frac{1}{3} + \frac{1}{4} = \frac{4}{12} + \frac{3}{12} = \frac{7}{12}$

$1 - \frac{7}{12} = \frac{12}{12} - \frac{7}{12} = \frac{5}{12}$

$\frac{5}{12}$ of the roofing job remains to be done.

Since $\frac{5}{12}$ is less than $\frac{1}{2}$, the roofing can be finished in another day.

151. Strategy   To find the amount earned:

→Add the times worked $\left(4\frac{1}{3}+5+3\frac{2}{3}\right)$.

→Multiply the hours worked by 9.

Solution $4\frac{1}{3}+5+3\frac{2}{3}=12\frac{3}{3}=13$

$13\cdot 9=117$

The student earned $117.

153. Strategy   To find course length, replace a, b, and c with $4\frac{3}{10}$, $3\frac{7}{10}$, and $2\frac{1}{2}$ in the given formula

and solve for P.

Solution $P=a+b+c$

$P=4\frac{3}{10}+3\frac{7}{10}+2\frac{1}{2}=4\frac{3}{10}+3\frac{7}{10}+2\frac{5}{10}=10\frac{1}{2}$

The total length of the course is $10\frac{1}{2}$ mi.

155. Strategy   To find the difference in heights, subtract the low 1970s $\left(\frac{3}{16}\right)$ from the high in 1960s $\left(\frac{7}{32}\right)$.

Solution $\frac{7}{32}-\frac{3}{16}=\frac{7}{32}-\frac{6}{32}=\frac{1}{32}$

The difference in heights between 1970s and 1960s was $\frac{1}{32}$ in.

157. Strategy   To find the amount of weight to gain:

→Add the amounts already gained $\left(4\frac{1}{2}+3\frac{3}{4}\right)$.

→Subtract the amount gained from the goal amount (15).

Solution $4\frac{1}{2}+3\frac{3}{4}=4\frac{2}{4}+3\frac{3}{4}=7\frac{5}{4}-8\frac{1}{4}$

$15-8\frac{1}{4}=14\frac{4}{4}-8\frac{1}{4}=6\frac{3}{4}$

The boxer has $6\frac{3}{4}$ lb left to gain.

Critical Thinking 3.4

159. Answers will vary. This is an example. Fractions with the same denominator are added by adding the numerators and placing the sum over the common denominator.

$\frac{2}{8}+\frac{3}{8}=\frac{2+3}{8}=\frac{5}{8}$

Projects or Group Activities 3.4

161. Not possible. Two positive numbers added together will result in a positive number.

163. Answers will vary. One example is:

$$\frac{3}{4} + \left(-\frac{1}{2}\right) = \frac{3}{4} + \left(-\frac{2}{4}\right) = \frac{1}{4}$$

165. Answers will vary. One example is:

$$\frac{1}{2} - \frac{3}{4} = \frac{2}{4} - \frac{3}{4} = -\frac{1}{4}$$

167. Not possible. Subtracting a positive number from a negative number will result in a negative number.

Section 3.5

Concept Check

1. 8

3. $\frac{1}{5}$

Objective A Exercises

5. $\frac{x}{4} = 9$

$4 \cdot \frac{x}{4} = 4 \cdot 9$

$x = 36$

The solution is 36.

7. $-3 = \frac{m}{4}$

$4(-3) = 4 \cdot \frac{m}{4}$

$-12 = m$

The solution is -12.

9. $\frac{2}{5}x = 10$

$\frac{5}{2} \cdot \frac{2}{5}x = \frac{5}{2} \cdot 10$

$x = 25$

The solution is 25.

11. $-\frac{5}{6}w = 10$

$-\frac{6}{5}\left(-\frac{5}{6}\right)w = -\frac{6}{5}(10)$

$w = -12$

The solution is -12.

13. $\frac{1}{4} + y = \frac{3}{4}$

$\frac{1}{4} - \frac{1}{4} + y = \frac{3}{4} - \frac{1}{4}$

$y = \frac{2}{4}$

$y = \frac{1}{2}$

The solution is $\frac{1}{2}$.

15. $x + \frac{1}{4} = \frac{5}{6}$

$x = \frac{1}{4} - \frac{1}{4} = \frac{5}{6} - \frac{1}{4}$

$x = \frac{10}{12} - \frac{3}{12}$

$x = \frac{7}{12}$

The solution is $\frac{7}{12}$.

17. $-\frac{2x}{3} = -\frac{1}{2}$

$-\frac{3}{2}\left(-\frac{2}{3}x\right) = -\frac{3}{2}\left(-\frac{1}{2}\right)$

$x = \frac{3}{4}$

The solution is $\frac{3}{4}$.

19. $\frac{5n}{6} = -\frac{2}{3}$

$\frac{6}{5}\left(\frac{5}{6}n\right) = \frac{6}{5}\left(-\frac{2}{3}\right)$

$n = -\frac{4}{5}$

The solution is $-\frac{4}{5}$.

21. $-\frac{3}{8}t = -\frac{1}{4}$

$-\frac{8}{3}\left(-\frac{3}{8}t\right) = -\frac{8}{3}\left(-\frac{1}{4}\right)$

$t = \frac{2}{3}$

The solution is $\frac{2}{3}$.

23. $4a = 6$

$\frac{4a}{4} = \frac{6}{4}$

$a = \frac{3}{2}$

$a = 1\frac{1}{2}$

The solution is $1\frac{1}{2}$.

25. $-9c = 12$

$$\frac{-9c}{-9} = \frac{12}{-9}$$

$$c = -\frac{4}{3}$$

$$c = -1\frac{1}{3}$$

The solution is $-1\frac{1}{3}$.

27. $-2x = \frac{8}{9}$

$$\frac{-2x}{-2} = \frac{8}{9} \div (-2)$$

$$x = -\left(\frac{8}{9} \cdot \frac{1}{2}\right)$$

$$x = -\frac{4}{9}$$

The solution is $-\frac{4}{9}$.

29. iii

Objective B Exercises

31. The unknown number: n

a number minus one-third	equals	one-half

$$n - \frac{1}{3} = \frac{1}{2}$$

$$n - \frac{1}{3} + \frac{1}{3} = \frac{1}{2} + \frac{1}{3}$$

$$n = \frac{3}{6} + \frac{2}{6}$$

$$n = \frac{5}{6}$$

The number is $\frac{5}{6}$.

33. The unknown number: n

three-fifths times a number	is	nine-tenths

$$\frac{3}{5}n = \frac{9}{10}$$

$$\frac{5}{3} \cdot \frac{3}{5}n = \frac{5}{3} \cdot \frac{9}{10}$$

$$n = \frac{3}{2}$$

$$n = 1\frac{1}{2}$$

The number is $1\frac{1}{2}$.

35. The unknown number: n

the quotient of a number and negative four	is	three-fourths

$$\frac{n}{-4} = \frac{3}{4}$$

$$-4\left(\frac{n}{-4}\right) = -4\left(\frac{3}{4}\right)$$

$$n = -3$$

The number is -3.

37. The unknown number: n

| negative three-fourths of a number | is equal to | one-sixth |

$$-\frac{3}{4}n = \frac{1}{6}$$

$$-\frac{4}{3}\left(-\frac{3}{4}n\right) = -\frac{4}{3}\left(\frac{1}{6}\right)$$

$$n = -\frac{2}{9}$$

The number is $-\frac{2}{9}$.

39. **Strategy**   To find the mechanic's monthly income, write and solve an equation using x to represent the monthly income.

Solution | the rent | is | two-fifths of the monthly income |

$$1,800 = \frac{2}{5}x$$

$$\frac{5}{2}(1,800) = \frac{5}{2}\left(\frac{2}{5}x\right)$$

$$4,500 = x$$

The mechanic's monthly income is $4,500.

41. **Strategy**   To find the worldwide gross income for *Avatar*, write and solve an equation using x to represent the worldwide gross income for *Avatar*.

Solution | income of *Twilight Saga:Eclipse* | is | one-fourth the income of *Avatar* |

$$698,000,000 = \frac{1}{4}x$$

$$4(698,000,000) = 4\left(\frac{1}{4}x\right)$$

$$2,790,000,000 = x$$

The worldwide gross income for Avatar is $2,790,000,000.

43. **Strategy**   To find the total number of quarts in the punch, write and solve an equation using x to represent the total number of quarts in the punch.

Solution | the number of quarts of orange juice | is | three-fifths of the total number of quarts |

$$15 = \frac{3}{5}x$$

$$\frac{5}{3}(15) = \frac{5}{3}\left(\frac{3}{5}x\right)$$

$$25 = x$$

There is a total of 25 quarts in the punch.

45. Strategy   To find the number of miles, replace a by 14 and g by 38 in the given formula and solve for m.

Solution $a = \dfrac{m}{g}$

$14 = \dfrac{m}{38}$

$38 \cdot 14 = 38 \cdot \dfrac{m}{38}$

$532 = m$

The truck can travel 532 mi on 38 gal of diesel fuel.

Critical Thinking 3.5

47. In explaining why dividing each side of $3x = 6$ by 3 is the same as multiplying each side of the equation by $\dfrac{1}{3}$, students should mention the relationship between multiplication and division by the reciprocal.

For example, they might state that 3 and $\dfrac{1}{3}$ are reciprocals, and division by 3 is the same as

multiplication by the reciprocal, $\dfrac{1}{3}$. $3 \div 3 = 3 \cdot \dfrac{1}{3} = 1$

Section 3.6

Concept Check

1. four

3. $\dfrac{1}{3}$; $\dfrac{5}{6}$

Objective A Exercises

5. $\left(\dfrac{3}{4}\right)^2 = \dfrac{3}{4} \cdot \dfrac{3}{4}$
$= \dfrac{3 \cdot 3}{4 \cdot 4} = \dfrac{9}{16}$

7. $\left(-\dfrac{1}{6}\right)^3 = \left(-\dfrac{1}{6}\right)\left(-\dfrac{1}{6}\right)\left(-\dfrac{1}{6}\right)$
$= -\left(\dfrac{1}{6} \cdot \dfrac{1}{6} \cdot \dfrac{1}{6}\right)$
$= -\dfrac{1 \cdot 1 \cdot 1}{6 \cdot 6 \cdot 6} = -\dfrac{1}{216}$

9. $\left(\dfrac{5}{8}\right)^3 \cdot \left(\dfrac{2}{5}\right)^2 = \dfrac{5}{8} \cdot \dfrac{5}{8} \cdot \dfrac{5}{8} \cdot \dfrac{2}{5} \cdot \dfrac{2}{5}$
$= \dfrac{5 \cdot 5 \cdot 5 \cdot 2 \cdot 2}{8 \cdot 8 \cdot 8 \cdot 5 \cdot 5}$
$= \dfrac{5}{128}$

11. $\left(\dfrac{4}{5}\right)^4 \cdot \left(-\dfrac{5}{8}\right)^3$
$= \dfrac{4}{5} \cdot \dfrac{4}{5} \cdot \dfrac{4}{5} \cdot \dfrac{4}{5} \cdot \left(-\dfrac{5}{8}\right)\left(-\dfrac{5}{8}\right)\left(-\dfrac{5}{8}\right)$
$= -\left(\dfrac{4}{5} \cdot \dfrac{4}{5} \cdot \dfrac{4}{5} \cdot \dfrac{4}{5} \cdot \dfrac{5}{8} \cdot \dfrac{5}{8} \cdot \dfrac{5}{8}\right)$
$= -\dfrac{4 \cdot 4 \cdot 4 \cdot 4 \cdot 5 \cdot 5 \cdot 5}{5 \cdot 5 \cdot 5 \cdot 5 \cdot 8 \cdot 8 \cdot 8} = -\dfrac{1}{10}$

13. $7^2 \cdot \left(\dfrac{2}{7}\right)^3 = \dfrac{7}{1} \cdot \dfrac{7}{1} \cdot \dfrac{2}{7} \cdot \dfrac{2}{7} \cdot \dfrac{2}{7}$
$= \dfrac{7 \cdot 7 \cdot 2 \cdot 2 \cdot 2}{1 \cdot 1 \cdot 7 \cdot 7 \cdot 7}$
$= \dfrac{8}{7} = 1\dfrac{1}{7}$

15. $4 \cdot \left(\dfrac{4}{7}\right)^2 \cdot \left(-\dfrac{3}{4}\right)^3$
$= \dfrac{4}{1} \cdot \dfrac{4}{7} \cdot \dfrac{4}{7} \cdot \left(-\dfrac{3}{4}\right)\left(-\dfrac{3}{4}\right)\left(-\dfrac{3}{4}\right)$
$= -\left(\dfrac{4}{1} \cdot \dfrac{4}{7} \cdot \dfrac{4}{7} \cdot \dfrac{3}{4} \cdot \dfrac{3}{4} \cdot \dfrac{3}{4}\right)$
$= -\dfrac{4 \cdot 4 \cdot 4 \cdot 3 \cdot 3 \cdot 3}{1 \cdot 7 \cdot 7 \cdot 4 \cdot 4 \cdot 4}$
$= -\dfrac{27}{49}$

17. x^4
$\left(\dfrac{2}{3}\right)^4 = \dfrac{2}{3} \cdot \dfrac{2}{3} \cdot \dfrac{2}{3} \cdot \dfrac{2}{3}$
$= \dfrac{2 \cdot 2 \cdot 2 \cdot 2}{3 \cdot 3 \cdot 3 \cdot 3}$
$= \dfrac{16}{81}$

19. $x^4 y^2$
$\left(\dfrac{5}{6}\right)^4 \cdot \left(-\dfrac{3}{5}\right)^2 = \dfrac{5}{6} \cdot \dfrac{5}{6} \cdot \dfrac{5}{6} \cdot \dfrac{5}{6}\left(-\dfrac{3}{5}\right)\left(-\dfrac{3}{5}\right)$
$= \dfrac{5}{6} \cdot \dfrac{5}{6} \cdot \dfrac{5}{6} \cdot \dfrac{5}{6} \cdot \dfrac{3}{5} \cdot \dfrac{3}{5}$
$= \dfrac{5 \cdot 5 \cdot 5 \cdot 5 \cdot 3 \cdot 3}{6 \cdot 6 \cdot 6 \cdot 6 \cdot 5 \cdot 5}$
$= \dfrac{25}{144}$

21. $x^3 y^2$
$\left(\dfrac{2}{3}\right)^3 \cdot \left(1\dfrac{1}{2}\right)^2 = \dfrac{2}{3} \cdot \dfrac{2}{3} \cdot \dfrac{2}{3} \cdot \dfrac{3}{2} \cdot \dfrac{3}{2}$
$= \dfrac{2 \cdot 2 \cdot 2 \cdot 3 \cdot 3}{3 \cdot 3 \cdot 3 \cdot 2 \cdot 2}$
$= \dfrac{2}{3}$

23. True

Objective B Exercises

25. $\dfrac{\frac{9}{16}}{\frac{3}{4}} = \dfrac{9}{16} \div \dfrac{3}{4}$
$= \dfrac{9}{16} \cdot \dfrac{4}{3} = \dfrac{3}{4}$

27. $\dfrac{-\frac{5}{6}}{\frac{15}{16}} = -\dfrac{5}{6} \div \dfrac{15}{16}$
$= -\left(\dfrac{5}{6} \cdot \dfrac{16}{15}\right) = -\dfrac{8}{9}$

29. $\dfrac{\frac{2}{3}+\frac{1}{2}}{7} = \dfrac{\frac{7}{6}}{\frac{7}{1}}$

$= \dfrac{7}{6} \div \dfrac{7}{1}$

$= \dfrac{7}{6} \cdot \dfrac{1}{7} = \dfrac{1}{6}$

31. $\dfrac{2+\frac{1}{4}}{\frac{3}{8}} = \dfrac{\frac{9}{4}}{\frac{3}{8}}$

$= \dfrac{9}{4} \div \dfrac{3}{8}$

$= \dfrac{9}{4} \cdot \dfrac{8}{3} = 6$

33. $\dfrac{\frac{9}{25}}{\frac{4}{5}-\frac{1}{10}} = \dfrac{\frac{9}{25}}{\frac{7}{10}}$

$= \dfrac{9}{25} \div \dfrac{7}{10}$

$= \dfrac{9}{25} \cdot \dfrac{10}{7} = \dfrac{18}{35}$

35. $\dfrac{\frac{1}{3}-\frac{3}{4}}{\frac{1}{6}+\frac{2}{3}} = \dfrac{-\frac{5}{12}}{\frac{5}{6}}$

$= -\dfrac{5}{12} \div \dfrac{5}{6}$

$= -\left(\dfrac{5}{12} \cdot \dfrac{6}{5}\right) = -\dfrac{1}{2}$

37. $\dfrac{3+2\frac{1}{3}}{5\frac{1}{6}-1} = \dfrac{5\frac{1}{3}}{4\frac{1}{6}}$

$= 5\dfrac{1}{3} \div 4\dfrac{1}{6}$

$= \dfrac{16}{3} \div \dfrac{25}{6}$

$= \dfrac{16}{3} \cdot \dfrac{6}{25} = \dfrac{32}{25} = 1\dfrac{7}{25}$

39. $\dfrac{5\frac{2}{3}-1\frac{1}{6}}{3\frac{5}{8}-2\frac{1}{4}} = \dfrac{4\frac{1}{2}}{1\frac{3}{8}}$

$= 4\dfrac{1}{2} \div 1\dfrac{3}{8}$

$= \dfrac{9}{2} \div \dfrac{11}{8}$

$= \dfrac{9}{2} \cdot \dfrac{8}{11} = \dfrac{36}{11} = 3\dfrac{3}{11}$

41. $\dfrac{x+y}{z}$

$\dfrac{\frac{2}{3}+\frac{3}{4}}{\frac{1}{12}} = \dfrac{\frac{17}{12}}{\frac{1}{12}}$

$= \dfrac{17}{12} \div \dfrac{1}{12}$

$= \dfrac{17}{12} \cdot \dfrac{12}{1} = 17$

43. $\dfrac{xy}{z}$

$\dfrac{\frac{3}{4} \cdot \left(-\frac{2}{3}\right)}{\frac{5}{8}} = \dfrac{-\frac{1}{2}}{\frac{5}{8}}$

$= -\dfrac{1}{2} \div \dfrac{5}{8}$

$= -\dfrac{1}{2} \cdot \dfrac{8}{5} = -\dfrac{4}{5}$

45. $\dfrac{x-y}{z}$

$\dfrac{2\frac{5}{8}-1\frac{1}{4}}{1\frac{3}{8}} = \dfrac{1\frac{3}{8}}{1\frac{3}{8}}$

$= 1\dfrac{3}{8} \div 1\dfrac{3}{8} = \dfrac{11}{8} \div \dfrac{11}{8}$

$= \dfrac{11}{8} \cdot \dfrac{8}{11} = 1$

47. 1

49. 0

Objective C Exercises

51. addition, division, subtraction

53. $-\dfrac{3}{7} \cdot \dfrac{14}{15} + \dfrac{4}{5} = -\dfrac{2}{5} + \dfrac{4}{5} = \dfrac{2}{5}$

55. $\left(\dfrac{5}{6}\right)^2 - \dfrac{5}{9} = \dfrac{25}{36} - \dfrac{5}{9}$

$= \dfrac{5}{36}$

57. $\dfrac{3}{4} \cdot \left(\dfrac{11}{12} - \dfrac{7}{8}\right) + \dfrac{5}{16} = \dfrac{3}{4} \cdot \dfrac{1}{24} + \dfrac{5}{16}$

$= \dfrac{1}{32} + \dfrac{5}{16}$

$= \dfrac{11}{32}$

59. $\dfrac{11}{16} - \left(\dfrac{3}{4}\right)^2 + \dfrac{7}{8} = \dfrac{11}{16} - \dfrac{9}{16} + \dfrac{7}{8}$

$= \dfrac{1}{8} + \dfrac{7}{8}$

$= 1$

61. $\left(1\dfrac{1}{3} - \dfrac{5}{6}\right) + \dfrac{7}{8} \div \left(-\dfrac{1}{2}\right)^2$

$= \dfrac{1}{2} + \dfrac{7}{8} \div \left(-\dfrac{1}{2}\right)^2$

$= \dfrac{1}{2} + \dfrac{7}{8} \div \dfrac{1}{4}$

$= \dfrac{1}{2} + \dfrac{7}{8} \cdot \dfrac{4}{1}$

$= \dfrac{1}{2} + \dfrac{7}{2} = 4$

63. $\left(\dfrac{2}{3}\right)^2 + \dfrac{8-7}{3-9} \div \dfrac{3}{8} = \left(\dfrac{2}{3}\right)^2 + \left(-\dfrac{1}{6}\right) \div \dfrac{3}{8}$

$= \dfrac{4}{9} + \left(-\dfrac{1}{6}\right) \div \dfrac{3}{8}$

$= \dfrac{4}{9} + \left(-\dfrac{1}{6}\right) \cdot \dfrac{8}{3}$

$= \dfrac{4}{9} + \left(\dfrac{-4}{9}\right) = 0$

65. $-\dfrac{1}{2} + \dfrac{\frac{13}{25}}{4 - \frac{3}{4}} \div \dfrac{1}{5} = -\dfrac{1}{2} + \dfrac{\frac{13}{25}}{\frac{13}{4}} \div \dfrac{1}{5}$

$= -\dfrac{1}{2} + \dfrac{13}{25} \div \dfrac{13}{4} \div \dfrac{1}{5}$

$= -\dfrac{1}{2} + \dfrac{13}{25} \cdot \dfrac{4}{13} \div \dfrac{1}{5}$

$= -\dfrac{1}{2} + \dfrac{4}{25} \div \dfrac{1}{5} = -\dfrac{1}{2} + \dfrac{4}{25} \cdot \dfrac{5}{1}$

$= -\dfrac{1}{2} + \dfrac{4}{5} = -\dfrac{5}{10} + \dfrac{8}{10} = \dfrac{3}{10}$

67. $\left(\dfrac{2}{3}\right)^2 + \dfrac{\frac{5}{8} - \frac{1}{4}}{\frac{2}{3} - \frac{1}{6}} \cdot \dfrac{8}{9} = \left(\dfrac{2}{3}\right)^2 + \dfrac{\frac{3}{8}}{\frac{1}{2}} \cdot \dfrac{8}{9}$

$= \left(\dfrac{2}{3}\right)^2 + \dfrac{3}{8} \div \dfrac{1}{2} \cdot \dfrac{8}{9}$

$= \dfrac{4}{9} + \dfrac{3}{8} \div \dfrac{1}{2} \cdot \dfrac{8}{9}$

$= \dfrac{4}{9} + \dfrac{3}{8} \cdot \dfrac{2}{1} \cdot \dfrac{8}{9}$

$= \dfrac{4}{9} + \dfrac{3}{4} \cdot \dfrac{8}{9}$

$= \dfrac{4}{9} + \dfrac{2}{3} = \dfrac{10}{9} = 1\dfrac{1}{9}$

69. $\dfrac{x}{y} - z^2$

$\dfrac{\frac{5}{6}}{\frac{1}{3}} - \left(-\dfrac{3}{4}\right)^2 = \dfrac{5}{6} \div \dfrac{1}{3} - \left(-\dfrac{3}{4}\right)^2$

$= \dfrac{5}{6} \div \dfrac{1}{3} - \dfrac{9}{16}$

$= \dfrac{5}{6} \cdot \dfrac{3}{1} - \dfrac{9}{16}$

$= \dfrac{5}{2} - \dfrac{9}{16}$

$= \dfrac{31}{16} = 1\dfrac{15}{16}$

71. $xy^3 + z$

$\dfrac{9}{10} \cdot \left(\dfrac{1}{3}\right)^3 + \dfrac{7}{15} = \dfrac{9}{10} \cdot \dfrac{1}{27} + \dfrac{7}{15}$

$= \dfrac{1}{30} + \dfrac{7}{15}$

$= \dfrac{1}{2}$

73. $\dfrac{w}{xy} - z$

$\dfrac{2\frac{1}{2}}{4 \cdot \frac{3}{8}} - \dfrac{2}{3} = \dfrac{\frac{5}{2}}{\frac{3}{2}} - \dfrac{2}{3}$

$= \dfrac{5}{2} \div \dfrac{3}{2} - \dfrac{2}{3}$

$= \dfrac{5}{2} \cdot \dfrac{2}{3} - \dfrac{2}{3}$

$\dfrac{5}{3} - \dfrac{2}{3} = \dfrac{3}{3} = 1$

Critical Thinking 3.6

75.

$\left(1 - \dfrac{1}{2^2}\right)\left(1 - \dfrac{1}{3^2}\right)\left(1 - \dfrac{1}{4^2}\right) \cdots \left(1 - \dfrac{1}{9^2}\right)\left(1 - \dfrac{1}{10^2}\right)$

$= \left(1 - \dfrac{1}{4}\right)\left(1 - \dfrac{1}{9}\right)\left(1 - \dfrac{1}{16}\right) \cdots \left(1 - \dfrac{1}{81}\right)\left(1 - \dfrac{1}{10^2}\right)$

$= \left(\dfrac{3}{4}\right)\left(\dfrac{8}{9}\right)\left(\dfrac{15}{16}\right)\left(\dfrac{24}{25}\right)\left(\dfrac{35}{36}\right)\left(\dfrac{48}{49}\right)\left(\dfrac{63}{64}\right)\left(\dfrac{80}{81}\right)\left(\dfrac{99}{100}\right)$

$= \left(\dfrac{3}{1}\right)\left(\dfrac{2}{9}\right)\left(\dfrac{15}{2}\right)\left(\dfrac{3}{25}\right)\left(\dfrac{35}{3}\right)\left(\dfrac{4}{49}\right)\left(\dfrac{63}{4}\right)\left(\dfrac{5}{81}\right)\left(\dfrac{99}{100}\right)$

$= \left(\dfrac{1}{1}\right)\left(\dfrac{2}{3}\right)\left(\dfrac{3}{2}\right)\left(\dfrac{3}{5}\right)\left(\dfrac{5}{3}\right)\left(\dfrac{4}{7}\right)\left(\dfrac{7}{4}\right)\left(\dfrac{5}{9}\right)\left(\dfrac{9}{20}\dfrac{11}{}\right)$

$= \dfrac{11}{20}$

Projects or Group Activities 3.6

77. Adding the diagonal from lower-left to upper-right, we see that the sum for each column, row, and diagonal must be:

$$\frac{1}{2}+\frac{5}{8}+\frac{3}{4}=\frac{4}{8}+\frac{5}{8}+\frac{6}{8}=\frac{15}{8}$$

To make the diagonal from upper-left to lower-right have a sum of $\frac{15}{8}$, we need to have the upper-left corner be $\frac{3}{8}$. The entry in the top row, middle column must be $\frac{3}{4}$. The entry in the second row, first column must be 1. The entry in the second row, last column must be $\frac{1}{4}$. The entry in the bottom row, middle column must be $\frac{1}{2}$

Chapter Review Exercises

1.
$$\begin{array}{r} 9 \\ 2\overline{)19} \\ \underline{-18} \\ 1 \end{array} \qquad \frac{19}{2}=9\frac{1}{2}$$

2. $6\frac{2}{9}-3\frac{7}{18}=6\frac{4}{18}-3\frac{7}{18}$
$$=5\frac{22}{18}-3\frac{7}{18}$$
$$=2\frac{15}{18}=2\frac{5}{6}$$

3. $x \div y$
$$2\frac{5}{8} \div 1\frac{3}{4}=\frac{21}{8} \div \frac{7}{4}$$
$$=\frac{21}{8} \cdot \frac{4}{7}$$
$$=\frac{3}{2}=1\frac{1}{2}$$

4. $\left(-2\frac{1}{3}\right) \cdot \frac{3}{7}=-\left(\frac{7}{3} \cdot \frac{3}{7}\right)$
$$=-\frac{7 \cdot 3}{3 \cdot 7}$$
$$=-1$$

5. $3\frac{3}{4} \div 1\frac{7}{8}=\frac{15}{4} \div \frac{15}{8}$
$$=\frac{15}{4} \cdot \frac{8}{15}$$
$$=\frac{15 \cdot 8}{4 \cdot 15}$$
$$=\frac{3 \cdot 5 \cdot 2 \cdot 2 \cdot 2}{2 \cdot 2 \cdot 3 \cdot 5}=2$$

6. $3 \cdot \frac{8}{9}=\frac{3}{1} \cdot \frac{8}{9}$
$$=\frac{3 \cdot 8}{1 \cdot 9}$$
$$=\frac{3 \cdot 2 \cdot 2 \cdot 2}{1 \cdot 3 \cdot 3}$$
$$=\frac{8}{3}=2\frac{2}{3}$$

7. $\dfrac{x}{y+z}$
$$\frac{\frac{7}{8}}{\frac{4}{5}+\left(-\frac{1}{2}\right)}=\frac{\frac{7}{8}}{\frac{3}{10}}$$
$$=\frac{7}{8} \div \frac{3}{10}$$
$$=\frac{7}{8} \cdot \frac{10}{3}=\frac{35}{12}=2\frac{11}{12}$$

8. $\frac{3}{5}=\frac{9}{15} \qquad \frac{7}{15}=\frac{7}{15}$
$$\frac{9}{15}>\frac{7}{15}$$
$$\frac{3}{5}>\frac{7}{15}$$

9. $50=2 \cdot 5^2$
$75=3 \cdot 5^2$
$\text{LCM}=2 \cdot 3 \cdot 5^2=150$

10. $6\frac{11}{15}+4\frac{7}{10}=6\frac{22}{30}+4\frac{21}{30}$
$$=10\frac{43}{30}$$
$$=11\frac{13}{30}$$

11. xy
$$8 \cdot \frac{5}{12}=\frac{8}{1} \cdot \frac{5}{12}$$
$$=\frac{8 \cdot 5}{1 \cdot 12}$$
$$=\frac{2 \cdot 2 \cdot 2 \cdot 5}{1 \cdot 2 \cdot 2 \cdot 3}$$
$$=\frac{10}{3}=3\frac{1}{3}$$

12. $\frac{10}{7}; 1\frac{3}{7}$

13. $\frac{7}{8} = \frac{35}{40} \qquad \frac{17}{20} = \frac{34}{40}$

$\qquad \frac{35}{49} > \frac{34}{40}$

$\qquad \frac{7}{8} > \frac{17}{20}$

14. $\dfrac{\frac{5}{8} - \frac{1}{4}}{\frac{1}{2} + \frac{1}{8}} = \dfrac{\frac{3}{8}}{\frac{5}{8}}$

$\qquad = \frac{3}{8} \div \frac{5}{8}$

$\qquad = \frac{3}{8} \cdot \frac{8}{5}$

$\qquad = \frac{3 \cdot 8}{8 \cdot 5} = \frac{3}{5}$

15. $72 \div 9 = 8$

$\qquad \frac{4}{9} = \frac{4 \cdot 8}{9 \cdot 8} = \frac{32}{72}$

$\qquad \frac{32}{72}$ is equivalent to $\frac{4}{9}$.

16. $x^2 y^3$

$\qquad \left(\frac{8}{9}\right)^2 \cdot \left(-\frac{3}{4}\right)^3$

$\qquad = \frac{8}{9} \cdot \frac{8}{9} \cdot \left(-\frac{3}{4}\right)\left(-\frac{3}{4}\right)\left(-\frac{3}{4}\right)$

$\qquad = -\left(\frac{8}{9} \cdot \frac{8}{9} \cdot \frac{3}{4} \cdot \frac{3}{4} \cdot \frac{3}{4}\right)$

$\qquad = -\frac{8 \cdot 8 \cdot 3 \cdot 3 \cdot 3}{9 \cdot 9 \cdot 4 \cdot 4 \cdot 4} = -\frac{1}{3}$

17. $ab^2 - c$

$\qquad 4 \cdot \left(\frac{1}{2}\right)^2 - \frac{5}{7} = 4 \cdot \frac{1}{4} - \frac{5}{7}$

$\qquad = \frac{4}{1} \cdot \frac{1}{4} - \frac{5}{7} = 1 - \frac{5}{7}$

$\qquad = \frac{7}{7} - \frac{5}{7}$

$\qquad = \frac{2}{7}$

18. $42 = 2 \cdot 3 \cdot 7$

$\qquad 63 = 3 \cdot 3 \cdot 7$

$\qquad \text{GCF} = 3 \cdot 7 = 21$

19. $2\frac{5}{14} = \frac{(14 \cdot 2) + 5}{14}$

$\qquad = \frac{28 + 5}{14} = \frac{33}{14}$

20. $x + y + z$

$\qquad \frac{5}{8} + \left(-\frac{3}{4}\right) + \frac{1}{2} = \frac{5}{8} + \frac{-3}{4} + \frac{1}{2}$

$\qquad = \frac{5}{8} + \frac{-6}{8} + \frac{4}{8}$

$\qquad = \frac{5 + (-6) + 4}{8}$

$\qquad = \frac{3}{8}$

21. $\frac{5}{9} \div \left(-\frac{2}{3}\right) = -\left(\frac{5}{9} \div \frac{2}{3}\right)$

$\qquad = -\left(\frac{5}{9} \cdot \frac{3}{2}\right)$

$\qquad = -\frac{5 \cdot 3}{9 \cdot 2}$

$\qquad = -\frac{5 \cdot 3}{3 \cdot 3 \cdot 2} = -\frac{5}{6}$

22. $\frac{2}{5} \div \frac{4}{7} + \frac{3}{8} = \frac{2}{5} \cdot \frac{7}{4} + \frac{3}{8}$

$\qquad = \frac{7}{10} + \frac{3}{8}$

$\qquad = \frac{28}{40} + \frac{15}{40}$

$\qquad = \frac{28 + 15}{40} = \frac{43}{40} = 1\frac{3}{40}$

23. $5\frac{1}{4} \cdot \frac{8}{9} \cdot (-3) = -\left(\frac{21}{4} \cdot \frac{8}{9} \cdot \frac{3}{1}\right)$

$\qquad = -\frac{21 \cdot 8 \cdot 3}{4 \cdot 9 \cdot 1}$

$\qquad = -\frac{3 \cdot 7 \cdot 2 \cdot 2 \cdot 2 \cdot 3}{2 \cdot 2 \cdot 3 \cdot 3 \cdot 1}$

$\qquad = -14$

24. $\frac{2}{3} - \frac{11}{18} = \frac{12}{18} - \frac{11}{18}$

$\qquad = \frac{12 - 11}{18}$

$\qquad = \frac{1}{18}$

25. $\frac{7}{8} - \left(-\frac{5}{6}\right) = \frac{7}{8} - \left(-\frac{5}{6}\right)$

$\qquad = \frac{21}{24} - \frac{-20}{24}$

$\qquad = \frac{21 - (-20)}{24} = \frac{21 + 20}{24}$

$\qquad = \frac{41}{24} = 1\frac{17}{24}$

26. $\left(-\dfrac{3}{8}\right)^2 \cdot 4^2 = \dfrac{9}{64} \cdot 16$

$\qquad = \dfrac{9}{64} \cdot \dfrac{16}{1}$

$\qquad = \dfrac{9}{4} = 2\dfrac{1}{4}$

27. $3\dfrac{7}{12} + 5\dfrac{1}{2} = 3\dfrac{7}{12} + 5\dfrac{6}{12}$

$\qquad = 8\dfrac{13}{12} = 9\dfrac{1}{12}$

28. $\dfrac{30}{105} = \dfrac{2 \cdot 3 \cdot 5}{3 \cdot 5 \cdot 7} = \dfrac{2}{7}$

29. $a - b$

$\quad 7 - 2\dfrac{3}{10} = 6\dfrac{10}{10} - 2\dfrac{3}{10}$

$\qquad\quad = 4\dfrac{7}{10}$

30. $-\dfrac{5}{9} = \dfrac{1}{6} + p$

$\quad -\dfrac{5}{9} - \dfrac{1}{6} = \dfrac{1}{6} - \dfrac{1}{6} + p$

$\quad -\dfrac{10}{18} - \dfrac{3}{18} = p$

$\qquad -\dfrac{13}{18} = p$

The solution is $-\dfrac{13}{18}$.

31. Strategy   To find the fraction, write a fraction with 40 in the numerator and the number of minutes in one hour (60) in the denominator. Write the fraction in simplest form.

     Solution $\dfrac{40}{60} = \dfrac{2 \cdot 2 \cdot 2 \cdot 5}{2 \cdot 2 \cdot 3 \cdot 5} = \dfrac{2}{3}$

            40 min is $\dfrac{2}{3}$ of an hour.

32. Strategy   To find the entire length, substitute $12\dfrac{1}{12}$, $29\dfrac{1}{3}$, and $26\dfrac{3}{4}$ for a, b, and c in the given formula and solve for P.

     Solution $P = a + b + c$

$\qquad P = 12\dfrac{1}{12} + 29\dfrac{1}{3} + 26\dfrac{3}{4}$

$\qquad\quad = 12\dfrac{1}{12} + 29\dfrac{4}{12} + 26\dfrac{9}{12}$

$\qquad\quad = 68\dfrac{2}{12} = 68\dfrac{1}{6}$

     The entire length is $68\dfrac{1}{6}$ yd.

33. Strategy   To find the amount of weight to gain:

→Add the amounts already gained $\left(3\dfrac{1}{2} + 2\dfrac{1}{4}\right)$

→Subtract the amount gained from the goal amount (12).

     Solution $3\dfrac{1}{2} + 2\dfrac{1}{4} = 3\dfrac{2}{4} + 2\dfrac{1}{4}$

$\qquad\qquad = 5\dfrac{3}{4}$

$\quad 12 - 5\dfrac{3}{4} = 11\dfrac{4}{4} - 5\dfrac{3}{4} = 6\dfrac{1}{4}$

     The wrestler has $6\dfrac{1}{4}$ lb left to gain.

34. Strategy   To find the number of units:

→Find the number of minutes in 8 h.

→Divide the number of minutes worked by the time to assemble one unit.

     Solution $8 \cdot 60 = 480$

$\quad 480 \div 2\dfrac{1}{2} = 480 \div \dfrac{5}{2}$

$\qquad\qquad = \dfrac{480}{1} \cdot \dfrac{2}{5}$

$\qquad\qquad = 192$

     The employee can assemble 192 units.

35. Strategy   To find the overtime pay, substitute $6\frac{1}{4}$ for H and 24 for R in the given formula and solve for P.

Solution $P = RH$

$P = 24 \cdot 6\frac{1}{4}$

$P = 24 \cdot \frac{25}{4}$

$P = \frac{24}{1} \cdot \frac{25}{4} = 150$

The employee is due $150 in overtime pay.

36. Strategy   To find the final velocity, substitute 0 for S and $15\frac{1}{2}$ for t in the given formula and solve for t.

Solution $V = S + 32t$

$V = 0 + 32 \cdot 15\frac{1}{2}$

$V = 32 \cdot \frac{31}{2}$

$V = \frac{32}{1} \cdot \frac{31}{2} = 496$

The final velocity is 496 ft/s.

Chapter Test

1.
$$7\overline{)\begin{array}{c} 2 \\ 18 \\ -14 \\ \hline 4 \end{array}} \qquad \frac{18}{7} = 2\frac{4}{7}$$

2. $7\frac{3}{4} - 3\frac{5}{6} = 7\frac{9}{12} - 3\frac{10}{12}$

$= 6\frac{21}{12} - 3\frac{10}{12}$

$= 3\frac{11}{12}$

3. xy

$6\frac{3}{7} \cdot 3\frac{1}{2} = \frac{45}{7} \cdot \frac{7}{2}$

$= \frac{45}{2}$

$= 22\frac{1}{2}$

4. $-\frac{2}{3} \cdot \left(\frac{-7}{8}\right) = \frac{2}{3} \cdot \frac{7}{8}$

$= \frac{2 \cdot 7}{3 \cdot 8}$

$= \frac{7}{12}$

5. $30 = 2 \cdot 3 \cdot 5$
$45 = 3 \cdot 3 \cdot 5$
$\text{LCM} = 2 \cdot 3^2 \cdot 5 = 90$

6. $\frac{11}{12} + \left(\frac{-3}{8}\right) = \frac{22}{24} + \frac{-9}{24}$

$= \frac{22 - 9}{24}$

$= \frac{13}{24}$

7. $x^3 y^2$

$\left(1\frac{1}{2}\right)^3 \cdot \left(\frac{5}{6}\right)^2 = \left(\frac{3}{2}\right)^3 \cdot \left(\frac{5}{6}\right)^2$

$= \frac{3}{2} \cdot \frac{3}{2} \cdot \frac{3}{2} \cdot \frac{5}{6} \cdot \frac{5}{6}$

$= \frac{3 \cdot 3 \cdot 3 \cdot 5 \cdot 5}{2 \cdot 2 \cdot 2 \cdot 6 \cdot 6}$

$= \frac{75}{32} = 2\frac{11}{32}$

8. $3\frac{4}{5} = \frac{(5 \cdot 3) + 4}{5}$

$= \frac{15 + 4}{5} = \frac{19}{5}$

9. $-\frac{7}{12} \div \left(\frac{-3}{4}\right) = \frac{7}{12} \cdot \frac{4}{3}$

$= \frac{7 \cdot 4}{12 \cdot 3} = \frac{7}{9}$

10. $\frac{2}{7} \div \frac{3}{14} + \frac{2}{3} = \frac{2}{7} \cdot \frac{14}{3} + \frac{2}{3}$

$= \frac{4}{3} + \frac{2}{3} = 2$

11. $\dfrac{x}{yz}$

$$\dfrac{\frac{7}{20}}{\frac{2}{15}\cdot\frac{3}{8}} = \dfrac{\frac{7}{20}}{\frac{1}{20}}$$

$$= \dfrac{7}{20} \div \dfrac{1}{20}$$

$$= \dfrac{7}{20} \cdot \dfrac{20}{1} = 7$$

12. $18 = 2 \cdot 3^2$
$54 = 2 \cdot 3^3$
$\text{GCF} = 2 \cdot 3^2 = 18$

13. $\dfrac{13}{14} - \dfrac{16}{21} = \dfrac{39}{42} - \dfrac{32}{42}$

$$= \dfrac{7}{42} = \dfrac{1}{6}$$

14. $\dfrac{60}{75} = \dfrac{2 \cdot 2 \cdot 3 \cdot 5}{3 \cdot 5 \cdot 5} = \dfrac{4}{5}$

15. $x + y + z$

$$1\dfrac{3}{8} + \dfrac{1}{2} + \dfrac{5}{6} = 1\dfrac{9}{24} + \dfrac{12}{24} + \dfrac{20}{24}$$

$$= 2\dfrac{17}{24}$$

16. $\dfrac{5}{6} = \dfrac{25}{30} \qquad \dfrac{11}{15} = \dfrac{22}{30}$

$$\dfrac{25}{30} > \dfrac{22}{30}$$

$$\dfrac{5}{6} > \dfrac{11}{15}$$

17. $a^2 b - c^2$

$$\left(\dfrac{2}{3}\right)^2 \cdot 9 - \left(\dfrac{3}{5}\right)^2 = \dfrac{2}{3} \cdot \dfrac{2}{3} \cdot 9 - \dfrac{3}{5} \cdot \dfrac{3}{5}$$

$$= 4 - \dfrac{9}{25}$$

$$= \dfrac{100}{25} - \dfrac{9}{25}$$

$$= \dfrac{91}{25} = 3\dfrac{16}{25}$$

18. $\dfrac{\frac{3}{4} - \frac{1}{3}}{\frac{1}{6} + \frac{1}{3}} = \dfrac{\frac{9}{12} - \frac{4}{12}}{\frac{1}{6} + \frac{2}{6}} = \dfrac{\frac{5}{12}}{\frac{3}{6}}$

$$= \dfrac{5}{12} \div \dfrac{3}{6}$$

$$= \dfrac{5}{12} \cdot \dfrac{6}{3} = \dfrac{5}{6}$$

19. $\dfrac{x - y}{z^3}$

$$\dfrac{\frac{4}{9} - \frac{10}{27}}{\left(\frac{1}{3}\right)^3} = \dfrac{\frac{12}{27} - \frac{10}{27}}{\frac{2}{3} \cdot \frac{2}{3} \cdot \frac{2}{3}} = \dfrac{\frac{2}{27}}{\frac{8}{27}}$$

$$= \dfrac{2}{27} \div \dfrac{8}{27}$$

$$= \dfrac{2}{27} \cdot \dfrac{27}{8} = \dfrac{1}{4}$$

20. $x \div y$

$$-\dfrac{8}{9} \div \dfrac{16}{27} = \dfrac{-8}{9} \cdot \dfrac{27}{16}$$

$$= -\dfrac{3}{2} = -1\dfrac{1}{2}$$

21. $\dfrac{3x}{5} = -\dfrac{3}{10}$

$$\dfrac{5}{3} \cdot \dfrac{3}{5} x = \dfrac{5}{3} \cdot \left(-\dfrac{3}{10}\right)$$

$$x = -\dfrac{1}{2}$$

22.

$$z + \dfrac{1}{5} = \dfrac{11}{20}$$

$-\dfrac{3}{4} + \dfrac{1}{5}$	$\dfrac{11}{20}$
$-\dfrac{15}{20} + \dfrac{4}{20}$	$\dfrac{11}{20}$
$-\dfrac{11}{20}$	$\dfrac{11}{20}$

$$-\dfrac{11}{20} \neq \dfrac{11}{20}$$

No, $-\dfrac{3}{4}$ is not a solution of the equation.

23. $2\dfrac{7}{8} \cdot \dfrac{2}{11} \cdot 4 = \dfrac{23}{8} \cdot \dfrac{2}{11} \cdot \dfrac{4}{1}$

$$= \dfrac{23}{11} = 2\dfrac{1}{11}$$

24. $x + \dfrac{1}{3} = \dfrac{5}{6}$

$x + \dfrac{1}{3} - \dfrac{1}{3} = \dfrac{5}{6} - \dfrac{1}{3}$

$x = \dfrac{5}{6} - \dfrac{2}{6}$

$x = \dfrac{3}{6} = \dfrac{1}{2}$

25. $28 \div 7 = 4$

$\dfrac{3}{7} = \dfrac{3 \cdot 4}{7 \cdot 4} = \dfrac{12}{28}$

$\dfrac{12}{28}$ is equivalent to $\dfrac{3}{7}$.

26. The unknown number: n

a number minus one-half	equals	one-third

$n - \dfrac{1}{2} = \dfrac{1}{3}$

$n - \dfrac{1}{2} + \dfrac{1}{2} = \dfrac{1}{3} + \dfrac{1}{2}$

$n = \dfrac{2}{6} + \dfrac{3}{6}$

$n = \dfrac{5}{6}$

The number is $\dfrac{5}{6}$.

27. $\dfrac{10}{24} = \dfrac{5}{12}$

28. Strategy To find the amount of weight to lose:
→Add the amounts already lost $\left(11\dfrac{1}{6} + 8\dfrac{5}{8} \right)$.
→Subtract the amount lost from the goal amount (30).

Solution $11\dfrac{1}{6} + 8\dfrac{5}{8} = 11\dfrac{4}{24} + 8\dfrac{15}{24}$

$= 19\dfrac{19}{24}$

$30 - 19\dfrac{19}{24} = 29\dfrac{24}{24} - 19\dfrac{19}{24}$
$= 10\dfrac{5}{24}$

The patient has $10\dfrac{5}{24}$ lbs left to lose.

29. Strategy To find the amount of hamburger meat, multiply the number of hamburgers per person (2) by the weight of each hamburger $\left(\dfrac{1}{4} \right)$ by the number of guests (35).

Solution $2 \cdot \dfrac{1}{4} \cdot 35 = \dfrac{2}{1} \cdot \dfrac{1}{4} \cdot \dfrac{35}{1} = 17\dfrac{1}{2}$

You should buy $17\dfrac{1}{2}$ lbs of hamburger.

30. Strategy To find the amount of felt needed, substitute 20 for b and 12 for h in the given formula and solve for A.

Solution $A = \dfrac{1}{2}bh$

$A = \dfrac{1}{2} \cdot 20 \cdot 12 = 120$

The amount of felt needed is 120 in^2.

31. Strategy To find the amount of hours still required:

→Add the hours already contributed $\left(7\frac{1}{4}+2\frac{3}{4}\right)$.

→Subtract the amount already contributed from the total hours required (20).

Solution $7\frac{1}{4}+2\frac{3}{4}=10$

$20-10=10$

You must still contribute 10 h of community service.

32. Strategy To find the number of units:

→Multiply the number of hours (6) by 60 minutes.

→Divide the minutes by the time to assemble one unit $\left(4\frac{1}{2}\right)$.

Solution $6\cdot 60=360$

$360\div 4\frac{1}{2}=\frac{360}{1}\div\frac{9}{2}$

$=\frac{360}{1}\cdot\frac{2}{9}=80$

The employee can assemble 80 units in 6 h.

33. Strategy To find the rate, substitute the distance walked $\left(4\frac{1}{2}\right)$ for d

and time walked $\left(1\frac{1}{2}\right)$ for t

in the given formula and solve for r.

Solution $r=\dfrac{d}{t}$

$r=\dfrac{4\frac{1}{2}}{1\frac{1}{2}}=\dfrac{\frac{9}{2}}{\frac{3}{2}}=\dfrac{9}{2}\cdot\dfrac{2}{3}=3$

The hiker walked at a rate of 3 mph.

Cumulative Review Exercises

1. $3a+(a-b)^3$

$3\cdot 4+(4-1)^3=3\cdot 4+3^3$

$=3\cdot 4+27$

$=12+27$

$=39$

2. $4\cdot\dfrac{7}{8}=\dfrac{4}{1}\cdot\dfrac{7}{8}$

$=\dfrac{4\cdot 7}{1\cdot 8}$

$=\dfrac{2\cdot 2\cdot 7}{1\cdot 2\cdot 2\cdot 2}$

$=\dfrac{7}{2}=3\dfrac{1}{2}$

3. $4\dfrac{7}{9}+3\dfrac{5}{6}=4\dfrac{14}{18}+3\dfrac{15}{18}$

$=7\dfrac{29}{18}$

$=8\dfrac{11}{18}$

4. $-42-(-27)=-42+27$

$=-15$

5. $72=2\cdot 2\cdot 2\cdot 3\cdot 3$

$108=2\cdot 2\cdot 3\cdot 3\cdot 3$

$GCF=2\cdot 2\cdot 3\cdot 3=36$

6. $3\dfrac{1}{13}\cdot 5\dfrac{1}{5}=\dfrac{40}{13}\cdot\dfrac{26}{5}=\dfrac{40\cdot 26}{13\cdot 5}$

$=\dfrac{2\cdot 2\cdot 2\cdot 5\cdot 2\cdot 13}{13\cdot 5}$

$=16$

7. $\dfrac{8}{9}\div\left(-\dfrac{4}{5}\right)=-\left(\dfrac{8}{9}\div\dfrac{4}{5}\right)$

$=-\left(\dfrac{8}{9}\cdot\dfrac{5}{4}\right)$

$=-\dfrac{8\cdot 5}{9\cdot 4}$

$=-\dfrac{2\cdot 2\cdot 2\cdot 5}{3\cdot 3\cdot 2\cdot 2}$

$=-\dfrac{10}{9}=-1\dfrac{1}{9}$

8. $-\dfrac{2}{3}-\left(-\dfrac{2}{5}\right)=\dfrac{-2}{3}-\left(\dfrac{-2}{5}\right)$

$=\dfrac{-10}{15}-\dfrac{-6}{15}=\dfrac{-10-(-6)}{15}$

$=\dfrac{-10+6}{15}=\dfrac{-4}{15}=-\dfrac{4}{15}$

9. $\dfrac{\frac{1}{5}+\frac{1}{4}}{\frac{1}{4}-\frac{1}{5}}=\dfrac{\frac{9}{20}}{\frac{1}{20}}$

$=\dfrac{9}{20}\div\dfrac{1}{20}$

$=\dfrac{9}{20}\cdot\dfrac{20}{1}=\dfrac{9\cdot20}{20\cdot1}$

$=9$

10. $\dfrac{7}{11}=\dfrac{35}{55}\qquad\dfrac{4}{5}=\dfrac{44}{55}$

$\dfrac{35}{55}<\dfrac{44}{55}$

$\dfrac{7}{11}<\dfrac{4}{5}$

11. $-2\dfrac{1}{3}\div1\dfrac{2}{7}=-\left(\dfrac{7}{3}\div\dfrac{9}{7}\right)$

$=-\left(\dfrac{7}{3}\cdot\dfrac{7}{9}\right)$

$=-\dfrac{7\cdot7}{3\cdot9}$

$=-\dfrac{7\cdot7}{3\cdot3\cdot3}=-\dfrac{49}{27}$

$=-1\dfrac{22}{27}$

12. $-\dfrac{3}{8}\cdot\dfrac{2}{5}\cdot\left(-\dfrac{4}{9}\right)=\dfrac{3}{8}\cdot\dfrac{2}{5}\cdot\dfrac{4}{9}$

$=\dfrac{3\cdot2\cdot4}{8\cdot5\cdot9}$

$=\dfrac{3\cdot2\cdot2\cdot2}{2\cdot2\cdot2\cdot5\cdot3\cdot3}$

$=\dfrac{1}{15}$

13. abc

$\dfrac{4}{7}\cdot1\dfrac{1}{6}\cdot3=\dfrac{4}{7}\cdot\dfrac{7}{6}\cdot\dfrac{3}{1}$

$=\dfrac{4\cdot7\cdot3}{7\cdot6\cdot1}$

$=\dfrac{2\cdot2\cdot7\cdot3}{7\cdot2\cdot3\cdot1}=2$

14. $8\dfrac{3}{4}-1\dfrac{5}{7}=8\dfrac{21}{28}-1\dfrac{20}{28}$

$=7\dfrac{1}{28}$

15. $\dfrac{7}{12}-\left(-\dfrac{3}{8}\right)=\dfrac{7}{12}-\dfrac{-3}{8}$

$=\dfrac{14}{24}-\dfrac{-9}{24}=\dfrac{14-(-9)}{24}$

$=\dfrac{14+9}{24}=\dfrac{23}{24}$

16. $\dfrac{2}{5}\div\dfrac{9-6}{3+7}+\left(-\dfrac{1}{2}\right)^2=\dfrac{2}{5}\div\dfrac{3}{10}+\left(-\dfrac{1}{2}\right)^2$

$=\dfrac{2}{5}\div\dfrac{3}{10}+\dfrac{1}{4}$

$=\dfrac{2}{5}\cdot\dfrac{10}{3}+\dfrac{1}{4}$

$=\dfrac{4}{3}+\dfrac{1}{4}=\dfrac{19}{12}$

$=1\dfrac{7}{12}$

17. $a-b$

$\dfrac{3}{4}-\left(-\dfrac{7}{8}\right)=\dfrac{3}{4}-\dfrac{-7}{8}$

$=\dfrac{6}{8}-\dfrac{-7}{8}$

$=\dfrac{6-(-7)}{8}$

$=\dfrac{6+7}{8}$

$=\dfrac{13}{8}=1\dfrac{5}{8}$

18. $1\dfrac{9}{16}+4\dfrac{5}{8}=1\dfrac{9}{16}+4\dfrac{10}{16}$

$=5\dfrac{19}{16}=6\dfrac{3}{16}$

19. $28=-7y$

$\dfrac{28}{-7}=\dfrac{-7y}{-7}$

$-4=y$

The solution is -4.

20. $9\overline{)41}\;\dfrac{4}{}\qquad\dfrac{41}{9}=4\dfrac{5}{9}$

$\underline{-36}$

5

21. $\dfrac{5}{14}-\dfrac{9}{42}=\dfrac{15}{42}-\dfrac{9}{42}$

$=\dfrac{6}{42}=\dfrac{1}{7}$

22. x^3y^4

$\left(\dfrac{7}{12}\right)^3\left(\dfrac{6}{7}\right)^4=\dfrac{7}{12}\cdot\dfrac{7}{12}\cdot\dfrac{7}{12}\cdot\dfrac{6}{7}\cdot\dfrac{6}{7}\cdot\dfrac{6}{7}\cdot\dfrac{6}{7}$

$=\dfrac{7\cdot7\cdot7\cdot6\cdot6\cdot6\cdot6}{12\cdot12\cdot12\cdot7\cdot7\cdot7\cdot7}$

$=\dfrac{3}{28}$

23. $2a - (b - a)^2$

$2 \cdot 2 - (-3 - 2)^2 = 2 \cdot 2 - (-5)^2$

$= 2 \cdot 2 - 25$

$= 4 - 25 = 4 + (-25)$

$= -21$

24.
$$\begin{array}{r} 6,847 \\ 3,501 \\ +924 \\ \hline 11,272 \end{array}$$

25. $(x - y)^3 + 5x$

$(8 - 6)^3 + 5 \cdot 8 = 2^3 + 5 \cdot 8$

$= 8 + 5 \cdot 8$

$= 8 + 40$

$= 48$

26. $x + \dfrac{4}{5} = \dfrac{1}{4}$

$x + \dfrac{4}{5} - \dfrac{4}{5} = \dfrac{1}{4} - \dfrac{4}{5}$

$x = \dfrac{5}{20} - \dfrac{16}{20}$

$x = \dfrac{5 - 16}{20}$

$x = -\dfrac{11}{20}$

The solution is $-\dfrac{11}{20}$.

27.
$$\begin{array}{r} 89,357 \rightarrow 90,000 \\ 66,042 \rightarrow -70,000 \\ \hline 20,000 \end{array}$$

28. $-8 - (-12) - (-15) - 32$

$= -8 + 12 + 15 + (-32)$

$= 4 + 15 + (-32)$

$= 19 + (-32)$

$= -13$

29. $7\dfrac{3}{4} = \dfrac{(4 \cdot 7) + 3}{4}$

$= \dfrac{28 + 3}{4}$

$= \dfrac{31}{4}$

30.
$$\begin{array}{r} 7 \\ 5{\overline{)35}} \end{array}$$
$$2{\overline{)70}}$$
$$2{\overline{)140}}$$

$140 = 2 \cdot 2 \cdot 5 \cdot 7 = 2^2 \cdot 5 \cdot 7$

31. Strategy   To determine how many more calories you would burn:

→Calculate the number of calories burned by bicycling at 12 mph for 4 h.

→Calculate the number of calories burned by walking at a rate of 3 mph for 5 h.

→Find the difference between the two calculations.

Solution   Bicycling: $4 \cdot 410 = 1,640$

Walking: $320 \cdot 5 = 1,600$

$1,640 - 1,600 = 40$

You would burn 40 more calories by bicycling at 12 mph for 4 h.

32. Strategy   To find the population increase, subtract the population in 2000 (13,922,517) from the population in 2010 (14,444,865).

Solution $14,444,865 - 13,922,517 = 522,348$

The projected increase in population is 522,348 people.

33. Strategy   To find the average life span of a $100 bill, multiply the average life span of a $1 $\left(1\dfrac{1}{2} \text{ years}\right)$ by 6.

Solution $1\dfrac{1}{2} \cdot 6 = \dfrac{3}{2} \cdot \dfrac{6}{1} = 9$

The average life span of a $100 bill is 9 years.

34. Strategy   To find the length of fencing, substitute $16\frac{1}{2}$ for s in the given formula and solve for P.

Solution $P = 4s$

$$P = 4 \cdot 16\frac{1}{2} = \frac{4}{1} \cdot \frac{33}{2}$$

$$P = \frac{132}{2} = 66$$

The length of fencing needed is 66 ft.

35. Strategy   To find the distance traveled, substitute $5\frac{1}{2}$ for r and $\frac{3}{4}$ for t in the given formula and solve for d.

Solution $d = rt$

$$d = 5\frac{1}{2} \cdot \frac{3}{4}$$

$$d = \frac{11}{2} \cdot \frac{3}{4}$$

$$d = \frac{33}{8} = 4\frac{1}{8}$$

The distance traveled is $4\frac{1}{8}$ miles.

36. Strategy   To find the pressure, substitute $14\frac{3}{4}$ for D in the given formula and solve for P.

Solution
$$P = 15 + \frac{1}{2}D$$

$$P = 15 + \frac{1}{2} \cdot 14\frac{3}{4}$$

$$P = 15 + \frac{1}{2} \cdot \frac{59}{4}$$

$$P = 15 + 7\frac{3}{8} = 22\frac{3}{8}$$

The pressure is $22\frac{3}{8}$ pounds per square inch.

Chapter 4: Decimals and Real Numbers

Prep Test

1. $\frac{3}{10}$

2. 36,900

3. four thousand seven hundred ninety-one

4. 6,842

5.

6. $-37 + 8,892 + 465 = 8,855 + 465 = 9,320$

7. $2,403 - (-765) = 2,403 + 765 = 3,168$

8. $-844(-91) = 76,804$

9.
$$
\begin{array}{r}
278\,\text{r}18 \\
23\overline{)6412} \\
\underline{-46} \\
181 \\
\underline{-161} \\
202 \\
\underline{-184} \\
18
\end{array}
$$

10. $8^2 = (8)(8) = 64$

Section 4.1

Concept Check

1. hundredths; thousandths; hundred-thousandths; millionths

3. and; thousandths

5. <; <

Objective A Exercises

7. The digit 5 is in the thousandths' place.

9. The digit 5 is in the ten-thousandths' place.

11. The digit 5 is in the hundredths' place.

13. $\frac{3}{10} = 0.3$ [three tenths]

15. $\frac{21}{100} = 0.21$ [twenty-one hundredths]

17. $\frac{461}{1,000} = 0.461$ [four hundred sixty-one thousandths]

19. $\frac{93}{1,000} = 0.093$ [ninety-three thousandths]

21. $0.1 = \frac{1}{10}$ [one tenth]

23. $0.47 = \frac{47}{100}$ [forty-seven hundredths]

25. $0.289 = \frac{289}{1,000}$ [two hundred eighty-nine thousandths]

27. $0.09 = \frac{9}{100}$ [nine hundredths]

29. thirty-seven hundredths

31. nine and four tenths

33. fifty-three ten-thousandths

35. forty-five thousandths

37. twenty-six and four hundredths

39. 3.0806

41. 407.03

43. 246.024

45. 73.02684

Objective B Exercises

47. 0.7 = 0.70
0.70 > 0.56
0.7 > 0.56

49. 3.605 > 3.065

51. 9.04 = 9.040
9.004 < 9.040
9.004 < 9.04

53. 9.31 = 9.310
9.310 > 9.031
9.31 > 9.031

55. 4.6 < 40.6

57. 0.07046 > 0.07036

59. 0.609, 0.660, 0.696, 0.699
0.609, 0.66, 0.696, 0.699

61. 1.237, 1.327, 1.372, 1.732

63. 21.780, 21.805, 21.870, 21.875
21.78, 21.805, 21.87, 21.875

65. 0.62 > 0.062

Objective C Exercises

67. 5.3̅9̅8̅ *Given place value*
9 > 5

5.398 rounded to the nearest tenth is 5.4.

69. 30.0̅0̅9̅2 *Given place value*
0 < 5

30.0092 rounded to the nearest tenth is 30.0.

71. 413.5̅9̅7̅2 *Given place value*
7 > 5

413.5972 rounded to the nearest hundredth is 413.60.

73. 6.06̅1̅7̅4̅5 *Given place value*
7 > 5

6.061745 rounded to the nearest thousandth is 6.062.

75. 96.8̅0̅2̅7 *Given place value*
8 > 5

96.8027 rounded to the nearest whole number is 97.

77. 5,439.8̅3 *Given place value*
8 > 5

5,439.83 rounded to the nearest whole number is 5,440.

79. 0.02̅3̅5̅9̅1 *Given place value*
9 > 5

0.023591 rounded to the nearest ten-thousandth is 0.0236.

Objective D Exercises

81. Strategy
a. To find if the cost to mint a penny is greater or less than the face value of a penny, compare the cost to mint a penny (1.67 cents) with the value of a penny (1 cent).

b. To find the cost to mint a penny to the nearest cent, round the cost to mint a penny (1.67 cents) to the nearest integer.

c. To find if the cost to mint a nickel is greater or less than the face value of a nickel, compare the cost to mint a nickel (7.7 cents) with the value of a nickel (5 cents).

d. To find the cost to mint a nickel to the nearest cent, round the cost to mint a nickel (7.7 cents) to the nearest integer.

Solution
a. 1.67 > 1
The cost to mint a penny is greater than the face value of a penny.

b. 1.67 rounded to the nearest integer is 2. The cost to mint a penny to the nearest cent is 2 cents.

c. 7.7 > 5
The cost to mint a nickel is greater than the face value of a nickel.

d. 7.7 rounded to the nearest integer is 8. The cost to mint a nickel to the nearest cent is 8 cents.

83. Strategy   To find the distance, round 26.21875 to the nearest tenth.

Solution   26.21875 rounded to the nearest tenth is 26.2.
To the nearest tenth, the Boston Marathon is 26.2 mi.

85. Strategy

To find the shipping and handling charges for each order, compare each amount ordered with the amounts in the table and read the corresponding shipping and handling charges from the table.

Solution

a. $10.01 < $12.42 < $20.00

The shipping and handling charge is $2.40.

b. $20.01 < $23.56 < $30.00

The shipping and handling charge is $3.60.

c. $40.01 < $47.80 < $50.00

The shipping and handling charge is $6.00.

d. $66.91 > $50.01

The shipping and handling charge is $7.00.

e. $30.01 < $35.75 < $40.00

The shipping and handling charge is $4.70.

f. $20.00 = $20.00

The shipping and handling charge is $2.40.

g. $10.01 < $18.25 < $20.00

The shipping and handling charge is $2.40.

Critical Thinking 4.1

87. a. Answers will vary.

For example, 0.11, 0.12, 0.13, 0.14, 0.15, 0.16, 0.17, 0.18, and 0.19 are numbers between 0.1 and 0.2. But any number of digits can be attached to 0.1, and the number will be between 0.1 and 0.2. For example, 0.123456789 is a number between 0.1 and 0.2.

b. Answers will vary.

For example, 1.01, 1.02, 1.03, 1.04, 1.05, 1.06, 1.07, 1.08, and 1.09 are numbers between 1 and 1.1. But any number of digits can be attached to 1.0, and the number will be between 1 and 1.1. For example, 1.0123456789 is a number between 1 and 1.1.

c. Answers will vary.

For example, 0.001, 0.002, 0.003, and 0.004 are numbers between 0 and 0.005. But any number of digits can be attached to 0.001, 0.002, 0.003, or 0.004, and the number will be between 0 and 0.005. For example, 0.00123456789 is a number between 0 and 0.005.

Section 4.2

Concept Check

1.

$$\begin{array}{r} \overset{1\ \ \ 1}{2.391} \\ 45 \\ + 13.0784 \\ \hline 60.4694 \end{array}$$

Objective A Exercises

3.

$$\begin{array}{r} \overset{1\ \ \ 11}{1.864} \\ 39 \\ + \ 25.0781 \\ \hline 65.9421 \end{array}$$

5.

$$\begin{array}{r} \overset{2\,1}{35.9} \\ 8.217 \\ + 146.74 \\ \hline 190.857 \end{array}$$

7.

$$\begin{array}{r} 36.47 \\ - 15.21 \\ \hline 21.26 \end{array}$$

9.

$$\begin{array}{r} \overset{7\ 9\ 10}{28.0\!\!\!/0\!\!\!/} \\ - \ 6.74 \\ \hline 21.26 \end{array}$$

11.

$$\begin{array}{r} \overset{5\ \ 9\,11\,10}{6.0\!\!\!/2\!\!\!/0\!\!\!/} \\ - 3.252 \\ \hline 2.768 \end{array}$$

13. $-42.1 - 8.6 = -42.1 + (-8.6)$
$= -50.7$

15. $5.73 - 9.042 = 5.73 + (-9.042)$
$= -3.312$

17. $-9.37 + 3.465 = -5.905$

19. $-19 - (-2.65) = -19 + 2.65$
$= -16.35$

21. $-12.3 - 4.07 + 6.82$
$= -12.3 + (-4.07) + 6.82$
$= -16.37 + 6.82$
$= -9.55$

23. $-5.6 - (-3.82) - 17.409$
$= -5.6 + 3.82 + (-17.409)$
$= -1.78 + (-17.409)$
$= -19.189$

25. $6.24 + 8.573 + 19.06 + 22.488$
$= 14.813 + 19.06 + 22.488$
$= 33.873 + 22.488$
$= 56.361$

27. $62.57 - 8.9 = 62.57 + (-8.9)$
$= 53.67$

29. $-65.47 + (-32.91) = -98.38$

31. $-138.72 - 510.64$
$= -138.72 + (-510.64) = -649.36$

33. $-31 - (-62.09) = -31 + 62.09$
$= 31.09$

35. $\begin{array}{r} 6.408 \\ +\, 5.917 \\ \hline 12.325 \end{array}$ $\rightarrow$ $\begin{array}{r} 6 \\ +\, 6 \\ \hline 12 \end{array}$

37. $\begin{array}{r} 56.87 \\ -\, 23.24 \\ \hline 33.63 \end{array}$ $\rightarrow$ $\begin{array}{r} 60 \\ -\, 20 \\ \hline 40 \end{array}$

39. $\begin{array}{r} 0.931 \\ -\, 0.628 \\ \hline 0.303 \end{array}$ $\rightarrow$ $\begin{array}{r} 0.9 \\ -\, 0.6 \\ \hline 0.3 \end{array}$

41. $\begin{array}{r} 87.65 \\ -\, 49.032 \\ \hline 38.618 \end{array}$ $\rightarrow$ $\begin{array}{r} 90 \\ -\, 50 \\ \hline 40 \end{array}$

43. **a.** $48.54 + 6.32 + 1.1 = 55.96$
The total children is
55.96 million.

   b. $48.54 - 6.32 = 42.22$
There are 42.22 million more children
in public school.

45. $x + y$
$5.904 + (-7.063) = -1.159$

47. $x + y$
$-6.175 + (-19.49) = -25.665$

49. $x + y + z$
$-6.059 + 3.884 + 15.71$
$= -2.175 + 15.71$
$= 13.535$

51. $x + y + z$
$-16.219 + 47 + (-2.3885)$
$= 30.781 + (-2.3885)$
$= 28.3925$

53. $x - y$
$6.029 - (-4.708) = 6.029 + 4.708$
$= 10.737$

55. $x - y$
$-21.073 - 6.48 = -21.073 + (-6.48)$
$= -27.553$

57. $x - y$
$-8.21 - (-6.798) = -8.21 + 6.798$
$= -1.412$

59. $\begin{array}{c} 0.8 - p = 3.6 \\ \hline \begin{array}{c|c} 0.8 - (-2.8) & 3.6 \\ 0.8 + 2.8 & 3.6 \end{array} \\ 3.6 = 3.6 \end{array}$
Yes, -2.8 is a solution of the equation.

61. $\begin{array}{c} 27.4 = y - 9.4 \\ \hline \begin{array}{c|c} 27.4 & 36.8 - 9.4 \end{array} \\ 27.4 = 27.4 \end{array}$
Yes, 36.8 is a solution of the equation.

63. iii

Objective B Exercises

65. **a.** The total amount of money deposited in
the account during the month

   b. The total amount of money withdrawn
from the account during the month

   c. The amount of money in the account at
the end of the month

67. Strategy   To find how many degrees the
temperature fell, subtract the
lower temperature
(-13.33) from the higher
temperature (12.78).

Solution $12.78 - (-13.33)$
$= 12.78 + 13.33 = 26.11$
The temperature fell $26.11°C$
during the 15-minute period.

69. Strategy   To find between which two years the difference in net income was the greatest, find the change in net income from 2006 to 2007, 2007 to 2008, 2008 to 2009 and 2009 to 2010, then compare the results.

Solution   2006 to 2007: $-2.88 - (-5.63)$
$= -2.88 + 5.63 = 2.75$
2007 to 2008: $-5.98 - (-2.88)$
$= -5.98 + 2.88 = -3.1$
2008 to 2009: $0.87 - (-5.98)$
$= 0.87 + 5.98 = 6.85$
2009 to 2010: $0.19 - 0.87 = -0.68$
The difference in net income was greatest between 2008 and 2009 with an increase of $6.85 billion.

71. Strategy   To find the new balance:
→Add the amount of the deposit (189.53) to the old balance (347.80).
→Subtract the amount of the check (62.89) from the sum

Solution $347.08 + 189.53 = 536.61$
$536.61 - 62.89 = 473.72$
The new balance is $473.72

73. Strategy   To estimate the bill, estimate the following items and then add the estimated values: soup (5.75), cheese sticks (8.25), swordfish (26.95), chicken divan (24.95), and carrot cake (7.25).

Solution
$5.75 \rightarrow 6.00$
$8.25 \rightarrow 8.00$
$26.95 \rightarrow 30.00$
$24.95 \rightarrow 20.00$
$7.25 \rightarrow + 7.00$
$\overline{71.00}$

The bill is approximately $71.00.

75. Strategy
a. To determine if life expectancy has increased for both males and females between each 10-year period shown in the graph, see if the numbers increase from left to right along each of the lines in the graph.

b. To determine whether males or females had a longer life expectancy, compare the life expectancy for males in 2000 with the life expectancy for females in 2000. To find the difference in the life expectancy, subtract the smaller of the two numbers from the larger.

c. To determine during which year the difference between male and female life expectancy was greatest, subtract the lower life expectancy from the higher for each year shown in the graph. Find the largest difference.

Solution
a. The numbers along each of the lines in the graph increase from left to right. Life expectancy has increased for both males and females between each 10-year period shown in the graph. Yes.

b. $79.4 > 73.6$
Females had a longer life expectancy.
$79.4 - 73.6 = 5.8$
Female life expectancy was longer by 5.8 years.

c. 1900: $49.6 - 49.1 = 0.5$
1910: $53.7 - 50.2 = 3.5$
1920: $56.3 - 54.6 = 1.7$
1930: $61.4 - 58.0 = 3.4$
1940: $65.3 - 60.9 = 4.4$
1950: $70.9 - 65.3 = 5.6$
1960: $73.2 - 66.6 = 6.6$
1970: $74.8 - 67.1 = 7.7$
1980: $77.5 - 69.9 = 7.6$
1990: $78.6 - 71.8 = 6.8$
2000: $79.4 - 73.6 = 5.6$
7.7 is the largest difference.
The difference between male and female life expectancy was greatest in 1970.

77. Strategy   To find the markup, substitute
2,231.81 for S and 1,653.19 for
C in the given formula and
solve for M.

Solution $M = S - C$
$M = 2,231.81 - 1,653.19$
$M = 578.62$
The markup is \$578.62.

79. Strategy   To find the equity, replace V by
225,000 and L by 167,853.25
in the given formula and solve
for E.

Solution $E = V - L$
$E = 225,000 - 167,853.25$
$E = 57,146.75$
The equity on the home is
\$57,146.75.

Critical Thinking 4.2

81. By trial and error, we see that
$3.45 + 6.78 + 9.01 = 19.24$

Projects or Group Activities 4.2

83. Theoretically, there is a temperature that is
the lowest possible temperature. Scientists
refer to the lowest possible temperature as
absolute zero. [Your students may not have
the scientific sophistication to understand
the definition of absolute zero as the
temperature at which a system has no
thermodynamic energy. An ideal gas under
constant pressure would reach zero volume
at absolute zero. In practice, molecular
motion continues at absolute zero, but no
energy is available for transfer to other
systems. When a material's temperature
approaches absolute zero, there is a
significant increase in conductivity.]

An English physicist, Lord William Kelvin,
devised the temperature scale called the
Kelvin scale. On the Kelvin scale, the zero
point is absolute zero and each degree
(which in the Kelvin scale is called a
Kelvin) is the same size as the Celsius
degree. The letter K denotes a temperature
on the Kelvin scale.

It is estimated that the temperature of
absolute zero is –273.15°C, or 273.15
degrees below 0°C. Thus, on the Kelvin
scale, the temperature at which water
freezes is 273.15 K, and the temperature at
which water boils is 373.15 K.

To convert temperatures from the Celsius
scale to the Kelvin scale, add 273.15 to the
Celsius temperature. To convert
temperatures from the Kelvin scale to the
Celsius scale, subtract 273.15 from the
Kelvin temperature. For example, –40°C =
$(-40 + 273.15)$K = 233.15 K, and
240 K = $(240 - 273.15) = -33.15$°C.

Section 4.3

Concept Check

1. 1; 2; 3; 1.113

3. two; whole; $304 \overline{)364.8}$ with quotient 1.2

5. 5; 4; terminating

7. multiplication

Objective A Exercises

9.
$$\begin{array}{r} 3.4 \\ \times\,0.5 \\ \hline 1.70 \end{array}$$

11.
$$\begin{array}{r} 8.29 \\ \times\,0.004 \\ \hline 0.03316 \end{array}$$

13. $(-6.3)(-2.4) = 15.12$

15. $-1.3(4.2) = -5.46$

17. $1.31(-0.006) = -0.00786$

19. $(-100)(4.73) = -473$

21. $1,000 \cdot 4.25 = 4,250$

23. $6.71 \cdot 10^4 = 67,100$

25. $(0.06)(-0.4)(-1.5) = (-0.024)(-1.5)$
$= 0.036$

27. 9.81 → 10
 0.77 → $\underline{\times 0.8}$
 8.0
 9.81 · 0.77 = 7.5537

29. 6.88 → 7
 9.97 → $\underline{\times 10}$
 70
 6.88 · 9.97 = 68.5936

31. 28.45 → 30
 1.13 → $\underline{\times 1}$
 30
 28.45 · 1.13 = 32.1485

33. 20,000(0.5266) = 10,532
 10,532 British pounds would be exchanged
 for 20,000 U.S. dollars.

35. *ab*
 6.27 · 8 = 50.16

37. 10*t*
 10(−4.8) = −48

39. *ab*
 (0.379)(−0.22) = −0.08338

41. *cd*
 (−2.537)(−9.1) = 23.0867

43. 1.6 = −0.2*z*
 $\overline{1.6\,|\,-0.2(-8)}$
 1.6 = 1.6
 Yes, −8 is a solution of the equation.

45. −83.25*r* = 8.325
 $\overline{-83.25(-10)\,|\,8.325}$
 832.5 ≠ 8.325
 No, −10 is not a solution of the equation.

47. two

Objective B Exercises

49. 32.3
 0.5$\overline{)\,16.15}$
 $\underline{-15}$
 11
 $\underline{-10}$
 15
 $\underline{-15}$
 0

51. 27.08 ÷ (−0.4) = −67.7

53. (−3.312) ÷ (−0.8) = 4.14

55. −2.501 ÷ 0.41 = −6.1

57. 55.63 ÷ 8.8 ≈ 6.3

59. (−52.8) ÷ (−9.1) ≈ 5.8

61. 6.457 ÷ 8 ≈ 0.81

63. 0.0416 ÷ (−0.53) ≈ −0.08

65. 52.78 ÷ 10 = 5.278

67. $48.05 \div 10^2 = 0.4805$

69. −19.04 ÷ 0.75 ≈ −25.4

71. 27.735 ÷ (−60.3) ≈ −0.5

73. 42.43 → 40
 3.8 → 4
 42.43 ÷ 3.8 ≈ 11.17
 40 ÷ 4 = 10

75. 6.398 → 6
 5.5 → 6
 6.398 ÷ 5.5 ≈ 1.16
 6 ÷ 6 = 1

77. 1.237 → 1
 0.021 → 0.02
 1.237 ÷ 0.021 ≈ 58.90
 1 ÷ 0.02 = 50

79. 33.14 → 30
 4.6 → 5
 33.14 ÷ 4.6 ≈ 7.20
 30 ÷ 5 = 6

81. Strategy To determine how many times
 greater sales were in 2010 than
 in 1997, divide the amount of
 sales in 2010 (28.6) by the
 amount of sales in 1997 (3.6).

 Solution 28.6 ÷ 3.6 ≈ 8
 Sales in 2010 were
 approximately 8 times greater
 than in 1997.

83. $\dfrac{x}{y}$

 $\dfrac{3.542}{0.7} = 3.542 \div 0.7 = 5.06$

85. $\dfrac{x}{y}$

$\dfrac{0.648}{-2.7} = 0.648 \div (-2.7) = -0.24$

87. $\dfrac{x}{y}$

$\dfrac{-8.034}{-3.9} = (-8.034) \div (-3.9) = 2.06$

89. $\dfrac{x}{y}$

$\dfrac{-2.501}{0.41} = -2.501 \div 0.41 = -6.1$

91. $\dfrac{q}{-8} = -3.1$

$\dfrac{24.8}{-8} \,\Big|\, -3.1$

$-3.1 = -3.1$

Yes, 24.8 is a solution of the equation.

93. $21 = \dfrac{t}{0.4}$

$21 \,\Big|\, \dfrac{-8.4}{0.4}$

$21 \approx -21$

No, -8.4 is not a solution of the equation.

95. three

Objective C Exercises

97. $8 \overline{)3.000}$ 0.375

$\dfrac{3}{8} = 0.375$

99. $11 \overline{)8.0000}$ 0.7272

$\dfrac{8}{11} = 0.\overline{72}$

101. $12 \overline{)7.00000}$ 0.58333

$\dfrac{7}{12} = 0.58\overline{3}$

103. $4 \overline{)7.00}$ 1.75

$\dfrac{7}{4} = 1.75$

105. Write $\dfrac{1}{2}$ as a decimal.

$2 \overline{)1.0}$ 0.5

$1\dfrac{1}{2} = 1.5$

107. Write $\dfrac{1}{6}$ as a decimal.

$6 \overline{)1.0000}$ 0.1666 $= 0.1\overline{6}$

$4\dfrac{1}{6} = 4.1\overline{6}$

109. Write $\dfrac{1}{4}$ as a decimal.

$4 \overline{)1.00}$ 0.25

$2\dfrac{1}{4} = 2.25$

111. Write $\dfrac{8}{9}$ as a decimal.

$9 \overline{)8.000}$ 0.888 $= 0.\overline{8}$

$3\dfrac{8}{9} = 3.\overline{8}$

113. $0.2 = \dfrac{2}{10} = \dfrac{1}{5}$

115. $0.75 = \dfrac{75}{100} = \dfrac{3}{4}$

117. $0.125 = \dfrac{125}{1,000} = \dfrac{1}{8}$

119. $2.5 = 2\dfrac{5}{10} = 2\dfrac{1}{2}$

121. $4.55 = 4\dfrac{55}{100} = 4\dfrac{11}{20}$

123. $1.72 = 1\dfrac{72}{100} = 1\dfrac{18}{25}$

125. $0.045 = \dfrac{45}{1,000} = \dfrac{9}{200}$

127. $\dfrac{9}{10} = 0.90$

$0.90 > 0.89$

$\dfrac{9}{10} > 0.89$

129. $\dfrac{4}{5} = 0.800$

$0.800 < 0.803$

$\dfrac{4}{5} < 0.803$

131. $\dfrac{4}{9} = 0.\overline{444}$

$0.444 < 0.\overline{444}$

$0.444 < \dfrac{4}{9}$

133. $\dfrac{3}{25} = 0.12$

$0.13 > 0.12$

$0.13 > \dfrac{3}{25}$

135. $\dfrac{5}{16} = 0.3125$

$0.3125 > 0.3120$

$\dfrac{5}{16} > 0.312$

137. $\dfrac{10}{11} = 0.9\overline{09}$

$0.9\overline{09} > 0.909$

$\dfrac{10}{11} > 0.909$

139. $\dfrac{3}{5}$

Objective D Exercises

141. a. The total cost of the sodas

b. The total cost of the meal

c. The total amount of money each friend will pay

143. Strategy

a. To determine the amount of the payments over 36 months, multiply the amount of each payment (499.50) by the number of payments (36).

b. To determine the total cost of the car, add the amount of the payments to the down payment (5000).

Solution

a. $499.50 \cdot 36 = 17{,}982$
The amount of the payments over 36 months is $17,982.

b. $17{,}982 + 5000 = 22{,}982$
The total cost of the car is $17,982.

145. Strategy To find Ramon's average yards per carry, divide the number of yards gained (162) by the number of carries (26). Round to the nearest hundredth.

Solution $162 \div 26 \approx 6.23$
Ramon averaged approximately 6.23 yards per carry.

147. Strategy To find the average annual cost of electricity, multiply the number of months in a year (12) times the average monthly bill (95.66).

Solution $12 \cdot 95.66 = 1147.92$
The average annual cost of electricity $1,147.92.

149. Strategy To find the amount received for the 400 cans, multiply the weight of the cans (18.75) by the amount paid per pound (0.75). Round to the nearest cent.

Solution $18.75 \cdot 0.75 \approx 14.06$
The amount received for the 400 cans is $14.06.

151. Strategy To find the amount of taxes paid, multiply the tax per gallon (0.466) by the number of gallons (12.5).

Solution $0.466 \cdot 12.5 \approx 5.83$
The amount of taxes paid is approximately $5.83.

153. Strategy   To find the added cost to the government for issuing paper checks find the cost of issuing paper check by multiplying the number of recipients (4,000,000) by the cost per paper check (0.89). Then find the cost of issuing electronic checks by multiplying the number of recipients (4,000,000) by the cost per electronic check (0.09). Finally, find the difference between the two costs.

Solution $4,000,000 \cdot 0.89 = 3,560,000$
$4,000,000 \cdot 0.09 = 360,000$
$3,560,000 - 360,000 = 3,200,000$
The added cost is $3,200,000.

155. Strategy   To find the amount of coal you will use in one year, divide the number of kilowatt-hours used per month (25) by the number of kilowatt-hours produced by one ton of coal (3000). Then multiply that amount by the number of months in a year (12).

Solution $25 \div 3000 = 0.0083333\ldots$
$0.008333 \cdot 12 = 0.1$
You will use 0.1 ton of coal in one year.

157. Strategy
a. To find the reduction in solid waste per month, divide the reduction in solid waste per year (1.6 billion) by the number of months in a year (12). Round to the nearest whole number.

b. To find the reduction in greenhouse gas emissions per month, divide the reduction in greenhouse gas emissions per year (2.1 million) by the number of months in a year (12). Round to the nearest whole number.

Solution
a. $1,600,000,000 \div 12 \approx 133,333,333$
The reduction in solid waste per month is approximately 133,333,333 tons.

b. $2,100,000 \div 12 = 175,000$
The reduction in greenhouse gas emissions per month is 175,000 tons.

159. Strategy   To find the profit:
→Convert 5 L to milliliters.
→Find the number of bottles by dividing the number of milliliters by 250.
→Find the total cost by multiplying the number of bottles by 0.75 and adding to 175.
→Find the income by multiplying the number of bottles by 15.89.
→Find the profit by subtracting the cost from the income.

Solution $5\ L = 5,000\ ml$
$5,000 \div 250 = 20$
$20 \cdot 0.75 + 175 = 15 + 175 = 190$
$20 \cdot 15.89 = 317.80$
$317.80 - 190 = 127.80$
The profit for the cough syrup was $127.80.

161. Strategy   To find the perimeter, substitute 4.5 for L and 3.25 for W in the formula below and solve for P.

Solution
$$P = 2L + 2W$$
$$P = 2 \cdot 4.5 + 2 \cdot 3.25$$
$$= 9 + 6.5 = 15.5$$
The perimeter is 15.5 in.

163. Strategy   To find the area, substitute 4.5 for L and 3.25 for W in the formula below and solve for A.

Solution
$$A = LW$$
$$A = 4.5 \cdot 3.25 = 14.625$$
The area is 14.625 in^2.

165. Strategy   To find the perimeter, substitute 2.8, 4.75, and 6.4 for a, b, and c in the formula below and solve for P.

Solution
$$P = a + b + c$$
$$P = 2.8 + 4.75 + 6.4 = 13.95$$
The perimeter is 13.95 m.

167. Strategy   To find the force on the falling object, substitute 4.25 for m and −9.80 for g in the given formula and solve for F.

Solution
$$F = ma$$
$$F = 4.25(-9.80)$$
$$F = -41.65$$
The force on the object is −41.65 newtons.

Critical Thinking 4.3

169. $1.3 \times 2.31 = \dfrac{13}{10} \times \dfrac{231}{100} = \dfrac{3003}{1000} = 3.003$

171. $3.46 \times 0.24 = 0.8304$

173. $0.064 \times 1.6 = 0.1024$

175. $3.0381 \div 1.23 = 2.47$

Projects or Group Activities 4.3

177. A batting average in baseball is calculated by dividing the number of hits by the number of times at bat. The quotient is rounded to the nearest thousandth. Some plate appearances are not included in the number of times at bat; for example, a walk is not counted as an "at bat."

Check Your Progress: Chapter 4

1. three and two hundred seventy-nine thousandths

2. 9.604

3. $5.809 > 5.089$

4. 512.68

5. $3.514 + 22.6981 + 145.78 = 171.9921$

6. $29.843 - 12.76 = 17.083$

7. $3.39 + 1.8 = 5.19$

8. $18.9174 - 8.82 = 10.0974$
$19 - 9 = 10$

9. $x + y + z$
$-19.55 + 7.448 + 26.1$
$= -12.102 + 26.1$
$= 13.998$

10. $3.75 = 2.13 - x$

$$\begin{array}{c|c} 3.75 & 2.13 - (-1.62) \end{array}$$

$3.75 = 3.75$
Yes, −1.62 is a solution of the equation.

11. $0.86(0.4) = 0.344$

12. $8.931 \div 1.3 = 6.87$

13. $5.628 \div 1.9 \approx 2.96$

14. $5.8(4.3) = 24.94$
$6(4) = 24$

15. $8.17 \cdot 10^4 = 81,700$

16. $325.7 \div 1000 = 0.3257$

17. $\dfrac{x}{y}$

$$\dfrac{60.9}{8.4} = 60.9 \div 8.4 = 7.25$$

18.
$$25\overline{)6.000} \;\; 0.240$$

$$\dfrac{6}{25} = 0.24$$

19. $1.2 = 1\dfrac{2}{10} = 1\dfrac{1}{5}$

20. $\dfrac{91}{100} = 0.91$

$0.90 < 0.91$

$0.90 < \dfrac{91}{100}$

21. Strategy To find the distance, round 42.195 to the nearest tenth.

Solution 42.195 rounded to the nearest tenth is 42.2.
To the nearest tenth, the New York City Marathon is 42.2 km.

22. Strategy To find the amount in your checking account, add the amount before the deposits (466.12) to the amount of the two deposits (49.90 and 67.34).

Solution $466.12 + 49.90 + 67.34 = 583.36$
The amount in your checking account is $583.36.

23. Strategy To find the amount in your checking account, subtract the amount of the check (61.57) from the amount in the account before writing the check (293.84).

Solution $293.84 - 61.57 = 232.27$
The balance in your checking account after writing the check is $232.27.

24. Strategy To find the cost of to operate the motor for 90 hours, multiply the cost per hour (0.038) times the number of hours (90).

Solution $90 \cdot 0.038 = 3.42$
The cost to operate the motor for 90 hours is $3.42.

25. Strategy To find the number of shelves, divide the total amount of wood (12) by the length of each shelf (3.4).

Solution $12 \div 3.4 \approx 3.529$
You can build 3 shelves.

Section 4.4

Concept Check

1. subtract 3.4 from each side

3. multiply each side by 3.4

Objective A Exercises

5. $y + 3.96 = 8.45$
$y + 3.96 - 3.96 = 8.45 - 3.96$
$y = 4.49$
The solution is 4.49.

7. $-9.3 = c - 15$
$-9.3 + 15 = c - 15 + 15$
$5.7 = c$
The solution is 5.7.

9. $7.3 = -\dfrac{n}{1.1}$

$-1.1(7.3) = -1.1\left(-\dfrac{n}{1.1}\right)$

$-8.03 = n$
The solution is -8.03.

11. $-7x = 8.4$
$\dfrac{-7x}{-7} = \dfrac{8.4}{-7}$
$x = -1.2$
The solution is -1.2.

13. $y - 0.234 = -0.09$
$y - 0.234 + 0.234 = -0.09 + 0.234$
$y = 0.144$
The solution is 0.144.

15. $6.21r = -1.863$

$$\frac{6.21r}{6.21} = \frac{-1.863}{6.21}$$

$r = -0.3$

17. $-0.001 = x + 0.009$

$-0.001 - 0.009 = x + 0.009 - 0.009$

$-0.01 = x$

The solution is -0.01.

19. $\frac{x}{2} = -0.93$

$2 \cdot \frac{x}{2} = 2(-0.93)$

$x = -1.86$

The solution is -1.86.

21. $-6v = 15$

$$\frac{-6v}{-6} = \frac{15}{-6}$$

$v = -2.5$

The solution is -2.5.

23. $0.908 = 2.913 + x$

$0.908 - 2.913 = 2.913 - 2.913 + x$

$-2.005 = x$

The solution is -2.005.

25. $n < 8.754$

Objective B Exercises

27. Strategy   To find the total cost, replace M by 0.42 and N by 25,000 in the given formula and solve for C.

Solution $M = \dfrac{C}{N}$

$0.42 = \dfrac{C}{25,000}$

$25,000(0.42) = 25,000 \cdot \dfrac{C}{25,000}$

$10,500 = C$

The total cost of operating the car is \$10,500.

29. Strategy   To find the stockholders' equity, replace A by 34.8 and L by 29.9 in the given formula and solve for S.

Solution $A = L + S$

$34.8 = 29.9 + S$

$34.8 - 29.9 = 29.9 - 29.9 + S$

$4.9 = S$

The stockholders' equity is \$4.9 million.

31. Strategy   To find the total consumer debt in 2001, write and solve an equation using x to represent the consumer debt in 2001.

Solution

Current debt	is	0.8 more than the 2001 debt

$2.5 = x + 0.8$
$2.5 - 0.8 = x + 0.8 - 0.8$
$1.7 = x$
The consumer debt in 2001 was $1.7 trillion.

33. Strategy   To find the monthly lease payment, write and solve an equation using x to represent the monthly lease payment.

Solution

the total of the monthly payments	is	the product of the number of months of the lease and the monthly lease payment

$21{,}387 = 60x$
$\dfrac{21{,}387}{60} = \dfrac{60x}{60}$
$356.45 = x$
The monthly lease payment is $356.45.

35. Strategy   To find the width, substitute 225 for A and 18 for L in the formula below and solve for W.

Solution $A = LW$
$225 = 18 \cdot W$
$\dfrac{225}{18} = \dfrac{18W}{18}$
$W = 12.5$
The width is 12.5 ft.

Critical Thinking 4.4

37. Yes, the solution shown is a proper method of solving the equation. Students should note that the decimals 0.375 and 0.6 were rewritten as fractions. 0.375 and $\dfrac{375}{1{,}000}$ represent the same number; they have the same value. The same is true of 0.6 and $\dfrac{6}{10}$. Rewriting $\dfrac{375}{1{,}000}$ as $\dfrac{3}{8}$ is a correct application of simplifying fractions, as is rewriting $\dfrac{6}{10}$ as $\dfrac{3}{5}$. $\dfrac{375}{1{,}000}$ and $\dfrac{3}{8}$ represent the same number, as do $\dfrac{6}{10}$ and $\dfrac{3}{5}$. In the third step, both sides of the equation are multiplied by the reciprocal of the coefficient of x. In the last step $\dfrac{8}{5}$ is correctly rewritten as the decimal 1.6.

Projects or Group Activities 4.4

39. Since x is divided by a, we can solve the equation for x by multiplying each side of the equation by a.

$$12 = \frac{x}{a}$$

$$a \cdot 12 = a \cdot \frac{x}{a}$$

$$a \cdot 12 = x$$

Let a be any positive number. Then find the corresponding value of x. We will use 1, 3, 5, 10, 100, and 1,000.

$a \cdot 12 = x$	$a \cdot 12 = x$	$a \cdot 12 = x$
$1 \cdot 12 = x$	$3 \cdot 12 = x$	$5 \cdot 12 = x$
$12 = x$	$36 = x$	$60 = x$

$a \cdot 12 = x$	$a \cdot 12 = x$	$a \cdot 12 = x$
$10 \cdot 12 = x$	$100 \cdot 12 = x$	$1{,}000 \cdot 12 = x$
$120 = x$	$1{,}200 = x$	$12{,}000 = x$

Students should note that as the value of a increases, the value of x increases.

Section 4.5

Concept Check

1. integer; 1, 4, 9, 49, 81, 100

3. **a.** 8; 8

   b. −8; −8

5. **a.** 25; 36; $\sqrt{25} = 5$; $\sqrt{36} = 6$

   b. $5 < \sqrt{33} < 6$

7. **a.** In their descriptions, students should paraphrase the definition of square root (a square root of a positive number x is a number whose square is x) and then relate it to finding the square root of a perfect square. A definition of a perfect square would be appropriate within the discussion.

   b. A student's description should include the idea of finding a factor of the radicand that is a perfect square. Be sure that the student indicates that the *largest* perfect square factor must be found. For example, 4 is a perfect square factor of 32, but it is not the largest perfect square factor. Therefore, if 32 is written as the product of 4 and 8, then $\sqrt{32} = \sqrt{4 \cdot 8} = 2\sqrt{8}$, which is not in simplest form. The largest square factor of 32 is 16, so 32 is written as the product of 16 and 2: $\sqrt{32} = \sqrt{16 \cdot 2} = 4\sqrt{2}$, which is in simplest form.

Objective A Exercises

9. Since $6^2 = 36$, $\sqrt{36} = 6$.

11. Since $3^2 = 9$, $-\sqrt{9} = -3$.

13. Since $13^2 = 169$, $\sqrt{169} = 13$.

15. Since $15^2 = 225$, $\sqrt{225} = 15$.

17. Since $5^2 = 25$, $-\sqrt{25} = -5$.

19. Since $10^2 = 100$, $-\sqrt{100} = -10$.

21. $\sqrt{8+17} = \sqrt{25}$
$= 5$

23. $\sqrt{49} + \sqrt{9} = 7 + 3$
$= 10$

25. $\sqrt{121} - \sqrt{4} = 11 - 2$
$= 9$

27. $3\sqrt{81} = 3 \cdot 9$
$= 27$

29. $-2\sqrt{49} = -2 \cdot 7$
$= -14$

31. $5\sqrt{16} - 4 = 5 \cdot 4 - 4$
$= 20 - 4$
$= 16$

33. $3 + 10\sqrt{1} = 3 + 10 \cdot 1$
$= 3 + 10$
$= 13$

35. $\sqrt{4} - 2\sqrt{16} = 2 - 2 \cdot 4$
$= 2 - 8$
$= 2 + (-8)$
$= -6$

37. $5\sqrt{25} + \sqrt{49} = 5 \cdot 5 + 7$
$= 25 + 7$
$= 32$

39. $\sqrt{\dfrac{1}{100}} = \dfrac{1}{10}$

41. $\sqrt{\dfrac{9}{16}} = \dfrac{3}{4}$

43. $\sqrt{\dfrac{1}{4}} + \sqrt{\dfrac{1}{64}} = \dfrac{1}{2} + \dfrac{1}{8}$
$= \dfrac{4}{8} + \dfrac{1}{8} = \dfrac{5}{8}$

45. $-4\sqrt{xy}$
$-4\sqrt{3 \cdot 12} = -4\sqrt{36}$
$= -4 \cdot 6$
$= -24$

47. $8\sqrt{x+y}$
$8\sqrt{19+6} = 8\sqrt{25}$
$= 8 \cdot 5$
$= 40$

49. $5 + 2\sqrt{ab}$
$5 + 2\sqrt{27 \cdot 3} = 5 + 2\sqrt{81}$
$= 5 + 2 \cdot 9$
$= 5 + 18$
$= 23$

51. $\sqrt{a^2 + b^2}$
$\sqrt{3^2 + 4^2} = \sqrt{9 + 16}$
$= \sqrt{25}$
$= 5$

53. $\sqrt{c^2 - b^2}$
$\sqrt{13^2 - 12^2} = \sqrt{169 - 144}$
$= \sqrt{25}$
$= 5$

55. $5 + \sqrt{9} = 5 + 3$
$= 8$

57. $6 - \sqrt{25} = 6 - 5$
$= 1$

59. $-4\sqrt{81} = -4 \cdot 9$
$= -36$

61. a. $\sqrt{\sqrt{16}} = \sqrt{4} = 2$

    b. $-\sqrt{\sqrt{81}} = -\sqrt{9} = -3$

Objective B Exercises

63. $\sqrt{3} \approx 1.7321$

65. $\sqrt{10} \approx 3.1623$

67. $2\sqrt{6} \approx 4.8990$

69. $3\sqrt{14} \approx 11.2250$

71. $-4\sqrt{2} \approx -5.6569$

73. $-8\sqrt{30} \approx -43.8178$

75. 23 is between the perfect squares 16 and 25.

$\sqrt{16} = 4$ and $\sqrt{25} = 5$

$4 < \sqrt{23} < 5$

77. 29 is between the perfect squares 25 and 36.

$\sqrt{25} = 5$ and $\sqrt{36} = 6$

$5 < \sqrt{29} < 6$

79. 62 is between the perfect squares 49 and 64.

$\sqrt{49} = 7$ and $\sqrt{64} = 8$

$7 < \sqrt{62} < 8$

81. 130 is between the perfect squares 121 and 144.

$\sqrt{121} = 11$ and $\sqrt{144} = 12$

$11 < \sqrt{130} < 12$

83. $\sqrt{8} = \sqrt{4 \cdot 2}$

$= \sqrt{4} \cdot \sqrt{2}$

$= 2\sqrt{2}$

85. $\sqrt{45} = \sqrt{9 \cdot 5}$

$= \sqrt{9} \cdot \sqrt{5}$

$= 3\sqrt{5}$

87. $\sqrt{20} = \sqrt{4 \cdot 5}$

$= \sqrt{4} \cdot \sqrt{5}$

$= 2\sqrt{5}$

89. $\sqrt{27} = \sqrt{9 \cdot 3}$

$= \sqrt{9} \cdot \sqrt{3}$

$= 3\sqrt{3}$

91. $\sqrt{48} = \sqrt{16 \cdot 3}$

$= \sqrt{16} \cdot \sqrt{3}$

$= 4\sqrt{3}$

93. $\sqrt{75} = \sqrt{25 \cdot 3}$

$= \sqrt{25} \cdot \sqrt{3}$

$= 5\sqrt{3}$

95. $\sqrt{63} = \sqrt{9 \cdot 7}$

$= \sqrt{9} \cdot \sqrt{7}$

$= 3\sqrt{7}$

97. $\sqrt{98} = \sqrt{49 \cdot 2}$

$= \sqrt{49} \cdot \sqrt{2}$

$= 7\sqrt{2}$

99. $\sqrt{112} = \sqrt{16 \cdot 7}$

$= \sqrt{16} \cdot \sqrt{7}$

$= 4\sqrt{7}$

101. $\sqrt{175} = \sqrt{25 \cdot 7}$

$= \sqrt{25} \cdot \sqrt{7}$

$= 5\sqrt{7}$

103. True

Objective C Exercises

105. Strategy   To find the velocity of the tsunami wave, substitute 100 for d in the given formula and solve for v.

Solution

$v = 3\sqrt{d}$

$v = 3\sqrt{100}$

$v = 3 \cdot 10 = 30$

The tsunami has a velocity of 30 ft per second.

107. Yes

109. Strategy   To find the time for the object to fall, substitute 64 for d in the given equation and solve for t.

Solution

$t = \sqrt{\dfrac{d}{16}}$

$t = \sqrt{\dfrac{64}{16}}$

$t = \sqrt{4} = 2$

It takes 2 s for the object to fall 64 ft.

111. Strategy   To find the distance, substitute 144 for E and 36 for S in the given formula and solve for d.

Solution

$d = 4{,}000\sqrt{\dfrac{E}{S}} - 4{,}000$

$d = 4{,}000\sqrt{\dfrac{144}{36}} - 4{,}000$

$d = 4{,}000\sqrt{4} - 4{,}000$

$d = 4{,}000 \cdot 2 - 4{,}000$

$d = 8{,}000 - 4{,}000$

$d = 4{,}000$

The space explorer is 4,000 mi above the surface.

Critical Thinking 4.5

113. a. Since $(0.9)^2 = 0.81$, $\sqrt{0.81} = 0.9$.

b. Since $(0.8)^2 = 0.64$, $-\sqrt{0.64} = -0.8$

c. $\sqrt{2\frac{7}{9}} = \sqrt{\frac{25}{9}}$

Since $\left(\frac{5}{3}\right)^2 = \frac{25}{9}$, $\sqrt{\frac{25}{9}} = \frac{5}{3} = 1\frac{2}{3}$.

$\sqrt{2\frac{7}{9}} = 1\frac{2}{3}$

d. $-\sqrt{3\frac{1}{16}} = -\sqrt{\frac{49}{16}}$

Since $\left(\frac{7}{4}\right)^2 = \frac{49}{16}$, $-\sqrt{\frac{49}{16}} = -\frac{7}{4} = -1\frac{3}{4}$.

$-\sqrt{3\frac{1}{16}} = -1\frac{3}{4}$

115. To simplify a principal square root, use the Product Property of Square Roots to simplify the radicand so that it contains no factor other than one that is a perfect square. To approximate a principal square root, we use a calculator to estimate the value of the square root to a desired level of accuracy.

Simplifying $\sqrt{20}$:

$$\sqrt{20} = \sqrt{4 \cdot 5} = \sqrt{4}\sqrt{5} = 2\sqrt{5}$$

Approximating $\sqrt{20}$:

$$\sqrt{20} \approx 4.4721$$

Projects or Group Activities 4.5

117. Different strategies for determining a perfect square between 350 and 400 are possible. One the students may describe is using a calculator to find the square root of 350 (which is approximately 18.708) and the square root of 400 (which is 20). The only whole number between 18.708 and 20 is 19, and $19^2 = 361$. Therefore, a perfect square between 350 and 400 is 361.

Section 4.6

Concept Check

1. -2; -1; left

3. 5; left

5. is greater than or equal to

7. $\leq$

9. $>$

Objective A Exercises

11. Draw a solid dot one half unit to the right of two on the number line.
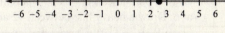

13. Draw solid dot one half unit to the left of negative three on the number line.
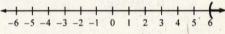

15. Draw a solid dot one half unit to the left of negative 4 on the number line.
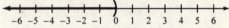

17. Draw a solid dot one half unit to the right of one on the number line.
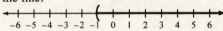

19. The real numbers greater than 6 are to the right of 6 on the number line. Draw a parenthesis at 6. Draw a heavy line to the right of 6. Draw an arrow at the right of the line.
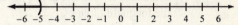

21. The real numbers less than 0 are to the left of 0 on the number line. Draw a parenthesis at 0. Draw a heavy line to the left of 0. Draw an arrow at the left of the line.

23. The real numbers greater than -1 are to the right of -1 on the number line. Draw a parenthesis at -1. Draw a heavy line to the right of -1. Draw an arrow at the right of the line.

25. The real numbers less than -5 are to the left of -5 on the number line. Draw a parenthesis at -5. Draw a heavy line to the left of -5. Draw an arrow at the left of the line.

27. Draw a parenthesis at 2 and a parenthesis at 5. Draw a heavy line between 2 and 5.

29. Draw a parenthesis at –4 and a parenthesis at 0. Draw a heavy line between –4 and 0.

31. Draw a parenthesis at –2 and a parenthesis at 6. Draw a heavy line between –2 and 6.

33. Draw a parenthesis at –6 and a parenthesis at 1. Draw a heavy line between –6 and 1.

35. both positive and negative numbers

37. **a.** The real numbers between 5 and 6

   b. Answers will vary; one example is 5.5.

Objective B Exercises

39. **a.** $x > 9$
   $-3.8 > 9$ False

   b. $x > 9$
   $0 > 9$ False

   c. $x > 9$
   $9 > 9$ False

   d. $x > 9$
   $\sqrt{101} > 9$ True
   The number $\sqrt{101}$ makes the inequality true.

41. **a.** $x \geq -2$
   $-6 \geq 2$ False

   b. $x \geq -2$
   $-2 \geq -2$ True

   c. $x \geq -2$
   $0.4 \geq -2$ True

   d. $x \geq -2$
   $\sqrt{17} \geq -2$ True
   The numbers $-2, 0.4,$ and $\sqrt{17}$ make the inequality true.

43. All real numbers less than 3 make the inequality true.

45. All real numbers greater than or equal to –1 make the inequality true.

47. Draw a parenthesis at –2. Draw a heavy line to the left of –2. Draw an arrow at the left of the line.

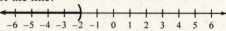

49. Draw a bracket at 0. Draw a heavy line to the right of 0. Draw an arrow at the right of the line.

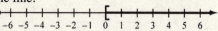

51. Draw a parenthesis at –5. Draw a heavy line to the right of –5. Draw an arrow at the right of the line.

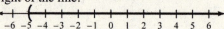

53. Draw a bracket at 2. Draw a heavy line to the left of 2. Draw an arrow at the left of the line.

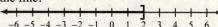

55. iv

Objective C Exercises

57. Strategy →To write the inequality, let h represent the number of credit hours allows per semester. Since h is a maximum, a part-time student carries credit hours that are less than or equal to 9.

→To determine if a student taking 8.5 credit hours is a part-time student, replace h in the inequality by 8.5 If the inequality is true the student is part-time. If the inequality is false, the student is a full-time student.

Solution $h \leq 9$
$8.5 \leq 9$ True
Yes, the student is a part-time student.

59. Strategy → To write the inequality, let s represent the number of sales. Since it is a minimum goal, the sales must be greater or equal to 50,000.
→ To determine if the sales goal has been reached, replace s in the inequality by 49,000. If the inequality is true, the sales quota has been reached. If the inequality is false, the sales quota has not been reached.

Solution $s \geq 50{,}000$
$49{,}000 \geq 50{,}000$ False
No, the sales representative has not met the minimum quota for sales.

61. Strategy → To write the inequality, let b represent the monthly budget. Since the budget is a maximum, b is less than or equal to 2,400.
→ To determine if the budget has been exceeded, replace b in the inequality by 2,380.50. If the inequality is true, the budget has not been exceeded. If the inequality is false, the monthly budget has been exceeded.

Solution $b \leq 2{,}400$
$2380.50 \leq 2{,}400$ True
Yes, the monthly budget has not been exceeded.

63. Strategy → To write the inequality, let p represent the points earned. Since the score is a minimum, let p be greater than 80.
→ To determine if the score is high enough to earn a B, substitute $80\frac{1}{2}$ in the inequality. If the inequality is true, then you will earn a B in the course. If the inequality is false, you will not earn a B.

Solution $p > 80$
$80\frac{1}{2} > 80$ True
Yes, you will receive a B in the history course.

65. Strategy → To write the inequality, let d represent the diameter of the ring on a basketball hoop. Since the diameter of the ring is a maximum, the diameter must be less than or equal to $\frac{5}{8}$.
→ To determine if the ring meets NCAA regulations, replace d in the inequality by $\frac{9}{16}$. If the inequality is true, the ring meets NCAA regulations. If the inequality if false, the ring does not meet NCAA regulations.

Solution $d \leq \frac{5}{8}$
$\frac{9}{16} \leq \frac{5}{8}$
$\frac{9}{16} \leq \frac{10}{16}$ True
Yes, the ring on the basketball hoop meets NCAA regulations.

Critical Thinking 4.6

67. **a.**

$|x| < 9$	$|x| < 9$	$|x| < 9$	$|x| < 9$
$|-2.5| < 9$	$|0| < 9$	$|9| < 9$	$|15.8| < 9$
$2.5 < 9$	$0 < 9$	$9 < 9$	$15.8 < 9$
True	True	False	False

The numbers -2.5 and 0 make the inequality $|x| < 9$ true.

b.

$|x| > -3$	$|x| > -3$	$|x| > -3$	$|x| > -3$
$|-6.3| > -3$	$|-3| > -3$	$|0| > -3$	$|6.7| > -3$
$6.3 > -3$	$3 > -3$	$0 > -3$	$6.7 > -3$
True	True	True	True

The numbers $-6.3, -3, 0,$ and 6.7 make the inequality $|x| > -3$ true.

c.

$|x| \geq 4$	$|x| \geq 4$	$|x| \geq 4$	$|x| \geq 4$
$|-1.5| \geq 4$	$|0| \geq 4$	$|4| \geq 4$	$|13.6| \geq 4$
$1.5 \geq 4$	$0 \geq 4$	$4 \geq 4$	$13.6 \geq 4$
False	False	True	True

The numbers 4 and 13.6 make the inequality $|x| \geq 4$ true.

d.

$|x| \leq 5$	$|x| \leq 5$	$|x| \leq 5$	$|x| \leq 5$
$|-4.9| \leq 5$	$|0| \leq 5$	$|2.1| \leq 5$	$|5| \leq 5$
$4.9 \leq 5$	$0 \leq 5$	$2.1 \leq 5$	$5 \leq 5$
True	True	True	True

The numbers $-4.9, 0, 2.1$ and 5 make the inequality $|x| \leq 5$ true.

Projects or Group Activities 4.6

69. The answer is a. For example:

    a. $1 + 2 < 3 + 4$
        $3 < 7$
        True

    b. $3 + 4 < 1 + 2$
        $7 < 3$
        False

    c. $1 + 4 < 3 + 2$
        $5 < 5$
        False

    d. $3 + 2 < 1 + 4$
        $5 < 5$
        False

Chapter Review Exercises

1. $3\sqrt{47} \approx 20.5670$

2. $0.918 \cdot 10^5 = 91,800$

3. $-\sqrt{121} = -11$

4. $-3.981 - 4.32 = -3.981 + (-4.32)$
      $= -8.301$

5. $a + b + c$
    $80.59 + (-3.647) + 12.3$
    $= 76.943 + 12.3$
    $= 89.243$

6. 5.034

7. $\sqrt{100} - 2\sqrt{49} = 10 - 2 \cdot 7$
                $= 10 - 14$
                $= 10 + (-14)$
                $= -4$

8. $14.2 \div 10^3 = 0.0142$

9. $4.2z = -1.428$
    $\dfrac{4.2z}{4.2} = \dfrac{-1.428}{4.2}$
    $z = -0.34$
    The solution is -0.34.

10. $8.31 = 8.310$
     $8.039 < 8.310$
     $8.039 < 8.31$

11. $\dfrac{x}{y}$
     $\dfrac{0.396}{3.6} = 0.11$

12. $\begin{array}{r} 9.47 \\ \times\, 0.26 \\ \hline 5682 \\ 1894 \\ \hline 2.4622 \end{array}$

13. **a.** $x \geq -1$
         $-6 \geq -1$
         False

    b. $x \geq -1$
        $-1 \geq -1$
        True

    c. $x \geq -1$
        $-0.5 \geq -1$
        True

    d. $x \geq -1$
        $\sqrt{10} \geq -1$
        True

The numbers -1, -0.5, and $\sqrt{10}$ make the inequality true.

14. $\dfrac{3}{7} \approx 0.4286$ $0.429 = 0.4290$
    $0.4286 < 0.4290$
    $\dfrac{3}{7} < 0.429$

15. $0.28 = \dfrac{28}{100} = \dfrac{7}{25}$

16. $-6.8 \div 47.92 \approx -0.1$

17. Strategy

    a. First find the cost of 5 lb of tomatoes today and 5 lb of tomatoes 30 years ago by multiplying the price for 1lb of tomatoes today (2.09) by 5 and the price of 1 lb of tomatoes 30 years ago (0.77) by 5. Then, to find the difference, subtract the cost of 5 lb of tomatoes 30 years ago from the cost of 5 lb of tomatoes today.

    b. To find how many times greater the cost of a dozen eggs today is than the cost 30 years ago, divide the cost of a dozen eggs today by the cost of a dozen eggs 30 years ago.

    Solution

    a. $5 \cdot 2.09 = 10.45$
    $5 \cdot 0.77 = 3.85$
    $10.45 - 3.85 = 6.60$
    Five lb of tomatoes today costs \$6.60 more than 5 lb of tomatoes 30 years ago.

    b. $1.73 \div 0.90 \approx 1.92$
    The cost of a dozen eggs today is approximately 2 times the cost of a dozen eggs 30 years ago..

18. Draw a parenthesis at −6 and a parenthesis at −2. Draw a heavy line between −6 and −2.

19. Draw a bracket at −3. Draw a heavy line to the right of −3. Draw an arrow at the right end of the line.

20. $-247.8 + (-193.4) = -441.2$

21. $614.3 \div 100 = 6.143$

22. $a - b$
    $80.32 - 29.577 = 80.32 + (-29.577)$
    $= 50.743$

23. $\sqrt{90} = \sqrt{9 \cdot 10}$
    $= \sqrt{9} \cdot \sqrt{10}$
    $= 3\sqrt{10}$

24. $60st$
    $60(5)(-3.7) = 300(-3.7)$
    $= -1,110$

25. 506.81 → 500
    64.1 →   <u>60</u>
           440

26. Strategy   →To write the inequality, let G represent the grade point average. Since it is a minimum qualification to qualify for a scholarship, the grade point average must be greater than or equal to 3.5
               →To determine if the grade point average qualifies the student for a scholarship, replace G by 3.48. If the inequality is true, the student qualifies. If the inequality is false, the student does not qualify for the scholarship.

     Solution $G \geq 3.5$
                   $3.48 \geq 3.5$ False
                   No, the student does not qualify for the scholarship.

27. Strategy   To find the difference, subtract the melting point of bromine (-7.2) from the boiling point of bromine (58.8).

     Solution $58.8 - (-7.2)$
                   $= 58.8 + 7.2$
                   $= 66$
                   The difference in temperature between the melting and boiling point of bromine is 66°C.

28. Strategy
    a. To find the difference, subtract the cost of World War I (0.38) from the cost of World War II (3.1).

    b. To find how many times greater the monetary cost of the Vietnam War was, divide the cost of the Vietnam War by the cost of World War I.

     Solution
    a. $3.1 - 0.38 = 2.72$
       The difference between the monetary costs of the two World Wars was $2.72 trillion.

    b. $0.57 \div 0.38 = 1.5$
       The cost of the Vietnam War was 1.5 times the cost of World War I.

29. Strategy   To find the cost per ounce, divide 11.78 by 7.

     Solution $\dfrac{11.78}{7} \approx 1.683$

                   To the nearest cent, the cost of instant coffee is $1.68 per ounce.

30. Strategy   To find the monthly lease payment, write and solve an equation using x to represent the monthly lease payment.

     Solution

the total of the monthly payments	is	the product of the number of months of the lease and the monthly lease payment

            $12,371.76 = 24x$
            $\dfrac{12,371.76}{24} = \dfrac{24x}{24}$
            $515.49 = x$
            The monthly lease payment is $515.49.

31. Strategy   To find the price of the treadmill, substitute 1,124.75 for C and 374.75 for M in the given formula and solve for P.

Solution $P = C + M$
$P = 1,124.75 + 374.75$
$P = 1,499.50$
The price of the treadmill is \$1,499.50.

32. Strategy   To find the velocity of the falling object, substitute 25 for d in the given formula and solve for v.

Solution $v = \sqrt{64d}$
$v = \sqrt{64 \cdot 25}$
$v = \sqrt{1,600}$
$v = 40$
The velocity of the falling object is 40 feet per second.

Chapter Test

1. 9.033

2. $4.003 < 4.009$

3. $6.05\overline{1}367$ *Given place value*
  $3 < 5$

6.051367 rounded to the nearest thousandth is 6.051.

4. $-30 - (-7.249) = -30 + 7.247$
  $= -22.753$

5. $x - y$
  $6.379 - (-8.28) = 6.379 + 8.28$
  $= 14.659$

6. $\begin{array}{rr} 92.34 \to & 90 \\ -17.95 \to & -20 \\ \hline 74.39 & 70 \end{array}$

7. $4.58 - 3.9 + 6.017$
  $= 4.58 + (-3.9) + 6.017$
  $= 0.68 + 6.017$
  $= 6.697$

8. $-2.5(7.36) = -18.4$

9. $-20cd$
  $-20 \cdot 0.5 \cdot (-6.4) = (-10) \cdot (-6.4)$
  $= 64$

10. $5.488 = -3.92p$
  $\dfrac{5.488}{-3.92} = \dfrac{-3.92p}{-3.92}$
  $p = -1.4$

11. $\sqrt{256} - 2\sqrt{121} = 16 - 2 \cdot 11$
  $= 16 - 22$
  $= 16 + (-22)$
  $= -6$

12. $84.96 \div 100 = 0.8496$

13. $\dfrac{x}{y}$
  $\dfrac{52.7}{-6.2} = -8.5$

14. $0.22 = 0.2200$ $\dfrac{2}{9} \approx 0.2222$
  $0.2200 < 0.2222$
  $0.22 < \dfrac{2}{9}$

15. $2\sqrt{46} \approx 13.5647$

16. $\sqrt{68} = \sqrt{4 \cdot 17}$
  $= \sqrt{4} \cdot \sqrt{17}$
  $= 2\sqrt{17}$

17. $63.6 - 22.8 = 40.8$
  The gross from *Thunderball* was \$40.8 million more than from *On Her Majesty's Secret Service*.

18. $8.4 = 5.9 + a$
  $8.4 \mid 5.9 + (-2.5)$
  $8.4 \neq 3.4$
  No, −2.5 is not a solution of the equation.

19. $8.973 \cdot 10^4 = 89,730$

20. Draw a parenthesis at –2 and a parenthesis at 2. Draw a heavy line between –2 and 2.

21. Draw a bracket at 3. Draw a heavy line to the right of 3. Draw an arrow at the right end of the line.

22. $x + y$
$-233.81 + 71.3 = -162.51$

23. $-8v = 26$
$$\frac{-8v}{-8} = \frac{26}{-8}$$
$p = -3.25$

24. Strategy — To find the difference, subtract the melting point of fluorine (–219.62) from the boiling point of fluorine (–188.14).

Solution — $-188.14 - (-219.62)$
$= -188.14 + 219.62 = 31.48$
The difference in temperature between the melting and boiling point of fluorine is 31.48°C.

25. Strategy — To find the velocity of the falling object, substitute 16 for d in the given formula and solve for v.

Solution — $v = \sqrt{64d}$
$v = \sqrt{64 \cdot 16}$
$v = \sqrt{1,024}$
$v = 32$
The velocity of the falling object is 32 feet per second.

26. Strategy — To find the stockholders' equity, substitute 48.2 for A and 27.6 for L in the given formula and solve for S.

Solution — $A = L + S$
$48.2 = 27.6 + S$
$48.2 - 27.6 = 27.6 - 27.6 + S$
$20.6 = S$
The stockholders' equity is $20.6 million.

27. Strategy — To find the perimeter, substitute 8.75, 5.25, and 4.5 for a, b, and c in the formula below and solve for P.

Solution — $P = a + b + c$
$P = 8.75 + 5.25 + 4.5 = 18.5$
The perimeter is 18.5 m.

28. Strategy — →To write the inequality, let x represent the units sold. Since it is a minimum to reach the goal, the sales must be greater or equal to 65,000 per year.
→To determine if the number of units sold has reached the goal, replace x by 57,000. If the inequality is true, the goal has been reached. If the inequality is false, the representative has not reached the sales goal.

Solution — $x \geq 65,000$
$57,000 \geq 65,000$ False
No, the representative has not reached the sales goal.

29. Strategy — To find the force on the falling object, substitute 5.75 for m and –9.80 for g in the given formula and solve for F.

Solution — $F = ma$
$F = 5.75(-9.80)$
$F = -56.35$
The force on the object is –56.35 newtons.

30. Strategy   To find the temperature fall, subtract the low temperature (-20.56) from the temperature 15 minutes later (2.78).

        Solution $2.78 - (-20.56) = 2.78 + 20.56$
$$= 23.34$$
The temperature rose 23.34°C in 15 minutes.

Cumulative Review Exercises

1. $387.9 \div 10^4 = 0.03879$

2. $(x+y)^2 - 2z$
$$(-3+2)^2 - 2(-5) = (-1)^2 - 2(-5)$$
$$= 1 - 2(-5)$$
$$= 1 - (-10)$$
$$= 1 + 10 = 11$$

3. $-9.8 = -0.49c$
$$\frac{-9.8}{-0.49} = \frac{-0.49c}{-0.49}$$
$$20 = c$$
The solution is 20.

4. 8,072,092

5. Draw a parenthesis at -4 and a parenthesis at 1. Draw a heavy line between -4 and 1.

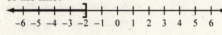

6. Draw a bracket at -2. Draw a heavy line to the left of -2. Draw an arrow at the left end of the line.

7. $-23 - (-19) = -23 + 19$
$$= -4$$

8. $372 \rightarrow 400$
$541 \rightarrow 500$
$608 \rightarrow 600$
$429 \rightarrow \underline{+\ 400}$
$ 1{,}900$

9. $\sqrt{192} = \sqrt{64 \cdot 3}$
$$= \sqrt{64} \cdot \sqrt{3}$$
$$= 8\sqrt{3}$$

10. $x \div y$
$$3\frac{2}{3} \div 2\frac{4}{9} = \frac{11}{3} \div \frac{22}{9}$$
$$= \frac{11}{3} \cdot \frac{9}{22}$$
$$= \frac{11 \cdot 9}{3 \cdot 22}$$
$$= \frac{11 \cdot 3 \cdot 3}{3 \cdot 2 \cdot 11}$$
$$= \frac{3}{2} = 1\frac{1}{2}$$

11. $-36.92 + 18.5 = -18.42$

12. $\left(\dfrac{5}{9}\right)\left(-\dfrac{3}{10}\right)\left(-\dfrac{6}{7}\right) = \left(\dfrac{5}{9} \cdot \dfrac{3}{10} \cdot \dfrac{6}{7}\right)$
$$= \frac{5 \cdot 3 \cdot 6}{9 \cdot 10 \cdot 7}$$
$$= \frac{5 \cdot 3 \cdot 2 \cdot 3}{3 \cdot 3 \cdot 2 \cdot 5 \cdot 7}$$
$$= \frac{1}{7}$$

13. $x^4 y^2$
$$2^4 \cdot 10^2 = (2 \cdot 2 \cdot 2 \cdot 2) \cdot (10 \cdot 10)$$
$$= 16 \cdot 100$$
$$= 1{,}600$$

14. $5\overline{)65}$
$\dfrac{13}{}$
$2\overline{)130}$
$2\overline{)260}$
$$260 = 2 \cdot 2 \cdot 5 \cdot 13 = 2^2 \cdot 5 \cdot 13$$

15. $25\overline{)19.00}$ $\dfrac{0.76}{}$
$$\frac{19}{25} = 0.76$$

16. $10\sqrt{19} \approx 95.3939$

17. a. $28 > 20$
Switzerland mandates more vacation days than Ireland.

    b. $35 \div 7 = 5$
Austria mandates 5 times more vacation days than Mexico.

18. $\dfrac{-8}{0}$
Division by zero is undefined.

19. $-\dfrac{5}{7} + \dfrac{4}{21} = \dfrac{-5}{7} + \dfrac{4}{21}$

$\qquad = \dfrac{-15}{21} + \dfrac{4}{21}$

$\qquad = \dfrac{-15+4}{21}$

$\qquad = \dfrac{-11}{21} = -\dfrac{11}{21}$

20. $4\sqrt{25} - \sqrt{81} = 4 \cdot 5 - 9$

$\qquad = 20 - 9$

$\qquad = 11$

21. $\begin{array}{r} 62.8 \to \quad 60 \\ 0.47 \to \times 0.5 \\ \hline 30 \end{array}$

22. $5(3-7) \div (-4) + 6(2)$

$\qquad = 5(-4) \div (-4) + 6(2)$

$\qquad = -20 \div (-4) + 6(2)$

$\qquad = 5 + 6(2)$

$\qquad = 5 + 12$

$\qquad = 17$

23. $\dfrac{a}{b+c}$

$\dfrac{\frac{3}{8}}{\frac{1}{2} + \frac{3}{4}} = \dfrac{\frac{3}{8}}{\frac{5}{4}}$

$\qquad = \dfrac{3}{8} \div \dfrac{5}{4}$

$\qquad = \dfrac{3}{8} \cdot \dfrac{4}{5}$

$\qquad = \dfrac{3 \cdot 4}{8 \cdot 5}$

$\qquad = \dfrac{3 \cdot 2 \cdot 2}{2 \cdot 2 \cdot 2 \cdot 5} = \dfrac{3}{10}$

24. $x - y + z$

$\dfrac{5}{12} - \left(-\dfrac{3}{8}\right) + \left(-\dfrac{3}{4}\right) = \dfrac{5}{12} + \dfrac{3}{8} + \dfrac{-3}{4}$

$\qquad = \dfrac{10}{24} + \dfrac{9}{24} + \dfrac{-18}{24}$

$\qquad = \dfrac{10 + 9 + (-18)}{24}$

$\qquad = \dfrac{1}{24}$

25. $2.617 \div 0.93 \approx 2.8$

26. Strategy   To calculate the copier service bill:

→Calculate the number of pages the service was used after the first 50 pages.

→Add the standard charge (6.95) to the product of 0.12 and the number of pages the service was used after the first 50 pages.

Solution $78 - 50 = 28$

The service was used for 28 pages after the first 50 pages.

$0.12(28) + 6.95$

$= 3.36 + 6.95$

$= 10.31$

The pager service bill is $10.31.

27. Strategy   To find the temperature fall, subtract the temperature at midnight (–29.4) from the temperature at noon (17.22).

Solution $17.22 - (-29.4) = 17.22 + 29.4$

$= 46.62$

The temperature fell 46.62°C in the 12-hour period.

28. Strategy   To find the cost per visit, substitute 515 for M and 125 for N in the given formula and solve for C.

Solution $C = \dfrac{M}{N}$

$C = \dfrac{515}{125}$

$C = 4.12$

The cost per visit is $4.12.

29. Strategy

    a. To find the number of hours worked per week, add the number of hours spent in all five categories.

    b. To determine which takes more time, compare the amount of time spent on face-to-face selling (13.9) with the sum of the time spent on administrative work (7.0) and placing service calls (5.6).

Solution

    a. $13.9 + 11.5 + 8.5 + 7.0 + 5.6 = 46.5$
        On average, a salesperson works 46.5 h per week.

    b. $7.0 + 5.6 = 12.6$
        $13.9 > 12.6$
        The average salesperson spends more time on face-to-face selling.

30. Strategy   To find the velocity, substitute 45 for d in the given equation and solve for v.

Solution $v = \sqrt{20d}$
             $v = \sqrt{20 \cdot 45}$
             $v = \sqrt{900}$
             $v = 30$
             The velocity of the car is 30 mph.

Chapter 5: Variable Expressions

Prep Test

1. $54 > 45$

2. $-19 + 8 = -11$

3. $26 - 38 = -12$

4. $-2(44) = -88$

5. $-\dfrac{3}{4}(-8) = \dfrac{24}{4} = 6$

6. $3.97 \cdot 10^4 = 3.97 \cdot 10,000 = 39,700$

7. $(-3)^2 = (-3)(-3) = 9$

8. $(8-6)^2 + 12 \div 4 \cdot 3^2 = (2)^2 + 12 \div 4 \cdot 9$
$$= 4 + 3 \cdot 9$$
$$= 4 + 27$$
$$= 31$$

Section 5.1

Concept Check

1. **a.** Associative

   b. -24

Objective A Exercises

3. The Associative Property of Multiplication

5. The Commutative Property of Addition

7. The Inverse Property of Addition

9. The Commutative Property of Multiplication

11. **a.** The Associative Property of Multiplication

    b. The Inverse Property of Multiplication

    c. The Multiplication Property of One

13. $x + (4 + y) = (x + 4) + y$

15. $5 \cdot \dfrac{1}{5} = 1$

17. $a \cdot 0 = 0$

19. $-\dfrac{2}{3}\left(-\dfrac{3}{2}\right) = 1$

21. $6(2x) = (6 \cdot 2)x$
$$= 12x$$

23. $-5(3x) = (-5 \cdot 3)x$
$$= -15x$$

25. $(3t) \cdot 7 = 7 \cdot (3t)$
$$= (7 \cdot 3)t$$
$$= 21t$$

27. $(-3p) \cdot 7 = 7 \cdot (-3p)$
$$= [7 \cdot (-3)]p$$
$$= -21p$$

29. $(-2)(-6q) = [(-2)(-6)]q$
$$= 12q$$

31. $\dfrac{1}{2}(4x) = \left(\dfrac{1}{2} \cdot 4\right)x$
$$= 2x$$

33. $-\dfrac{5}{3}(9w) = \left(-\dfrac{5}{3} \cdot 9\right)w$
$$= -15w$$

35. $-\dfrac{1}{2}(-2x) = \left[\left(-\dfrac{1}{2}\right)(-2)\right]x$
$$= 1 \cdot x = x$$

37. $(2x)(3x) = (2 \cdot 3)(x \cdot x)$
$$= 6x^2$$

39. $(-3x)(9x) = (-3 \cdot 9)(x \cdot x)$
$$= -27x^2$$

41. $\left(\dfrac{1}{2}x\right)(2x) = \left(\dfrac{1}{2} \cdot 2\right)(x \cdot x)$
$$= 1 \cdot x^2 = x^2$$

43. $\left(-\dfrac{2}{3}\right)(x)\left(-\dfrac{3}{2}\right) = \left[\left(-\dfrac{2}{3}\right)\left(-\dfrac{3}{2}\right)\right]x$
$$= 1 \cdot x = x$$

45. $6\left(\dfrac{1}{6}c\right) = \left(6 \cdot \dfrac{1}{6}\right)c$
$$= 1 \cdot c = c$$

47. $-5\left(-\dfrac{1}{5}a\right) = \left(-5 \cdot -\dfrac{1}{5}\right)a$
$$= 1 \cdot a = a$$

49. $\dfrac{4}{5}w \cdot 15 = \left(\dfrac{4}{5} \cdot 15\right)w$
$\qquad = 12w$

51. $2v \cdot 8w = (2 \cdot 8)(v \cdot w)$
$\qquad = 16vw$

53. $(-4b)(7c) = (-4 \cdot 7)(b \cdot c)$
$\qquad\qquad = -28bc$

55. $3x + (-3x) = 0$

57. $-12h + 12h = 0$

59. $9 + 2m + (-2m)$
$\quad = 9 + 0$
$\quad = 9$

61. $8x + 7 + (-8x)$
$\quad = 8x + (-8x) + 7$
$\quad = 0 + 7$
$\quad = 7$

63. $6t - 15 + (-6t)$
$\quad = -15 + 6t + (-6t\,)$
$\quad = -15 + 0$
$\quad = -15$

65. $8 + (-8) - 5y$
$\quad = 0 - 5y$
$\quad = -5y$

67. $(-4) + 4 + 13b$
$\quad = 0 + 13b$
$\quad = 13b$

69. negative

71. positive

Objective B Exercises

73. $2(5z + 2)$
$\quad = 2(5z) + 2(2)$
$\quad = 10z + 4$

75. $6(2y + 5z)$
$\quad = 6(2y) + 6(5z)$
$\quad = 12y + 30z$

77. $3(7x - 9)$
$\quad = 3(7x) - 3(9)$
$\quad = 21x - 27$

79. $-(2x - 7) = -2x + 7$

81. $-(-4x - 9) = 4x + 9$

83. $-5(y + 3)$
$\quad = -5(y) + (-5)(3)$
$\quad = -5y - 15$

85. $-6(2x - 3)$
$\quad = -6(2x) - (-6)(3)$
$\quad = -12x + 18$

87. $-5(4n - 8)$
$\quad = -5(4n) - (-5)(8)$
$\quad = -20n + 40$

89. $-8(-6z + 3)$
$\quad = -8(-6z) + (-8)(3)$
$\quad = 48z - 24$

91. $-6(-4p - 7)$
$\quad = -6(-4p) - (-6)(7)$
$\quad = 24p + 42$

93. $5(2a + 3b + 1)$
$\quad = 5(2a) + 5(3b) + 5(1)$
$\quad = 10a + 15b + 5$

95. $4(3x - y - 1)$
$\quad = 4(3x) - 4(y) - 4(1)$
$\quad = 12x - 4y - 4$

97. $9(4m - n + 2)$
$\quad = 9(4m) - 9(n) + 9(2)$
$\quad = 36m - 9n + 18$

99. $-6(-2v + 3w + 7)$
$\quad = -6(-2v) + (-6)(3w) + (-6)(7)$
$\quad = 12v - 18w - 42$

101. $-4(-5x - 1)$
$\quad = -4(-5x) - (-4)(1)$
$\quad = 20x + 4$

103. $5(4a - 5b + c)$
$\quad = 5(4a) - 5(5b) + 5(c)$
$\quad = 20a - 25b + 5c$

105. $-6(3p - 2r - 9)$
$\quad = -6(3p) - (-6)(2r) - (-6)(9)$
$\quad = -18p + 12r + 54$

107. $-(5a - 9b + 7) = -5a + 9b - 7$

109. $-(11p - 2q - r) = -11p + 2q + r$

111. (i) $7 - 4(x + y) = 7 - 4x - 4y$
(ii) $(7 - 4)(x + y)$
$\quad = (3)(x + y)$
$\quad = 3x + 3y$
The answer is (ii).

Critical Thinking 5.1

113. False

115. False

117. True

119. Students will provide different examples of two operations in everyday experience that are not commutative. These might include getting into a car and starting a car, turning on a computer and using a word processor, or going up a chair lift and skiing down a mountain.

Section 5.2

Concept Check

  1. $3x$; $5x$

  3. $-2x$; -9; $5x^2$; $-2x$; -9

  5. 12; 3; 5; 10

Objective A Exercises

  7. $3x^2$, $4x$, $\underline{-9}$

  9. b, $\underline{5}$

 11. $9\underline{a}^2$, $-12\underline{a}$, $4\underline{b}^2$

 13. ③$\underline{x}^2$

 15. $7a + 9a = 16a$

 17. $12x + 15x = 27x$

 19. $9z - 6z = 3z$

 21. $9x - x = 8x$

 23. $8z - 15z = -7z$

 25. $w - 7w = -6w$

 27. $12v - 12v = 0$

 29. $9s - 8s = s$

 31. $\dfrac{n}{5} + \dfrac{3n}{5} = \dfrac{n + 3n}{5}$
$$= \dfrac{4n}{5}$$

33. $\dfrac{x}{4} + \dfrac{x}{4} = \dfrac{x + x}{4}$
$$= \dfrac{2x}{4}$$
$$= \dfrac{x}{2}$$

35. $\dfrac{8y}{7} - \dfrac{4y}{7} = \dfrac{8y - 4y}{7}$
$$= \dfrac{4y}{7}$$

37. $\dfrac{5c}{6} - \dfrac{c}{6} = \dfrac{5c - c}{6}$
$$= \dfrac{4c}{6}$$
$$= \dfrac{2c}{3}$$

39. $4x - 3y + 2x$
$$= 4x + 2x - 3y$$
$$= 6x - 3y$$

41. $4r + 8p - 2r + 5p$
$$= 4r - 2r + 8p + 5p$$
$$= 2r + 13p$$

43. $9w - 5v - 12w + 7v$
$$= 9w - 12w - 5v + 7v$$
$$= -3w + 2v$$

45. $-4p + 9 - 5p + 2$
$$= -4p - 5p + 9 + 2$$
$$= -9p + 11$$

47. $8p + 7 - 6p - 7$
$$= 8p - 6p + 7 - 7$$
$$= 2p + 0$$
$$= 2p$$

49. $7h + 15 - 7h - 9$
$$= 7h - 7h + 15 - 9$$
$$= 0 + 6$$
$$= 6$$

51. $9y^2 - 8 + 4y^2 + 9 = 9y^2 + 4y^2 - 8 + 9$
$$= 13y^2 + 1$$

53. $3w^2 - 7 - 9 + 9w^2 = 3w^2 + 9w^2 - 7 - 9$
$$= 12w^2 - 16$$

55. $9w^2 - 15w + w - 9w^2$
$= 9w^2 - 9w^2 - 15w + w$
$= 0 - 14w$
$= -14w$

57. $7a^2b + 5ab^2 - 2a^2b + 3ab^2$
$= 7a^2b - 2a^2b + 5ab^2 + 3ab^2$
$= 5a^2b + 8ab^2$

59. $8a - 9b + 2 - 8a + 9b + 3$
$= 8a - 8a - 9b + 9b + 2 + 3$
$= 0 + 0 + 5$
$= 5$

61. $4y^2 + 7y + 1 + y^2 - 10y + 9$
$= 4y^2 + y^2 + 7y - 10y + 1 + 9$
$= 5y^2 - 3y + 10$

63. $-4a^2b - 4ab^2 + 4 + 4ab^2 - 4a^2b - 4$
$= -4a^2b - 4a^2b - 4ab^2 + 4ab^2 + 4 - 4$
$= -8a^2b$
The answer is (ii) and (iii).

Objective B Exercises

65. $5x + 2(x + 1)$
$= 5x + 2x + 2$
$= 7x + 2$

67. $9n - 3(2n - 1)$
$= 9n - 6n + 3$
$= 3n + 3$

69. $7a - (3a - 4)$
$= 7a - 3a + 4$
$= 4a + 4$

71. $7 + 2(2a - 3)$
$= 7 + 4a - 6$
$= 4a + 1$

73. $6 + 4(2x + 9)$
$= 6 + 8x + 36$
$= 8x + 42$

75. $8 - 4(3x - 5)$
$= 8 - 12x + 20$
$= -12x + 28$

77. $2 - 9(2m + 6)$
$= 2 - 18m - 54$
$= -18m - 52$

79. $3(6c + 5) + 2(c + 4)$
$= 18c + 15 + 2c + 8$
$= 20c + 23$

81. $2(a - 2b) + 3(2a + 3b)$
$= 2a - 4b + 6a + 9b$
$= 8a + 5b$

83. $6(7z - 5) - 3(9z - 6)$
$= 42z - 30 - 27z + 18$
$= 15z - 12$

85. $-2(6y + 2) + 3(4y - 5)$
$= -12y - 4 + 12y - 15$
$= -19$

87. $-5(x - 2y) - 4(2x + 3y)$
$= -5x + 10y - 8x - 12y$
$= -13x - 2y$

89. $2 - 3(2v - 1) + 2(2v + 4)$
$= 2 - 6v + 3 + 4v + 8$
$= -2v + 13$

91. $2c - 3(c + 4) - 2(2c - 3)$
$= 2c - 3c - 12 - 4c + 6$
$= -5c - 6$

93. $8a + 3(2a - 1) + 6(4 - 2a)$
$= 8a + 6a - 3 + 24 - 12a$
$= 2a + 21$

95. $3n - 2[5 - 2(2n - 4)]$
$= 3n - 2[5 - 4n + 8]$
$= 3n - 2[13 - 4n]$
$= 3n - 26 + 8n$
$= 11n - 26$

97. $9x - 3[8 - 2(5 - 3x)]$
$= 9x - 3[8 - 10 + 6x]$
$= 9x - 3[-2 + 6x]$
$= 9x + 6 - 18x$
$= -9x + 6$

99. $21r - 4[3(4 - 5r) - 3(2 - 7r)]$
$= 21r - 4[12 - 15r - 6 + 21r]$
$= 21r - 4[6 + 6r]$
$= 21r - 24 - 24r$
$= -3r - 24$

101. $5[2a + 3(a - 4)]$
$= 10a + 15(a - 4)$
The answer is (i).

Critical Thinking 5.2

103. The simplification of $2 + 3(2x + 4)$ as $5(2x + 4)$ is incorrect. By the Order of Operations Agreement, multiplication is performed before addition. Therefore, the multiplication $3(2x + 4)$ must be performed before the addition $2 + 3$. Therefore, the correct simplification is:
$2 + 3(2x + 4) = 2 + 6x + 12 = 6x + 14$

Section 5.3

Concept Check

1. Yes

3. No, variable in denominator

5. Yes

7. No, variable under square root

9. Yes

11. No, variable under square root

13. 3 terms

15. 1 term

17. binomial

19. monomial

21. $3x^3 + 8x^2 - 2x - 6$

23. $5a^3 - 3a^2 + 2a + 1$

25. $-b^2 + 4$

27. a. $-2y^3$; $-6y$; 4

    b. 3; -4; -3

    c. $3y^3 - 4y - 3$

29. opposite

31. $-8x^3 - 5x^2 + 3x + 6$

33. $9a^3 - a^2 + 2a - 9$

Objective A Exercises

35. $\left(7m^2 - 9m - 8\right) + \left(5m^2 + 10m + 4\right) = \left(7m^2 + 5m^2\right) + \left(-9m + 10m\right) + \left(-8 + 4\right)$
$$= 12m^2 + m - 4$$

37. $\left(-8x^2 - 11x - 15\right) + \left(4x^2 - 12x + 13\right) = \left(-8x^2 + 4x^2\right) + \left(-11x - 12x\right) + \left(-15 + 13\right)$
$$= -4x^2 - 23x - 2$$

39. $\left(11p^3-9p^2-6p\right)+\left(10p^2-8p+4\right)=11p^3+\left(-9p^2+10p^2\right)+\left(-6p-8p\right)+4$
$$=11p^3+p^2-14p+4$$

41. $\left(7x^3-8x^2+9x-12\right)+\left(-3x^3-7x^2+5x-9\right)$
$$=\left(7x^3-3x^3\right)+\left(-8x^2-7x^2\right)+\left(9x+5x\right)+\left(-12-9\right)$$
$$=4x^3-15x^2+14x-21$$

43. $(8y^2-3y-1)+(-6y^3-8y^2+3y-1)$
$$=-6y^3+(8y^2-8y^2)+(-3y+3y)+(-1-1)$$
$$=-6y^3-2$$

45. $(8y^2-3y-1)+(-6y^2+3y-1)$
$$=(8y^2-6y^2)+(-3y+3y)+(-1-1)$$
$$=2y^2-2$$

47.
$$8v^2-\ 9v+12$$
$$\underline{12v^2-11v-2}$$
$$20v^2-20v+10$$

49.
$$13z^3-\ 7z^2+4z$$
$$\underline{\qquad 10z^2+5z-9}$$
$$13z^3+\ 3z^2+9z-9$$

51.
$$5a^3-a^2+4a-19$$
$$\underline{-a^3+a^2\ -7a+19}$$
$$4a^3\qquad -3a$$

53.
$$12c^3\qquad +9c$$
$$\underline{\quad -7c^2\ -c-8}$$
$$12c^3-7c^2+8c-8$$

55.
$$12x^2+7x+8$$
$$\underline{3x^3-12x^2-7x-11}$$
$$3x^3\qquad\qquad -3$$

57.
$$8z^3+5z^2-4z+7$$
$$\underline{-3z^3\ -z^2+6z-2}$$
$$5z^3+4z^2+2z+5$$

59. (i) and (ii)

Objective B Exercises

61. $\left(7y^2-8y-10\right)-\left(3y^2+2y-9\right)=\left(7y^2-8y-10\right)+\left(-3y^2-2y+9\right)$
$$=4y^2-10y-1$$

63. $\left(13w^3+3w^2-9\right)-\left(7w^3-9w+10\right)=\left(13w^3+3w^2-9\right)+\left(-7w^3+9w-10\right)$
$$=6w^3+3w^2+9w-19$$

65. $\left(15t^3 - 9t^2 + 8t + 11\right) - \left(17t^3 - 9t^2 - 8t + 6\right) = \left(15t^3 - 9t^2 + 8t + 11\right) + \left(-17t^3 + 9t^2 + 8t - 6\right)$

$\qquad\qquad\qquad\qquad\qquad\qquad\quad = -2t^3 + 16t + 5$

67. $\left(8p^3 + 14p\right) - \left(9p^2 - 12\right) = \left(8p^3 + 14p\right) + \left(-9p^2 + 12\right)$

$\qquad\qquad\qquad\qquad\quad = 8p^3 - 9p^2 + 14p + 12$

69. $\left(-4v^2 + 8v - 2\right) - \left(6v^3 - 13v^2 + 7v + 1\right) = \left(-4v^2 + 8v - 2\right) + \left(-6v^3 + 13v^2 - 7v - 1\right)$

$\qquad\qquad\qquad\qquad\qquad\qquad\qquad = -6v^3 + 9v^2 + v - 3$

71. $\left(7m^2 - 3m - 6\right) - \left(2m^2 - m + 5\right) = \left(7m^2 - 3m - 6\right) + \left(-2m^2 + m - 5\right)$

$\qquad\qquad\qquad\qquad\qquad\qquad = 5m^2 - 2m - 11$

73.
$$\begin{array}{r} 8b^2 - 7b - 6 \\ -5b^2 - 8b - 12 \\ \hline 3b^2 - 15b - 18 \end{array}$$

75.
$$\begin{array}{r} 10y^3 \qquad\;\; - 8y - 13 \\ -6y^2 - 2y - 7 \\ \hline 10y^3 - 6y^2 - 10y - 20 \end{array}$$

77.
$$\begin{array}{r} 4a^2 + 8a + 12 \\ -3a^3 - 4a^2 - 7a + 12 \\ \hline -3a^3 \qquad\; + a + 24 \end{array}$$

79.
$$\begin{array}{r} 7m - 6 \\ -2m^3 + m^2 \\ \hline -2m^3 + m^2 + 7m - 6 \end{array}$$

81.
$$\begin{array}{r} 4q^3 + 7q^2 + 8q - 9 \\ -14q^3 - 7q^2 + 8q + 9 \\ \hline -10q^3 \qquad\quad + 16q \end{array}$$

83.
$$\begin{array}{r} 7x^4 + 3x^2 - 11 \\ 5x^4 + 8x^2 - 6 \\ \hline 12x^4 + 11x^2 - 17 \end{array}$$

85. $(-x + a) - (-x - a)$

$\quad = -x + x + a + a$

$\quad = 2a$

True

Objective C Exercises

87. Strategy   To find the distance from Haley to Bedford, add the distance from Haley to Lincoln to the distance from Lincoln to Bedford.

Solution

$$2y^2 + y - 4$$
$$\underline{5y^2 - y + 3}$$
$$7y^2 \quad\;\; -1$$

The distance from Haley to Bedford is $(7y^2 - 1)$ km.

89. Strategy   To find the perimeter, substitute $(n^2 + 3)$ for a and $(n^2 - 2)$ for b and $(n^2 + 5)$ for c in the formula below and solve for P.

Solution

$$P = a + b + c$$
$$P = (n^2 + 3) + (n^2 - 2) + (n^2 + 5)$$
$$= 3n^2 + 6$$

The perimeter is $(3n^2 + 6)$ m.

91. Strategy   To find the monthly profit, substitute $(50n + 4000)$ for C and $(-0.6n^2 + 250n)$ for R in the formula and solve for P.

Solution

$$P = R - C$$
$$P = -0.6n^2 + 250n - (50n + 4000)$$
$$= -0.6n^2 + 250n - 50n - 4000$$
$$= -0.6n^2 + 200n - 4000$$

The monthly profit is $(-0.6n^2 + 200n - 4000)$ dollars.

93. Strategy   To find the monthly profit, substitute $(100n + 1500)$ for C and $(-n^2 + 800n)$ for R in the formula and solve for P.

Solution

$$P = R - C$$
$$P = -n^2 + 800n - (100n + 1500)$$
$$= -n^2 + 800n - 100n - 1500$$
$$= -n^2 + 700n - 1500$$

The monthly profit is $(-n^2 + 700n - 1500)$ dollars.

Critical Thinking 5.3

95. In explaining the meaning of monomial, binomial, and trinomial, a student should provide the number of terms in each of these polynomials. A student description of a polynomial should paraphrase the definition "a variable expression in which the terms are monomials." Be sure that the student has not defined a monomial in terms of a polynomial and a polynomial in terms of a monomial; one of these terms should be defined independently of the other. Their examples will vary.

Check Your Progress: Chapter 5

1. The Addition Property of Zero

2. $y \cdot z = z \cdot y$

3. $-6(4y) = (-6 \cdot 4)y$
$= -24z$

4. $(3z) \cdot 7 = 7 \cdot (3z)$
$= (7 \cdot 3)z$
$= 21z$

5. $(-2b)(-8b) = (-2 \cdot -8)(b \cdot b)$
$= 16b^2$

6. $(-3)(-c) = [(-3)(-1)]c$
$= 3c$

7. $-8t + 5 + 8t$
$= -8t + 8t + 5$
$= 0 + 5$
$= 5$

8. $6(3x - 4y)$
$= 6(3x) - 6(4y)$
$= 18x - 24y$

9. $-7(2d - 9)$
$= -7(2d) - 7(-9)$
$= -14d + 63$

10. $5(3a - 4b + 6c)$
$= 5(3a) - 5(4b) + 5(6c)$
$= 15a - 20b + 30c$

11. $-(8x + 2y - 1) = -8x - 2y + 1$

12. $15b - b = 14a$

13. $7x + 5 - 8x - 2$
$= 7x - 8x + 5 - 2$
$= -x + 3$

14. $\dfrac{4x}{5} + \dfrac{2x}{5} = \dfrac{4x + 2x}{5}$
$= \dfrac{6x}{5}$

15. $3y^2 - 4y + y^2 - 6y = 3y^2 + y^2 - 4y - 6y$
$= 4y^2 - 10y$

16. $6s + 9t - 4 + 3s - 10t + 8$
$= 6s + 3s + 9t - 10t - 4 + 8$
$= 9s - t + 4$

17. $7 + 4(8c - 2)$
$= 7 + 32c - 8$
$= 32c - 1$

18. $5(3a - 2b) - 4(6a - b)$
$= 15a - 10b - 24a + 4b$
$= -9a - 6b$

19. $3x - 2[5 - 4(x + 7)]$
$= 3x - 2[5 - 4x - 28]$
$= 3x - 2[-23 - 4x]$
$= 3x + 46 + 8x$
$= 11x + 46$

20. $9 - 8(4x - y) + 3(5x - 1)$
$= 9 - 32x + 8y + 15x - 3$
$= -17x + 8y + 6$

21. $\begin{array}{r} 3x^2 - 4x + 7 \\ 5x^2 - 3x - 9 \\ \hline 8x^2 - 7x - 2 \end{array}$

22. $\begin{array}{r} 2x^3 \qquad\ -3x + 4 \\ 5x^2 - 7x - 1 \\ \hline 2x^3 + 5x^2 - 10x + 3 \end{array}$

23. $\begin{array}{r} 4x^2 - 7x + 6 \\ -9x^2 + 12x - 8 \\ \hline -5x^2 + 5x - 2 \end{array}$

24. $\begin{array}{r} -7x^2 + 4x - 2 \\ -6x^2 + \ x + 1 \\ \hline -13x^2 + 5x - 1 \end{array}$

25. Strategy   To find the distance from Marion to Palmer, add the distance from Marion to Newburg to the distance from Newburg to Palmer.

      Solution
$$3x^2 + 5x - 4$$
$$\underline{6x^2 + x - 1}$$
$$9x^2 + 6x - 5$$

The distance from Marion to Palmer is ($9x^2 + 6x - 5$) km.

Section 5.4

Concept Check

1. 6; 3; x^9

3. 3; 4; x^{12}

Objective A Exercises

5. No

7. No

9. $a^4 \cdot a^5 = a^9$

11. $x^9 \cdot x^7 = x^{16}$

13. $n^4 \cdot n^2 = n^6$

15. $z^3 \cdot z \cdot z^4 = z^8$

17. $\left(a^3 b^2\right)\left(a^5 b\right) = \left(a^3 \cdot a^5\right)\left(b^2 \cdot b\right)$
$$= a^8 b^3$$

19. $\left(-m^3 n\right)\left(m^6 n^2\right) = -\left(m^3 \cdot m^6\right)\left(n \cdot n^2\right)$
$$= -m^9 n^3$$

21. $\left(2x^3\right)\left(5x^4\right) = (2 \cdot 5)\left(x^3 \cdot x^4\right)$
$$= 10x^7$$

23. $\left(8x^2 y\right)\left(xy^5\right) = 8\left(x^2 \cdot x\right)\left(y \cdot y^5\right)$
$$= 8x^3 y^6$$

25. $\left(-4m^3\right)\left(3m^4\right) = (-4 \cdot 3)\left(m^3 \cdot m^4\right)$
$$= -12m^7$$

27. $\left(7v^3\right)(-2w) = \left[7 \cdot (-2)\right]\left(v^3\right) \cdot w$
$$= -14v^3 w$$

29. $\left(ab^2 c^3\right)\left(-2b^3 c^2\right) = -2 \cdot a\left(b^2 \cdot b^3\right)\left(c^3 \cdot c^2\right)$
$$= -2ab^5 c^5$$

31. $\left(4b^4 c^2\right)\left(6a^3 b\right) = (4 \cdot 6)\left(a^3\right)\left(b^4 \cdot b\right)\left(c^2\right)$
$$= 24a^3 b^5 c^2$$

33. $\left(-8r^2 t^3\right)\left(-5rt^4 v\right)$
$$= \left[(-8)(-5)\right]\left(r^2 \cdot r\right)\left(t^3 \cdot t^4\right)(v)$$
$$= 40r^3 t^7 v$$

35. $\left(9mn^4 p\right)\left(-3mp^2\right)$
$$= \left[9 \cdot (-3)\right](m \cdot m)\left(n^4\right)\left(p \cdot p^2\right)$$
$$= -27m^2 n^4 p^3$$

37. $(2x)\left(3x^2\right)\left(4x^4\right)$
$$= (2 \cdot 3 \cdot 4)\left(x \cdot x^2 \cdot x^4\right)$$
$$= 24x^7$$

39. $(3ab)\left(2a^2 b^3\right)\left(a^3 b\right)$
$$= (3 \cdot 2)\left(a \cdot a^2 \cdot a^3\right)\left(b \cdot b^3 \cdot b\right)$$
$$= 6a^6 b^5$$

Objective B Exercises

41. Yes

43. No

45. $\left(p^3\right)^5 = p^{3 \cdot 5} = p^{15}$

47. $\left(b^2\right)^4 = b^{2 \cdot 4} = b^8$

49. $\left(p^4\right)^7 = p^{4 \cdot 7} = p^{28}$

51. $\left(c^7\right)^4 = c^{7 \cdot 4} = c^{28}$

53. $(3x)^2 = 3^{1 \cdot 2} x^{1 \cdot 2}$
$$= 3^2 x^2 = 9x^2$$

55. $\left(x^2 y^3\right)^6 = x^{2 \cdot 6} y^{3 \cdot 6} = x^{12} y^{18}$

57. $\left(r^3 t\right)^4 = r^{3 \cdot 4} t^{1 \cdot 4} = r^{12} t^4$

59. $\left(-y^2\right)^2 = (-1)^{1 \cdot 2} y^{2 \cdot 2} = y^4$

61. $\left(2x^4\right)^3 = 2^{1\cdot3}\,x^{4\cdot3}$
$= 2^3\,x^{12}$
$= 8x^{12}$

63. $\left(-2a^2\right)^3 = (-2)^{1\cdot3}\,a^{2\cdot3}$
$= (-2)^3\,a^6$
$= -8a^6$

65. $\left(3x^2y\right)^2 = 3^{1\cdot2}\,x^{2\cdot2}\,y^{1\cdot2}$
$= 3^2\,x^4\,y^2$
$= 9x^4\,y^2$

67. $\left(2a^3bc^2\right)^3 = 2^{1\cdot3}\,a^{3\cdot3}b^{1\cdot3}c^{2\cdot3}$
$= 2^3\,a^9b^3c^6$
$= 8a^9b^3c^6$

69. $\left(-mn^5p^3\right)^4$
$= (-1)^{1\cdot4}\,m^{1\cdot4}n^{5\cdot4}\,p^{3\cdot4}$
$= m^4n^{20}p^{12}$

Critical Thinking 5.4

71. Strategy   To find the area, substitute $(7y^5)$ for s in the formula below and solve for A.

Solution $A = s^2$
$A = \left(7y^5\right)^2 = 7^2 \cdot y^{5\cdot2}$
$= 49y^{10}$
The area is $(49y^{10})$ cm^2.

Projects or Group Activities 5.4

73. a. $\left(2^3\right)^2 = 2^6 = 64$

$2^{\left(3^2\right)} = 2^9 = 512$
The results are not the same.
$2^{\left(3^2\right)}$ is the larger number.

b.   The order of operations is $x^{\left(m^n\right)}$.

Section 5.5

Concept Check

1. $x^2 + 2x - 10$;
$5x\left(x^2\right) + 5x(2x) - 5x(10) = 5x^3 + 10x^2 - 50x$

Objective A Exercises

3. $x\left(x^2 - 3x - 4\right)$
$= x\left(x^2\right) - x(3x) - x(4)$
$= x^3 - 3x^2 - 4x$

5. $4a\left(2a^2 + 3a - 6\right)$
$= 4a\left(2a^2\right) + 4a(3a) - 4a(6)$
$= 8a^3 + 12a^2 - 24a$

7. $-2a\left(3a^2 + 9a - 7\right)$
$= -2a\left(3a^2\right) + (-2a)(9a) - (-2a)(7)$
$= -6a^3 - 18a^2 + 14a$

9. $m^3\left(4m - 9\right)$
$= m^3\left(4m\right) - m^3\left(9\right)$
$= 4m^4 - 9m^3$

11. $2x^3\left(5x^2 - 6xy + 2y^2\right)$
$= 2x^3\left(5x^2\right) - 2x^3\left(6xy\right) + 2x^3\left(2y^2\right)$
$= 10x^5 - 12x^4y + 4x^3y^2$

13. $-6r^5\left(r^2 - 2r - 6\right)$
$= -6r^5\left(r^2\right) - \left(-6r^5\right)(2r) - \left(-6r^5\right)(6)$
$= -6r^7 + 12r^6 + 36r^5$

15. $4a^2\left(3a^2 + 6a - 7\right)$
$= 4a^2\left(3a^2\right) + 4a^2\left(6a\right) - 4a^2\left(7\right)$
$= 12a^4 + 24a^3 - 28a^2$

17. $-2n^2\left(3 - 4n^3 - 5n^5\right)$
$= -2n^2\left(3\right) - \left(-2n^2\right)\left(4n^3\right) - \left(-2n^2\right)\left(5n^5\right)$
$= -6n^2 + 8n^5 + 10n^7$

19. $ab^2\left(3a^2 - 4ab + b^2\right)$
$= ab^2\left(3a^2\right) - ab^2\left(4ab\right) + \left(ab^2\right)\left(b^2\right)$
$= 3a^3b^2 - 4a^2b^3 + ab^4$

21. $-x^2y^3\left(4x^5y^2 - 5x^3y - 7x\right)$
$= -x^2y^3(4x^5y^2) - (-x^2y^3)(5x^3y) - (-x^2y^3)(7x)$
$= -4x^7y^5 + 5x^5y^4 + 7x^3y^3$

23. $6r^2t^3\left(1 - rt - r^3t^3\right)$
$= 6r^2t^3\left(1\right) - 6r^2t^3\left(rt\right) - 6r^2t^3\left(r^3t^3\right)$
$= 6r^2t^3 - 6r^3t^4 - 6r^5t^6$

Objective B Exercises

25. $(x + 4)(x + 6)$
$= (x)(x) + (x)(6) + 4(x) + 4(6)$
$= x^2 + 6x + 4x + 24$
$= x^2 + 10x + 24$

27. $(a - 6)(a - 7)$
$= (a)(a) + a(-7) + (-6)(a) + (-6)(-7)$
$= a^2 - 7a - 6a + 42$
$= a^2 - 13a + 42$

29. $(y + 4)(y + 3)$
$= (y)(y) + (y)(3) + 4(y) + 4(3)$
$= y^2 + 3y + 4y + 12$
$= y^2 + 7y + 12$

31. $(3c + 4)(2c + 3)$
$= (3c)(2c) + (3c)(3) + 4(2c) + 4(3)$
$= 6c^2 + 9c + 8c + 12$
$= 6c^2 + 17c + 12$

33. $(3v - 7)(4v + 3)$
$= (3v)(4v) + (3v)(3) + (-7)(4v) + (-7)(3)$
$= 12v^2 + 9v - 28v - 21$
$= 12v^2 - 19v - 21$

35. $(8x - 3)(5x - 4)$
$= (8x)(5x) + (8x)(-4) + (-3)(5x) + (-3)(-4)$
$= 40x^2 - 32x - 15x + 12$
$= 40x^2 - 47x + 12$

37. $(4n - 9)(4n - 5)$
$= (4n)(4n) + (4n)(-5) + (-9)(4n) + (-9)(-5)$
$= 16n^2 - 20n - 36n + 45$
$= 16n^2 - 56n + 45$

39. $(3y - 4)(4y + 7)$
$= (3y)(4y) + (3y)(7) + (-4)(4y) + (-4)(7)$
$= 12y^2 + 21y - 16y - 28$
$= 12y^2 + 5y - 28$

41. $(4a - 5)(4a + 5)$
$= (4a)(4a) + (4a)(5) + (-5)(4a) + (-5)(5)$
$= 16a^2 + 20a - 20a - 25$
$= 16a^2 - 25$

43. No. A negative number times a negative number is a positive number.

Critical Thinking 5.5

45. Strategy   To find the area, substitute $(2x + 3)$ for L and $(x - 6)$ for W in the formula below and solve for A.

Solution $A = LW$
$A = (2x + 3) \cdot (x - 6)$
$= (2x)(x) + (2x)(-6) + 3(x) + 3(-6)$
$= 2x^2 - 12x + 3x - 18$
$= 2x^2 - 9x - 18$
The area is
$(2x^2 - 9x - 18)$ mi^2.

Section 5.6

Concept Check

1. 1; 1; 1

3. 1; 10; 10

5. 9; right; –9

Objective A Exercises

7. No

9. No

11. $27^0 = 1$

13. $-(17)^0 = -1$

15. $3^{-2} = \dfrac{1}{3^2} = \dfrac{1}{9}$

17. $2^{-3} = \dfrac{1}{2^3} = \dfrac{1}{8}$

19. $x^{-5} = \dfrac{1}{x^5}$

21. $w^{-8} = \dfrac{1}{w^8}$

23. $y^{-1} = \dfrac{1}{y}$

25. $\dfrac{1}{a^{-5}} = a^5$

27. $\dfrac{1}{b^{-3}} = b^3$

29. $\dfrac{a^8}{a^2} = a^{8-2} = a^6$

31. $\dfrac{q^5}{q} = q^{5-1} = q^4$

33. $\dfrac{m^4 n^7}{m^3 n^5} = m^{4-3} n^{7-5} = mn^2$

35. $\dfrac{t^4 u^8}{t^2 u^5} = t^{4-2} u^{8-5} = t^2 u^3$

37. $\dfrac{x^4}{x^9} = x^{4-9} = x^{-5} = \dfrac{1}{x^5}$

39. $\dfrac{b}{b^5} = b^{1-5} = b^{-4} = \dfrac{1}{b^4}$

Objective B Exercises

41. $2{,}370{,}000 = 2.37 \times 10^6$

43. $0.00045 = 4.5 \times 10^{-4}$

45. $309{,}000 = 3.09 \times 10^5$

47. $0.000000601 = 6.01 \times 10^{-7}$

49. $57{,}000{,}000{,}000 = 5.7 \times 10^{10}$

51. $0.000000017 = 1.7 \times 10^{-8}$

53. $7.1 \times 10^5 = 710{,}000$

55. $4.3 \times 10^{-5} = 0.000043$

57. $6.71 \times 10^8 = 671{,}000{,}000$

59. $7.13 \times 10^{-6} = 0.00000713$

61. $5 \times 10^{12} = 5{,}000{,}000{,}000{,}000$

63. $8.01 \times 10^{-3} = 0.00801$

65. No. 84.3 is not a number between 1 and 10.

67. No. 2.5 is not an integer.

69. 0.000000015 m
$= 1.5 \times 10^{-8}$ m

71. $5{,}980{,}000{,}000{,}000{,}000{,}000{,}000{,}000$ kg
$= 5.98 \times 10^{24}$ kg

73. 0.000000000000665
$= 6.65 \times 10^{-13}$ g

75. $2{,}450{,}000{,}000$ hertz
$= 2.45 \times 10^9$ hertz

77. 0.0000000000000000039
$= 3.9 \times 10^{-18}$ g

79. 0.0000037 m $= 3.7 \times 10^{-6}$ m

81. $m \times 10^{-5} < n \times 10^{-3}$

Critical Thinking 5.6

83. a. $3^{-(-2)} = 3^2 = 9,$
$3^{-(-1)} = 3^1 = 3,$
$3^{-0} = 3^0 = 1,$
$3^{-1} = \dfrac{1}{3},$
$3^{-2} = \dfrac{1}{3^2} = \dfrac{1}{9}$

   b. $2^{-(-2)} = 2^2 = 4,$
$2^{-(-1)} = 2^1 = 2,$
$2^{-0} = 2^0 = 1,$
$2^{-1} = \dfrac{1}{2},$
$2^{-2} = \dfrac{1}{2^2} = \dfrac{1}{4}$

85. $9^{-2} + 3^{-3} = \dfrac{1}{9^2} + \dfrac{1}{3^3}$
$= \dfrac{1}{81} + \dfrac{1}{27} = \dfrac{1}{81} + \dfrac{3}{81} = \dfrac{4}{81}$

87. $25^{-2} = \dfrac{1}{25^2} = \dfrac{1}{625} = 0.0016$

89. If $m = n+1$, then $\dfrac{a^m}{a^n} = \dfrac{a^{n+1}}{a^n} = a$.

91. If $x = 0$, then $-6.3^x = -6.3^0 = -1$.

Section 5.7

Concept Check

1. sum of; times

3. difference between; divided by

5. $\dfrac{1}{5}$; $\dfrac{3}{5}$; $\dfrac{4}{5}n$

7. $6W$

Objective A Exercises

9. Three <u>more than</u> t
   $t + 3$

11. the fourth <u>power</u> of q
    q^4

13. the <u>sum</u> of negative two and z
    $-2 + z$

15. the <u>total of twice</u> q and five
    $2q + 5$

17. seven <u>subtracted</u> from the <u>product</u> of eight
    and d
    $8d - 7$

19. the <u>difference</u> between six <u>times</u> c and
    twelve
    $6c - 12$

21. <u>twice</u> the <u>sum</u> of three and w
    $2(3 + w)$

23. four <u>times</u> the <u>difference</u> between <u>twice</u> r
    and five
    $4(2r - 5)$

25. the <u>quotient</u> of v and the <u>difference</u> between
    v and 4
    $\dfrac{v}{v - 4}$

27. The <u>sum</u> of the <u>square</u> of m and the <u>cube</u> of
    m
    $m^2 + m^3$

29. smaller numbers: s
    larger number: $31 - s$
    five <u>more than</u> the larger number
    $(31 - s) + 5$

31. iii

Objective B Exercises

33. Let the number be x.
    a number <u>decreased</u> by the <u>total of</u> the
    number and twelve
    $x - (x + 12)$
    $x - x - 12$
    -12

35. Let the number be x.
    the <u>difference</u> between two thirds of a
    number and three eighths of the number
    $\dfrac{2}{3}x - \dfrac{3}{8}x$
    $\dfrac{16}{24}x - \dfrac{9}{24}x$
    $\dfrac{7}{24}x$

37. Let the number be x.
    <u>twice</u> the <u>sum</u> of seven <u>times</u> a number and
    six
    $2(7x + 6)$
    $14x + 12$

39. Let the number be x.
    the <u>sum</u> of eleven <u>times</u> a number and the
    <u>product</u> of three and the number
    $11x + 3x$
    $14x$

41. Let the number be x.
    nine <u>times</u> the <u>sum</u> of a number and seven
    $9(x + 7)$
    $9x + 63$

43. Let the number be x.
    seven <u>more than</u> the <u>sum</u> of a number and
    five
    $(x + 5) + 7$
    $x + 12$

45. Let the number be x.
    the <u>product</u> of seven and the <u>difference</u>
    between a number and four
    $7(x - 4)$
    $7x - 28$

47. Let the number be x.
    the <u>difference</u> between ten <u>times</u> a number
    and the <u>product</u> of three and the number
    $10x - 3x$
    $7x$

49. Let the number be x.
a number <u>increased</u> by the <u>difference</u>
between seven <u>times</u> the number and eight
$x + (7x - 8)$
$x + 7x - 8$
$8x - 8$

51. Let the number be x.
fourteen <u>decreased</u> by the <u>sum</u> of a number
and thirteen
$14 - (x + 13)$
$14 - x - 13$
$-x + 1$

53. Let the number be x.
the <u>product</u> of eight <u>times</u> a number and
two
$8x(2)$
$16x$

55. Let the number be x.
a number <u>plus</u> nine <u>added</u> to the <u>difference</u>
between the number and three
$(x + 9) + (x - 3)$
$x + 9 + x - 3$
$2x + 6$

57. Let the smaller number be y.
The larger number is $9 - y$.
five <u>times</u> the larger number
$5(9 - y)$
$-5y + 45$

59. Let the larger number be m.
The smaller number is $17 - m$.
nine <u>less than</u> the smaller number
$(17 - m) - 9$
$-m + 8$

Objective C Exercises

61. the number of emails: A

the amount of spam: $\dfrac{1}{2} A$

63. the amount of cashews: A
the amount of peanuts: $3A$

65. the number of bones in your body: N

the number of bones in your foot: $\dfrac{1}{4} N$

67. the speed of the slower cyclist: r
the speed of the faster cyclist: $r + 6$

69. the length of the shorter piece: L
the length of the longer piece: $12 - L$

Critical Thinking 5.7

71. There are twice as many hydrogen atoms as
oxygen atoms in the water. Thus if there are
x oxygen atoms, there will be $2x$ hydrogen
atoms.

Projects of Group Activities 5.7

73. Student translations of the expression
$5x + 8$ might include:
8 more than the product of 5 and x.
The sum of 5 times x and 8.
The total of 5 times x and 8.

Student translations of the expression
$5(x + 8)$ might include:
The product of 5 and the sum of x and 8.
5 times the sum of x and 8.
The product of 5 and 8 more than x.

Look for ambiguities within a student
translation. For example, 5 times a number
plus 8 could be either $5x + 8$ or $5(x + 8)$.

Chapter Review Exercises

1. $4z^2 + 3z - 9z + 2z^2$
$= (4z^2 + 2z^2) + (3z - 9z)$
$= 6z^2 - 6z$

2. $-2(9z + 1) = -18z - 2$

3. $\left(3z^2 + 4z - 7\right) + \left(7z^2 - 5z - 8\right)$
$= \left(3z^2 + 7z^2\right) + (4z - 5z) + (-7 - 8)$
$= 10z^2 - z - 15$

4. $\left(2m^3 n\right)\left(-4m^2 n\right)$
$= \left[2(-4)\right]\left(m^3 \cdot m^2\right)(n \cdot n)$
$= -8m^5 n^2$

5. $3^{-5} = \dfrac{1}{3^5} = \dfrac{1}{243}$

6. The additive inverse of $\dfrac{3}{7}$ is $-\dfrac{3}{7}$.

7. $\dfrac{2}{3}\left(\dfrac{3}{2} x\right) = x$

8. $-5(2s - 5t) + 6(3t + s)$
$= -10s + 25t + 18t + 6s$
$= -4s + 43t$

9. $(-5xy^4)(-3x^2y^3)$
$= [(-5)(-3)](x \cdot x^2)(y^4 \cdot y^3)$
$= 15x^3y^7$

10. $(7a + 6)(3a - 4)$
$= 21a^2 - 28a + 18a - 24$
$= 21a^2 - 10a - 24$

11. $\quad 6b^3 - 7b^2 + 5b - 9$
$\quad \underline{-9b^3 + 7b^2 - b - 9}$
$\quad -3b^3 \qquad + 4b - 18$

12. $(2z^4)^5 = 2^{1 \cdot 5} z^{4 \cdot 5} = 2^5 z^{20} = 32z^{20}$

13. $-\dfrac{3}{4}(-8w) = 6w$

14. $5xyz^2(-3x^2z + 6yz^2 - x^3y^4)$
$= 5xyz^2(-3x^2z) + 5xyz^2(6yz^2)$
$\quad - 5xyz^2(x^3y^4)$
$= -15x^3yz^3 + 30xy^2z^4 - 5x^4y^5z^2$

15. The multiplicative inverse of $-\dfrac{9}{4}$ is $-\dfrac{4}{9}$.

16. $-4(3c - 8) = -12c + 32$

17. $2m - 6n + 7 - 4m + 6n + 9$
$= (2m - 4m) + (-6n + 6n) + (7 + 9)$
$= -2m + 16$

18. $(4a^3b^8)(-3a^2b^7)$
$= [4(-3)](a^3 \cdot a^2)(b^8 \cdot b^7)$
$= -12a^5b^{15}$

19. The Distributive Property

20. $(p^2q^3)^3 = p^{2 \cdot 3}q^{3 \cdot 3}$
$\qquad\qquad = p^6q^9$

21. $\dfrac{a^4}{a^{11}} = a^{4-11}$
$\qquad = a^{-7}$
$\qquad = \dfrac{1}{a^7}$

22. $0.0000397 = 3.97 \times 10^{-5}$

23. The Commutative Property of Addition

24. $(9y^3 + 8y^2 - 10) + (-6y^3 + 8y - 9)$
$= (9y^3 - 6y^3) + 8y^2 + 8y + (-10 - 9)$
$= 3y^3 + 8y^2 + 8y - 19$

25. $8(2c - 3d) - 4(c - 5d)$
$= 16c - 24d - 4c + 20d$
$= 12c - 4d$

26. $7(2m - 6) = 14m - 42$

27. $\dfrac{x^3y^5}{xy} = x^{3-1}y^{5-1} = x^2y^4$

28. $7a^2 + 9 - 12a^2 + 3a$
$= (7a^2 - 12a^2) + 3a - 9$
$= -5a^2 + 3a + 9$

29. $(3p - 9)(4p + 7)$
$= 12p^2 + 21p - 36p - 63$
$= 12p^2 - 15p - 63$

30. $-2a^2b(4a^3 - 5ab^2 + 3b^4)$
$= -2a^2b(4a^3) - (-2a^2b)(5ab^2)$
$\quad + (-2a^2b) + (3b^4)$
$= -8a^5b + 10a^3b^3 - 6a^2b^5$

31. $-12x + 7y + 15x - 11y$
$= (-12x + 15x) + (7y - 11y)$
$= 3x - 4y$

32. $-7(3a - 4b) - 5(3b - 4a)$
$= -21a + 28b - 15b + 20a$
$= -a + 13b$

33. $c^{-5} = \dfrac{1}{c^5}$

34. $\quad 12x^3 + 9x^2 - 5x - 1$
$\quad \underline{-6x^3 - 9x^2 - 5x + 1}$
$\quad 6x^3 \qquad\quad -10x$

35. $2.4 \times 10^5 = 240,000$

36. Strategy   To find the perimeter, substitute $(b^2 - 4)$ for a, $(b^2 + 2)$ for b and $(b^2 + 5)$ for c in the formula below and solve for P.

      Solution

$$P = a + b + c$$
$$= (b^2 - 4) + (b^2 + 2) + (b^2 + 5)$$
$$= 3b^2 + 3$$

The perimeter is $(3b^2 + 3)$ ft.

37. nine <u>less than</u> the quotient of four <u>times</u> a number and seven

$$\frac{4x}{7} - 9$$

38. the <u>sum</u> of three <u>times</u> a number and the <u>difference</u> between the number and seven

$$3x + (x - 7)$$
$$4x - 7$$

39. $602,300,000,000,000,000,000,000$
$$= 6.023 \times 10^{23}$$

40. the number of pounds of mocha java beans: p

the number of pounds of espresso beans: $30 - p$

Chapter Test

1. $\dfrac{2}{3}\left(-\dfrac{3}{2}r\right) = -r$

2. $-3(5y - 7) = -15y + 21$

3. $7y - 3 - 4y + 6$
$$= (7y - 4y) + (-3 + 6)$$
$$= 3y + 3$$

4. $4x^2 - 2z + 7z - 8x^2$
$$= (4x^2 - 8x^2) + (-2z + 7z)$$
$$= -4x^2 + 5z$$

5. $2a - 4b + 12 - 5a - 2b + 6$
$$= (2a - 5a) + (-4b - 2b) + (12 + 6)$$
$$= -3a - 6b + 18$$

6. The multiplicative inverse of $\dfrac{5}{4}$ is $\dfrac{4}{5}$.

7. $-2(3x - 4y) + 5(2x + y)$
$$= -6x + 8y + 10x + 5y$$
$$= 4x + 13y$$

8. $9 - 2(4b - a) + 3(3b - 4a)$
$$= 9 - 8b + 2a + 9b - 12a$$
$$= -10a + b + 9$$

9. $0.00000079 = 7.9 \times 10^{-7}$

10. $4.9 \times 10^6 = 4,900,000$

11. $(4x^2 - 2x - 2) + (2x^2 - 3x + 7)$
$$= (4x^2 + 2x^2) + (-2x - 3x) + (-2 + 7)$$
$$= 6x^2 - 5x + 5$$

12. $\left(v^2 w^5\right)^4 = v^{2\cdot4} w^{5\cdot4} = v^8 w^{20}$

13. $\left(3m^2 n^3\right)^3 = 3^{1\cdot3} m^{2\cdot3} n^{3\cdot3}$
$$= 3^3 m^6 n^9 = 27 m^6 n^9$$

14. $(-5v^2 z)(2v^3 z^2)$
$$= (-5)(2)(v^2 \cdot v^3)(z \cdot z^2)$$
$$= -10 v^5 z^3$$

15. $(3p - 8)(2p + 5)$
$$= 6p^2 + 15p - 16p - 40$$
$$= 6p^2 - p - 40$$

16. $(2m^2 n^2)(-4mn^3 + 2m^3 - 3n^4)$
$$= 2m^2 n^2(-4mn^3) + 2m^2 n^2(2m^3) + 2m^2 n^2(-3n^4)$$
$$= -8m^3 n^5 + 4m^5 n^2 - 6m^2 n^6$$

17. $3z + 4w = 4w + 3z$

18. $\dfrac{x^2 y^5}{xy^2} = x^{2-1} y^{5-2} = xy^3$

19. $a^{-5} = \dfrac{1}{a^5}$

20. The Associative Property of Multiplication

21.
$$\begin{array}{r} 5a^3 - 6a^2 + 4a - 8 \\ -8a^3 + 7a^2 - 4a - 2 \\ \hline -3a^3 + a^2 \qquad -10 \end{array}$$

22. $\dfrac{1}{c^{-6}} = c^6$

23. The Distributive Property

24. $6w \cdot 0 = 0$

25. $(3x - 7y)(3x + 7y)$
$$= 9x^2 + 21xy - 21xy - 49y^2$$
$$= 9x^2 - 49y^2$$

26. The additive inverse of $-\dfrac{4}{7}$ is $\dfrac{4}{7}$.

27. $720,000,000 = 7.2 \times 10^8$

28. $(3a - 6)(4a + 2)$
$= 12a^2 + 6a - 24a - 12$
$= 12a^2 - 18a - 12$

29. $2(4a - 3b) + 3(5a - 2b)$
$= 8a - 6b + 15a - 6b$
$= 23a - 12b$

30. $\dfrac{m^4 n^2}{m^2 n^5} = m^{4-2} n^{2-5} = \dfrac{m^2}{n^3}$

31. five <u>more than</u> three <u>times</u> a number
$3x + 5$

32. the <u>sum</u> of a number and the <u>difference</u>
between the number and six
$x + (x - 6)$
$2x - 6$

33. the number of cups of sugar in the batter: s
the number of cups of flour in the
batter: $s + 3$

Cumulative Review Exercises

1. $\dfrac{4.712}{-0.38} = -12.4$

2. $9v - 10 + 5v + 8$
$= (9v + 5v) + (-10 + 8)$
$= 14v - 2$

3. $(3x - 5)(2x + 4)$
$= 6x^2 + 12x - 10x - 20$
$= 6x^2 + 2x - 20$

4. $-a - b,\ a = \dfrac{11}{24}$ and $b = -\dfrac{5}{6}$
$-\dfrac{11}{24} - \left(-\dfrac{5}{6}\right) = \dfrac{-11}{24} - \dfrac{-5}{6}$
$= \dfrac{-11}{24} - \dfrac{-20}{24}$
$= \dfrac{-11 - (-20)}{24} = \dfrac{-11 + 20}{24}$
$= \dfrac{9}{24} = \dfrac{3 \cdot 3}{3 \cdot 8} = \dfrac{3}{8}$

5. $\sqrt{81} + 3\sqrt{25} = 9 + 3 \cdot 5$
$= 9 + 15$
$= 24$

6. Draw a parenthesis at –3. Draw a line to the
right of –3. Draw an arrow at the right end
of the line.

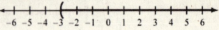

7. $\dfrac{1}{x^{-7}} = x^7$

8. $-4t = 36$
$\dfrac{-4t}{-4} = \dfrac{36}{-4}$
$t = -9$
The solution is –9.

9. $0.00000084 = 8.4 \times 10^{-7}$

10. $\left(5x^2 - 3x + 2\right) + \left(4x^2 + x - 6\right)$
$= \left(5x^2 + 4x^2\right) + \left(-3x + x\right) + (2 - 6)$
$= 9x^2 - 2x - 4$

11. $-5\sqrt{x + y},\ x = 18$ and $y = 31$
$-5\sqrt{18 + 31} = -5\sqrt{49}$
$= -5 \cdot 7 = -35$

12. $\dfrac{\frac{5}{8} + \frac{3}{4}}{3 - \frac{1}{2}} = \dfrac{\frac{11}{8}}{\frac{5}{2}}$
$= \dfrac{11}{8} \div \dfrac{5}{2}$
$= \dfrac{11}{8} \cdot \dfrac{2}{5}$
$= \dfrac{11 \cdot 2}{8 \cdot 5} = \dfrac{11 \cdot 2}{2 \cdot 2 \cdot 2 \cdot 5} = \dfrac{11}{20}$

13. $\left(-3a^2 b\right)\left(4a^5 b^8\right)$
$= (-3 \cdot 4)\left(a^2 \cdot a^5\right)\left(b \cdot b^8\right)$
$= -12a^7 b^9$

14. $\dfrac{x^3}{x^5} = x^{3-5}$
$= x^{-2}$
$= \dfrac{1}{x^2}$

15. $x^3 y^2,\ x = \dfrac{2}{5}$ and $y = 2\dfrac{1}{2}$
$\left(\dfrac{2}{5}\right)^3 \left(2\dfrac{1}{2}\right)^2 = \left(\dfrac{2}{5} \cdot \dfrac{2}{5} \cdot \dfrac{2}{5}\right) \cdot \left(\dfrac{5}{2} \cdot \dfrac{5}{2}\right)$
$= \dfrac{2 \cdot 2 \cdot 2 \cdot 5 \cdot 5}{5 \cdot 5 \cdot 5 \cdot 2 \cdot 2} = \dfrac{2}{5}$

16. $-8p(6) = (6)(-8p) = [(6)(-8)]p = -48p$

17.
$$\begin{array}{r} 829.43 \rightarrow 800 \\ 567.109 \rightarrow -600 \\ \hline 200 \end{array}$$

18. $-3ab^2\left(4a^2b + 5ab - 2ab^2\right)$
$$= -3ab^2(4a^2b) + (-3ab^2)(5ab) - (-3ab^2)(-2ab^2)$$
$$= -12a^3b^3 - 15a^2b^3 + 6a^2b^4$$

19. $6(5x - 4y) - 12(x - 2y)$
$$= 30x - 24y - 12x + 24y$$
$$= 18x$$

20. $\dfrac{a}{-b}$, $a = -56$ and $b = -8$
$$\frac{-56}{-(-8)} = \frac{-56}{8} = -7$$

21. $0.5625 = \dfrac{5625}{10,000} = \dfrac{625 \cdot 9}{625 \cdot 16} = \dfrac{9}{16}$

22. $6 \cdot (-2)^3 \div 12 - (-8) = 6 \cdot (-8) \div 12 - (-8)$
$$= -48 \div 12 - (-8)$$
$$= -4 - (-8)$$
$$= -4 + 8 = 4$$

23. $\sqrt{300} = \sqrt{100 \cdot 3}$
$$= \sqrt{100} \cdot \sqrt{3}$$
$$= 10\sqrt{3}$$

24. $\left(8y^2 - 7y + 4\right) - \left(3y^2 - 5y + 9\right)$
$$= \left(8y^2 - 7y + 4\right) + \left(-3y^2 + 5y - 9\right)$$
$$= 5y^2 - 2y - 5$$

25. $-6cd$, $c = -\dfrac{2}{9}$ and $d = \dfrac{3}{4}$
$$-6\left(-\frac{2}{9}\right)\left(\frac{3}{4}\right) = \frac{6}{1} \cdot \frac{2}{9} \cdot \frac{3}{4}$$
$$= \frac{6 \cdot 2 \cdot 3}{1 \cdot 9 \cdot 4}$$
$$= \frac{2 \cdot 3 \cdot 2 \cdot 3}{1 \cdot 3 \cdot 3 \cdot 2 \cdot 2} = 1$$

26. $-\left(3a^2\right)^0 = -1$

27. $\left(2a^4b^3\right)^5 = 2^{1 \cdot 5}a^{4 \cdot 5}b^{3 \cdot 5}$
$$= 2^5 a^{20}b^{15}$$
$$= 32a^{20}b^{15}$$

28. $(a - b)^2 + 5c$; $a = -4$, $b = 6$, $c = -2$
$$(-4 - 6)^2 + 5(-2) = (-10)^2 + 5(-2)$$
$$= 100 + 5(-2)$$
$$= 100 + (-10)$$
$$= 90$$

29. $2\dfrac{4}{5} \cdot \dfrac{6}{7} = \dfrac{14}{5} \cdot \dfrac{6}{7}$
$$= \frac{14 \cdot 6}{5 \cdot 7}$$
$$= \frac{2 \cdot 7 \cdot 2 \cdot 3}{5 \cdot 7}$$
$$= \frac{12}{5} = 2\frac{2}{5}$$

30. $6.23 \times 10^{-5} = 0.0000623$

31. Let the unknown number be x.
the <u>quotient</u> of ten and the <u>difference</u> between a number and nine
$$\frac{10}{x - 9}$$

32. Let the unknown number be x.
two <u>less than</u> twice the <u>sum</u> of a number and four
$$2(x + 4) - 2$$
$$2x + 8 - 2$$
$$2x + 6$$

33. Strategy — To find the difference, subtract the average rainfall in El Paso (7.82) from the average rainfall in Seattle (38.6).

Solution — $38.6 - 7.82 = 30.78$
The difference between the average annual rainfall in Seattle and El Paso is 30.78 in.

34. Strategy — To find the difference, subtract the amount of trash thrown away by a person in 1960 (2.7) from the amount of trash thrown away by a person in the 2009 (4.3). Multiply the difference by the number of days in a year (365). Round your answer.

Solution — $4.3 - 2.7 = 1.6$
$1.6 \cdot 365 = 584$
The person in 2009 throws away 584 lb more trash than a person in 1960.

35. the distance from Earth to the sun: d
    the distance from Neptune to the sun: $30d$

36. Strategy   To find the cost, substitute
    $15.375 for S and 200 for N in
    the given formula and solve for
    C.

    Solution $C = SN$
    $C = 15.375 \cdot 200$
    $C = 3,075$
    The cost of the stock was
    $3,075.

Chapter 6: First-Degree Equations

Prep Test

1. $8 - 12 = -4$

2. $-\dfrac{3}{4}\left(-\dfrac{4}{3}\right) = 1$

3. $-\dfrac{5}{8}(16) = -5(2) = -10$

4. $\dfrac{-3}{-3} = 1$

5. $-16 + 7y + 16 = 7y$

6. $8x - 9 - 8x = -9$

7. $2x + 3$
   $2(-4) + 3 = -8 + 3 = -5$

8. $y = -4x + 5$
   $y = -4(-2) + 5$
   $= 8 + 5$
   $= 13$

Section 6.1

Concept Check

1. The solution of the equation $x = 9$ is 9 because when we substitute 9 for the variable x, the result is the true equation $9 = 9$. So the solution to the equation $x = 9$ is the number x is equal to. Therefore, if, when solving an equation, we get the variable alone on one side, the solution of the equation is the number of the other side of the equal sign.

3. 10

5. $y = \dfrac{2}{7}$

Objective A Exercises

7. $x + 3 = 9$
   $x + 3 - 3 = 9 - 3$
   $x = 6$
   The solution is 6.

9. $4 + x = 13$
   $4 - 4 + x = 13 - 4$
   $x = 9$
   The solution is 9.

11. $m - 12 = 5$
    $m - 12 + 12 = 5 + 12$
    $m = 17$
    The solution is 17.

13. $x - 3 = -2$
    $x - 3 + 3 = -2 + 3$
    $x = 1$
    The solution is 1.

15. $a + 5 = -2$
    $a + 5 - 5 = -2 - 5$
    $a = -7$
    The solution is -7.

17. $3 + m = -6$
    $3 - 3 + m = -6 - 3$
    $m = -9$
    The solution is -9.

19. $8 = x + 3$
    $8 - 3 = x + 3 - 3$
    $5 = x$
    The solution is 5.

21. $3 = w - 6$
    $3 + 6 = w - 6 + 6$
    $9 = w$
    The solution is 9.

23. $-7 = -7 + m$
    $-7 + 7 = -7 + 7 + m$
    $0 = m$
    The solution is 0.

25. $-3 = v + 5$
    $-3 - 5 + v + 5 - 5$
    $-8 = v$
    The solution is -8.

27. $-5 = 1 + x$
    $-5 - 1 = 1 - 1 + x$
    $-6 = x$
    The solution is -6.

29. $3 = -9 + m$
    $3 + 9 = -9 + 9 + m$
    $12 = m$
    The solution is 12.

31. $4 + x - 7 = 3$
    $x - 3 = 3$
    $x - 3 + 3 = 3 + 3$
    $x = 6$
    The solution is 6.

33. $8t + 6 - 7t = -6$
$t + 6 = -6$
$t + 6 - 6 = -6 - 6$
$t = -12$
The solution is -12.

35. $y + \dfrac{4}{7} = \dfrac{6}{7}$

$y + \dfrac{4}{7} - \dfrac{4}{7} = \dfrac{6}{7} - \dfrac{4}{7}$

$y = \dfrac{2}{7}$

The solution is $\dfrac{2}{7}$.

37. $x - \dfrac{3}{8} = \dfrac{1}{8}$

$x - \dfrac{3}{8} + \dfrac{3}{8} = \dfrac{1}{8} + \dfrac{3}{8}$

$x = \dfrac{4}{8}$

$x = \dfrac{1}{2}$

The solution is $\dfrac{1}{2}$.

39. x would be less than $-\dfrac{21}{43}$. In order to solve this equation, we would subtract $\dfrac{13}{15}$ from $-\dfrac{21}{43}$, which would give us a number less than $-\dfrac{21}{43}$.

Objective B Exercises

41. $3x = 9$
$\dfrac{3x}{3} = \dfrac{9}{3}$
$x = 3$
The solution is 3.

43. $4c = -12$
$\dfrac{4c}{4} = \dfrac{-12}{4}$
$c = -3$
The solution is -3.

45. $-2r = 16$
$\dfrac{-2r}{-2} = \dfrac{16}{-2}$
$r = -8$
The solution is -8.

47. $-4m = -28$
$\dfrac{-4m}{-4} = \dfrac{-28}{-4}$
$m = 7$
The solution is 7.

49. $-3y = 0$
$\dfrac{-3y}{-3} = \dfrac{0}{-3}$
$y = 0$
The solution is 0.

51. $12 = 2c$
$\dfrac{12}{2} = \dfrac{2c}{2}$
$6 = c$
The solution is 6.

53. $-72 = 18v$
$\dfrac{-72}{18} = \dfrac{18v}{18}$
$-4 = v$
The solution is -4.

55. $-68 = -17t$
$\dfrac{-68}{-17} = \dfrac{-17t}{-17}$
$4 = t$
The solution is 4.

57. $12x = 30$
$\dfrac{12x}{12} = \dfrac{30}{12}$
$x = \dfrac{5}{2}$
The solution is $\dfrac{5}{2}$.

59. $-6a = 21$
$\dfrac{-6a}{-6} = \dfrac{21}{-6}$
$a = -\dfrac{7}{2}$
The solution is $-\dfrac{7}{2}$.

61. $\dfrac{2}{3}x = 4$

$\dfrac{3}{2} \cdot \dfrac{2}{3}x = \dfrac{3}{2}(4)$

$x = 6$
The solution is 6.

63. $-\dfrac{4c}{7} = 16$

$-\dfrac{7}{4}\left(-\dfrac{4}{7}c\right) = -\dfrac{7}{4}(16)$

$c = -28$
The solution is -28.

65. $8 = \dfrac{4}{5}y$

$\dfrac{5}{4}(8) = \dfrac{5}{4} \cdot \dfrac{4}{5}y$

$10 = y$

The solution is 10.

67. $\dfrac{5y}{6} = \dfrac{7}{12}$

$\dfrac{6}{5}\left(\dfrac{5}{6}y\right) = \dfrac{6}{5}\left(\dfrac{7}{12}\right)$

$y = \dfrac{7}{10}$

The solution is $\dfrac{7}{10}$.

69. $7y - 9y = 10$

$-2y = 10$

$\dfrac{-2y}{-2} = \dfrac{10}{-2}$

$y = -5$

The solution is –5.

71. $m - 4m = 21$

$-3m = 21$

$\dfrac{-3m}{-3} = \dfrac{21}{-3}$

$m = -7$

The solution is –7.

73. No. In order for $15x$ to be positive, x must be positive.

Critical Thinking 6.1

75. $\dfrac{1}{\frac{1}{x}} + 8 = -19$

$\dfrac{1}{\frac{1}{x}} = -27$

$1 \div \dfrac{1}{x} = -27$

$1 \cdot \dfrac{x}{1} = -27$

$x = -27$

77. $\dfrac{5}{\frac{7}{a}} - \dfrac{3}{\frac{7}{a}} = 6$

$\dfrac{2}{\frac{7}{a}} = 6$

$2 \div \dfrac{7}{a} = 6$

$2 \cdot \dfrac{a}{7} = 6$

$7\left(\dfrac{2a}{7}\right) = 7(6)$

$2a = 42$

$\dfrac{2a}{2} = \dfrac{42}{2}$

$a = 21$

Projects or Groups Activities 6.1

79. Answers will vary. For example, $5x = -10$.

Section 6.2

Concept Check

1. $3; 5$

Objective A Exercises

3. $5y + 1 = 11$

$5y + 1 - 1 = 11 - 1$

$5y = 10$

$\dfrac{5y}{5} = \dfrac{10}{5}$

$y = 2$

The solution is 2.

5. $2z - 9 = 11$

$2z - 9 + 9 = 11 + 9$

$2z = 20$

$\dfrac{2z}{2} = \dfrac{20}{2}$

$z = 10$

The solution is 10.

7. $12 = 2 + 5a$
$12 - 2 = 2 - 2 + 5a$
$10 = 5a$
$\dfrac{10}{5} = \dfrac{5a}{5}$
$2 = a$
The solution is 2.

9. $-5y + 8 = 13$
$-5y + 8 - 8 = 13 - 8$
$-5y = 5$
$\dfrac{-5y}{-5} = \dfrac{5}{-5}$
$y = -1$
The solution is -1.

11. $-12a - 1 = 23$
$-12a - 1 + 1 = 23 + 1$
$-12a = 24$
$\dfrac{-12a}{-12} = \dfrac{24}{-12}$
$a = -2$
The solution is -2.

13. $10 - c = 14$
$10 - 10 - c = 14 - 10$
$-c = 4$
$\dfrac{-1c}{-1} = \dfrac{4}{-1}$
$c = -4$
The solution is -4.

15. $4 - 3x = -5$
$4 - 4 - 3x = -5 - 4$
$-3x = -9$
$\dfrac{-3x}{-3} = \dfrac{-9}{-3}$
$x = 3$
The solution is 3.

17. $-33 = 3 - 4z$
$-33 - 3 = 3 - 3 - 4z$
$-36 = -4z$
$\dfrac{-36}{-4} = \dfrac{-4z}{-4}$
$9 = z$
The solution is 9.

19. $-4t + 16 = 0$
$-4t + 16 - 16 = 0 - 16$
$-4t = -16$
$\dfrac{-4t}{-4} = \dfrac{-16}{-4}$
$t = 4$
The solution is 4.

21. $5a + 9 = 12$
$5a + 9 - 9 = 12 - 9$
$5a = 3$
$\dfrac{5a}{5} = \dfrac{3}{5}$
$a = \dfrac{3}{5}$
The solution is $\dfrac{3}{5}$.

23. $2t - 5 = 2$
$2t - 5 + 5 = 2 + 5$
$2t = 7$
$\dfrac{2t}{2} = \dfrac{7}{2}$
$t = \dfrac{7}{2}$
The solution is $\dfrac{7}{2}$.

25. $8x + 1 = 7$
$8x + 1 - 1 = 7 - 1$
$8x = 6$
$\dfrac{8x}{8} = \dfrac{6}{8}$
$x = \dfrac{3}{4}$
The solution is $\dfrac{3}{4}$.

27. $4z - 5 = 1$
$4z - 5 + 5 = 1 + 5$
$4z = 6$
$\dfrac{4z}{4} = \dfrac{6}{4}$
$z = \dfrac{3}{2}$
The solution is $\dfrac{3}{2}$.

29. $25 = 11 + 8v$
$25 - 11 = 11 - 11 + 8v$
$14 = 8v$
$\dfrac{14}{8} = \dfrac{8v}{8}$
$\dfrac{7}{4} = v$
The solution is $\dfrac{7}{4}$.

31. $-3 = 7 + 4y$
$-3 - 7 = 7 - 7 + 4y$
$-10 = 4y$
$\dfrac{-10}{4} = \dfrac{4y}{4}$
$-\dfrac{5}{2} = y$
The solution is $-\dfrac{5}{2}$.

33. $8a - 5 = 31$

$8a - 5 + 5 = 31 + 5$

$8a = 36$

$\dfrac{8a}{8} = \dfrac{36}{8}$

$a = \dfrac{9}{2}$

The solution is $\dfrac{9}{2}$.

35. $7 - 12y = 7$

$7 - 7 - 12y = 7 - 7$

$-12y = 0$

$\dfrac{-12y}{-12} = \dfrac{0}{-12}$

$y = 0$

The solution is 0.

37. $-9 - 12y = 5$

$-9 + 9 - 12y = 5 + 9$

$-12y = 14$

$\dfrac{-12y}{-12} = \dfrac{14}{-12}$

$y = -\dfrac{7}{6}$.

39. $6z - \dfrac{1}{3} = \dfrac{5}{3}$

$6z - \dfrac{1}{3} + \dfrac{1}{3} = \dfrac{5}{3} + \dfrac{1}{3}$

$6z = \dfrac{6}{3}$

$6z = 2$

$\dfrac{6z}{6} = \dfrac{2}{6}$

$z = \dfrac{1}{3}$

The solution is $\dfrac{1}{3}$.

41. $3p - \dfrac{5}{8} = \dfrac{19}{8}$

$3p - \dfrac{5}{8} + \dfrac{5}{8} = \dfrac{19}{8} + \dfrac{5}{8}$

$3p = \dfrac{24}{8}$

$3p = 3$

$\dfrac{3p}{3} = \dfrac{3}{3}$

$p = 1$

The solution is 1.

43. $\dfrac{4}{5}y + 3 = 11$

$\dfrac{4}{5}y + 3 - 3 = 11 - 3$

$\dfrac{4}{5}y = 8$

$\dfrac{5}{4}\left(\dfrac{4}{5}y\right) = \dfrac{5}{4} \cdot 8$

$y = 10$

The solution is 10.

45. $\dfrac{3y}{7} - 2 = 10$

$\dfrac{3y}{7} - 2 + 2 = 10 + 2$

$\dfrac{3}{7}v = 12$

$\dfrac{7}{3}\left(\dfrac{3}{7}v\right) = \dfrac{7}{3} \cdot 12$

$v = 28$

The solution is 28.

47. $\dfrac{4z}{9} + 23 = 3$

$\dfrac{4z}{9} + 23 - 23 = 3 - 23$

$\dfrac{4z}{9} = -20$

$\dfrac{9}{4}\left(\dfrac{4}{9}z\right) = \dfrac{9}{4}(-20)$

$z = -45$

The solution is −45.

49. $\dfrac{y}{4} + 5 = 2$

$\dfrac{y}{4} + 5 - 5 = 2 - 5$

$\dfrac{y}{4} = -3$

$\dfrac{4}{1}\left(\dfrac{1}{4}y\right) = \dfrac{4}{1}(-3)$

$y = -12$

The solution is −12.

51. $\dfrac{2}{5}y - 3 = 1$

$\dfrac{2}{5}y - 3 + 3 = 1 + 3$

$\dfrac{2}{5}y = 4$

$\dfrac{5}{2}\left(\dfrac{2}{5}y\right) = \dfrac{5}{2}(4)$

$y = 10$

The solution is 10.

53. $5 - \dfrac{7}{8}y = 2$

$5 - 5 - \dfrac{7}{8}y = 2 - 5$

$-\dfrac{7}{8}y = -3$

$\left(-\dfrac{8}{7}\right)\left(-\dfrac{7}{8}y\right) = \left(-\dfrac{8}{7}\right)(-3)$

$y = \dfrac{24}{7}$

The solution is $\dfrac{24}{7}$.

55. $\dfrac{3}{5}y + \dfrac{1}{4} = \dfrac{3}{4}$

$\dfrac{3}{5}y + \dfrac{1}{4} - \dfrac{1}{4} = \dfrac{3}{4} - \dfrac{1}{4}$

$\dfrac{3}{5}y = \dfrac{1}{2}$

$\dfrac{5}{3}\left(\dfrac{3}{5}y\right) = \dfrac{5}{3}\left(\dfrac{1}{2}\right)$

$y = \dfrac{5}{6}$

The solution is $\dfrac{5}{6}$.

57. $\dfrac{3}{5} = \dfrac{2}{7}t + \dfrac{1}{5}$

$\dfrac{3}{5} - \dfrac{1}{5} = \dfrac{2}{7}t + \dfrac{1}{5} - \dfrac{1}{5}$

$\dfrac{2}{5} = \dfrac{2}{7}t$

$\dfrac{7}{2}\left(\dfrac{2}{5}\right) = \dfrac{7}{2}\left(\dfrac{2}{7}t\right)$

$\dfrac{7}{5} = t$

The solution is $\dfrac{7}{5}$.

59. $\dfrac{z}{3} - \dfrac{1}{2} = \dfrac{1}{4}$

$\dfrac{z}{3} - \dfrac{1}{2} + \dfrac{1}{2} = \dfrac{1}{4} + \dfrac{1}{2}$

$\dfrac{z}{3} = \dfrac{1}{4} + \dfrac{2}{4}$

$\dfrac{z}{3} = \dfrac{3}{4}$

$\dfrac{3}{1}\left(\dfrac{1}{3}z\right) = \dfrac{3}{1}\left(\dfrac{3}{4}\right)$

$z = \dfrac{9}{4}$

The solution is $\dfrac{9}{4}$.

61. $5.6t - 5.1 = 1.06$

$5.6t - 5.1 + 5.1 = 1.06 + 5.1$

$5.6t = 6.16$

$\dfrac{5.6t}{5.6} = \dfrac{6.16}{5.6}$

$t = 1.1$

The solution is 1.1.

63. $6.2 - 3.3t = -12.94$

$6.2 - 6.2 - 3.3t = -12.94 - 6.2$

$-3.3t = -19.14$

$\dfrac{-3.3t}{-3.3} = \dfrac{-19.14}{-3.3}$

$t = 5.8$

The solution is 5.8.

65. $6c - 2 - 3c = 10$

$3c - 2 = 10$

$3c - 2 + 2 = 10 + 2$

$3c = 12$

$\dfrac{3c}{3} = \dfrac{12}{3}$

$c = 4$

The solution is 4.

67. $4y + 5 - 12y = -3$

$-8y + 5 = -3$

$-8y + 5 - 5 = -3 - 5$

$-8y = -8$

$\dfrac{-8y}{-8} = \dfrac{-8}{-8}$

$y = 1$

The solution is 1.

69. $17 = 12p - 5 - 6p$

$17 = 6p - 5$

$17 + 5 = 6p - 5 + 5$

$22 = 6p$

$\dfrac{22}{6} = \dfrac{6p}{6}$

$\dfrac{11}{3} = p$

The solution is $\dfrac{11}{3}$.

71. Negative. -347 minus -73 will result in a negative number. That negative number divided by 15 will result in a negative number.

Objective B Exercises

73. Strategy To find the number of years, substitute 48,000 for V and 70,000 for C in the given equation and solve for t.

Solution
$$V = C - 5{,}500t$$
$$48{,}000 = 70{,}000 - 5{,}500t$$
$$48{,}000 - 70{,}000 = 70{,}000 - 70{,}000 - 5{,}500t$$
$$-22{,}000 = -5{,}500t$$
$$\frac{-22{,}000}{-5{,}500} = \frac{-5{,}500t}{-5{,}500}$$
$$4 = t$$
In 4 years, the X-ray machine will have a value of \$48,000.

75. Strategy To find the average crown spread, substitute $557\frac{1}{4}$ for P, 425 for c, and 118 for h in the given equation and solve for s.

Solution
$$P = c + h + \tfrac{1}{4}s$$
$$557\tfrac{1}{4} = 425 + 118 + \tfrac{1}{4}s$$
$$\tfrac{2229}{4} = 543 + \tfrac{1}{4}s$$
$$4\left(\tfrac{2229}{4}\right) = 4\left(543 + \tfrac{1}{4}s\right)$$
$$2229 = 2172 + s$$
$$57 = s$$
The average crown spread is 57 ft.

77. Strategy To find the number of grams of protein, substitute 174 for C, 2 for f, and 30 for c.

Solution
$$C = 9f + 4p + 4c$$
$$174 = 9(2) + 4p + 4(30)$$
$$174 = 18 + 4p + 120$$
$$174 = 4p + 138$$
$$36 = 4p$$
$$\frac{36}{4} = \frac{4p}{4}$$
$$9 = p$$
There are 9 grams of protein in an 8-ounce serving of vanilla yogurt.

79. Strategy To find the initial velocity, substitute 80 for s and 2 for t.

Solution
$$s = 16t^2 + vt$$
$$80 = 16(2)^2 + v(2)$$
$$80 = 16(4) + 2v$$
$$80 = 64 + 2v$$
$$16 = 2v$$
$$\frac{16}{2} = \frac{2v}{2}$$
$$8 = v$$
The initial velocity is 8 ft/s.

81. Strategy   To find the distance, substitute −3 for C in the given equation and solve for D.

Solution $C = \frac{1}{4}D - 45$

$-3 = \frac{1}{4}D - 45$

$-3 + 45 = \frac{1}{4}D - 45 + 45$

$42 = \frac{1}{4}D$

$\frac{4}{1}(42) = \frac{4}{1}\left(\frac{1}{4}D\right)$

$168 = D$

The car will slide 168 ft.

83. ii

Critical Thinking 6.2

85. $2x - 3y = 8$
$2x - 4(0) = 8$
$2x - 0 = 8$
$2x = 8$
$\dfrac{2x}{2} = \dfrac{8}{2}$
$x = 4$

87. $5x - 2y = -3$
$5(-3) - 2y = -3$
$-15 - 2y = -3$
$-2y = 12$
$\dfrac{-2y}{-2} = \dfrac{12}{-2}$
$y = -6$

89. $3x + 5 = -4$
$3x = -9$
$\dfrac{3x}{3} = \dfrac{-9}{3}$
$x = -3$
Substituting -3 for x:
$2x - 5$
$= 2(-3) - 5$
$= -6 - 5$
$= -11$

91. $2 - 3x = 11$
$-3x = 9$
$\dfrac{-3x}{-3} = \dfrac{9}{-3}$
$x = -3$
Substituting -3 for x:
$x^2 + 2x - 3$
$= (-3)^2 + 2(-3) - 3$
$= 9 - 6 - 3$
$= 0$

93. The sentence "Solve $3x + 4(x - 3)$" does not make sense. $3x + 4(x - 3)$ is an expression, which can be simplified, but not solved.

Projects or Group Activities 6.2

95. Answers will vary. For example,
$2x + 5 = -1$

Section 6.3

Concept Check

1. a. $2x$

b. 3

Objective A Exercises

3. $4x + 3 = 2x + 9$
$4x - 2x + 3 = 2x - 2x + 9$
$2x + 3 = 9$
$2x + 3 - 3 = 9 - 3$
$2x = 6$
$\dfrac{2x}{2} = \dfrac{6}{2}$
$x = 3$
The solution is 3.

5. $7y - 6 = 3y + 6$
$7y - 3y - 6 = 3y - 3y + 6$
$4y - 6 = 6$
$4y - 6 + 6 = 6 + 6$
$4y = 12$
$\dfrac{4y}{4} = \dfrac{12}{4}$
$y = 3$
The solution is 3.

7. $12m + 11 = 5m + 4$
$12m - 5m + 11 = 5m - 5m + 4$
$7m + 11 = 4$
$7m + 11 - 11 = 4 - 11$
$7m = -7$
$\dfrac{7m}{7} = \dfrac{-7}{7}$
$m = -1$
The solution is -1.

9. $7c - 5 = 2c - 25$
$7c - 2c - 5 = 2c - 2c - 25$
$5c - 5 = -25$
$5c - 5 + 5 = -25 + 5$
$5c = -20$
$\dfrac{5c}{5} = \dfrac{-20}{5}$
$c = -4$
The solution is -4.

11. $2n - 3 = 5n - 18$
$2n - 5n - 3 = 5n - 5n - 18$
$-3n - 3 = -18$
$-3n - 3 + 3 = -18 + 3$
$-3n = -15$
$\dfrac{-3n}{-3} = \dfrac{-15}{-3}$
$n = 5$
The solution is 5

13. $3z + 5 = 19 - 4z$
$3z + 4z + 5 = 19 - 4z + 4z$
$7z + 5 = 19$
$7z + 5 - 5 = 19 - 5$
$7z = 14$
$\dfrac{7z}{7} = \dfrac{14}{7}$
$z = 2$
The solution is 2.

15. $5v - 3 = 4 - 2v$
$5v + 2v - 3 = 4 - 2v + 2v$
$7v - 3 = 4$
$7v - 3 + 3 = 4 + 3$
$7v = 7$
$\dfrac{7v}{7} = \dfrac{7}{7}$
$v = 1$
The solution is 1.

17. $7 - 4a = 2a$
$7 - 4a + 4a = 2a + 4a$
$7 = 6a$
$\dfrac{7}{6} = \dfrac{6a}{6}$
$\dfrac{7}{6} = a$
The solution is $\dfrac{7}{6}$.

19. $12 - 5y = 3y - 12$
$12 - 5y - 3y = 3y - 3y - 12$
$12 - 8y = -12$
$12 - 12 - 8y = -12 - 12$
$-8y = -24$
$\dfrac{-8y}{-8} = \dfrac{-24}{-8}$
$y = 3$
The solution is 3.

21. $7r = 8 + 2r$
$7r - 2r = 8 + 2r - 2r$
$5r = 8$
$\dfrac{5r}{5} = \dfrac{8}{5}$
$r = \dfrac{8}{5}$
The solution is $\dfrac{8}{5}$.

23. $5a + 3 = 3a + 10$
$5a - 3a + 3 = 3a - 3a + 10$
$2a + 3 = 10$
$2a + 3 - 3 = 10 - 3$
$2a = 7$
$\dfrac{2a}{2} = \dfrac{7}{2}$
The solution is $\dfrac{7}{2}$.

25. $x - 7 = 5x - 21$
$x - 5x - 7 = 5x - 5x - 21$
$-4x - 7 = -21$
$-4x - 7 + 7 = -21 + 7$
$-4x = -14$
$\dfrac{-4x}{-4} = \dfrac{-14}{-4}$
$x = \dfrac{7}{2}$
The solution is $\dfrac{7}{2}$.

27. $5n - 1 + 2n = 4n + 8$
$7n - 1 = 4n + 8$
$7n - 4n - 1 = 4n - 4n + 8$
$3n - 1 = 8$
$3n - 1 + 1 = 8 + 1$
$3n = 9$
$\dfrac{3n}{3} = \dfrac{9}{3}$
$n = 3$
The solution is 3.

29. $3z - 2 - 7z = 4z + 6$
$-4z - 2 = 4z + 6$
$-4z - 4z - 2 = 4z - 4z + 6$
$-8z - 2 = 6$
$-8z - 2 + 2 = 6 + 2$
$-8z = 8$
$\dfrac{-8z}{-8} = \dfrac{8}{-8}$
$z = -1$
The solution is -1.

31. $4t - 8 + 12t = 3 - 4t - 11$
$16t - 8 = -8 - 4t$
$16t + 4t - 8 = -8 - 4t + 4t$
$20t - 8 = -8$
$20t - 8 + 8 = -8 + 8$
$20t = 0$
$\dfrac{20t}{20} = \dfrac{0}{20}$
$t = 0$
The solution is 0.

33. Negative

Objective B Exercises

35. iv

37. $3(4y + 5) = 25$
$12y + 15 = 25$
$12y + 15 - 15 = 25 - 15$
$12y = 10$
$\dfrac{12y}{12} = \dfrac{10}{12}$
$y = \dfrac{5}{6}$
The solution is $\dfrac{5}{6}$.

39. $-2(4x + 1) = 22$
$-8x - 2 = 22$
$-8x - 2 + 2 = 22 + 2$
$-8x = 24$
$\dfrac{-8x}{-8} = \dfrac{24}{-8}$
$x = -3$
The solution is -3.

41. $5(2k + 1) - 7 = 28$
$10k + 5 - 7 = 28$
$10k - 2 = 28$
$10k - 2 + 2 = 28 + 2$
$10k = 30$
$\dfrac{10k}{10} = \dfrac{30}{10}$
$k = 3$
The solution is 3.

43. $3(3v - 4) + 2v = 10$
$9v - 12 + 2v = 10$
$11v - 12 = 10$
$11v - 12 + 12 = 10 + 12$
$11v = 22$
$\dfrac{11v}{11} = \dfrac{22}{11}$
$v = 2$
The solution is 2.

45. $3y + 2(y + 1) = 12$
$3y + 2y + 2 = 12$
$5y + 2 = 12$
$5y + 2 - 2 = 12 - 2$
$5y = 10$
$\dfrac{5y}{5} = \dfrac{10}{5}$
$y = 2$
The solution is 2.

47. $7v - 3(v - 4) = 20$
$7v - 3v + 12 = 20$
$4v + 12 = 20$
$4v + 12 - 12 = 20 - 12$
$4v = 8$
$\dfrac{4v}{4} = \dfrac{8}{4}$
$v = 2$
The solution is 2.

49. $6 + 3(3x - 3) = 24$
$6 + 9x - 9 = 24$
$9x - 3 = 24$
$9x - 3 + 3 = 24 + 3$
$9x = 27$
$\dfrac{9x}{9} = \dfrac{27}{9}$
$x = 3$
The solution is 3.

51. $9 - 3(4a - 2) = 9$
$9 - 12a + 6 = 9$
$-12a + 15 = 9$
$-12a + 15 - 15 = 9 - 15$
$-12a = -6$
$\dfrac{-12a}{-12} = \dfrac{-6}{-12}$
$a = \dfrac{1}{2}$
The solution is $\dfrac{1}{2}$.

53. $3(2z - 5) = 4z + 1$
$6z - 15 = 4z + 1$
$6z - 4z - 15 = 4z - 4z + 1$
$2z - 15 = 1$
$2z - 15 + 15 = 1 + 15$
$2z = 16$
$\dfrac{2z}{2} = \dfrac{16}{2}$
$z = 8$
The solution is 8.

55. $2 - 3(5x + 2) = 2(3 - 5x)$

$2 - 15x - 6 = 6 - 10x$

$-15x - 4 = 6 - 10x$

$-15x + 10x - 4 = 6 - 10x + 10x$

$-5x - 4 = 6$

$-5x - 4 + 4 = 6 + 4$

$-5x = 10$

$\dfrac{-5x}{-5} = \dfrac{10}{-5}$

$x = -2$

The solution is -2.

57. $4r + 11 = 5 - 2(3r + 3)$

$4r + 11 = 5 - 6r - 6$

$4r + 11 = -6r - 1$

$4r + 6r + 11 = -6r + 6r - 1$

$10r + 11 = -1$

$10r + 11 - 11 = -1 - 11$

$10r = -12$

$\dfrac{10r}{10} = \dfrac{-12}{10}$

$r = -\dfrac{6}{5}$

The solution is $-\dfrac{6}{5}$.

Objective C Exercises

59. Strategy   To find the break-even point, substitute 1,600 for P, 950 for C, and 211,250 for F in the given equation and solve for x.

     Solution $Px = Cx + F$

$1,600x = 950x + 211,250$

$650x = 211,250$

$\dfrac{650x}{650} = \dfrac{211,250}{650}$

$x = 325$

325 units must be sold to break even.

61. Strategy   To find the break-even point, substitute 99 for P, 38 for C, and 24,400 for F in the given equation and solve for x.

     Solution $Px = Cx + F$

$99x = 38x + 24,400$

$61x = 24,400$

$\dfrac{61x}{61} = \dfrac{24,400}{61}$

$x = 400$

400 units must be sold to break even.

63. Strategy   To find the oxygen consumption, substitute 10.4 for m

     Solution $m = \frac{1}{6}(C - 5)$

$10.4 = \frac{1}{6}(C - 5)$

$\frac{6}{1}(10.4) = \frac{6}{1}[\frac{1}{6}(C - 5)]$

$62.4 = C - 5$

$67.4 = C$

The oxygen consumption is 67.4 ml/min

65. a. 5 ft

b. The person who is 3 ft from the fulcrum

c. No

67. Strategy   The distance of the fulcrum from the 175-pound adult: x
The distance of the fulcrum from the 70-pound child: $14 - x$
To find the placement of the fulcrum, replace the variables F_1, F_2, and d by the given values and solve for x.

     Solution $F_1 \cdot x = F_2 \cdot (d - x)$

$70 \cdot x = 175(14 - x)$

$70x = 2,450 - 175x$

$245x = 2,450$

$\dfrac{245x}{245} = \dfrac{2,450}{245}$

$x = 10$

The 70-pound child should be 10 ft from the fulcrum.

69.   Strategy   The distance of the fulcrum from the 90-pound child: x
The distance of the fulcrum from the 60-pound child: $12 - x$
To find the placement of the fulcrum, replace the variables F_1, F_2, and d by the given values and solve for x.

      Solution

$$F_1 \cdot x = F_2 \cdot (d - x)$$
$$90 \cdot x = 60(12 - x)$$
$$90x = 720 - 60x$$
$$150x = 720$$
$$\frac{150x}{150} = \frac{720}{150}$$
$$x = 4.8$$

The 90-pound child should be placed 4.8 ft from the fulcrum.

71.   Strategy   To find the force, substitute 30 for F_2, 9 for d, and 0.15 for x in the given equation and solve for F_1.

      Solution

$$F_1 \cdot x = F_2 \cdot (d - x)$$
$$F_1 \cdot 0.15 = 30(9 - 0.15)$$
$$0.15F_1 = 265.5$$
$$\frac{0.15F_1}{0.15} = \frac{265.5}{0.15}$$
$$F_1 = 1{,}770$$

The force on the lip of the can is 1,770 lb.

73.   Strategy   To find the number of miles, substitute 20.90 for F in the given equation

      Solution

$$F = 2.50 + 2.30(m - 1)$$
$$20.90 = 2.50 + 2.30(m - 1)$$
$$20.90 = 2.50 + 2.30m - 2.30$$
$$20.90 = 2.30m + 0.20$$
$$20.70 = 2.30\,m$$
$$\frac{20.70}{2.30} = \frac{2.30m}{2.30}$$
$$9 = m$$

The passenger traveled 9 miles.

Critical Thinking 6.3

75. $\dfrac{1}{5}(25 - 10b) + 4 = \dfrac{1}{3}(9b - 15) - 6$
$$5 - 2b + 4 = 3b - 5 - 6$$
$$-2b + 9 = 3b - 11$$
$$20 = 5b$$
$$\frac{20}{5} = \frac{5b}{5}$$
$$4 = b$$

77. $\dfrac{2(5x - 6) - 3(x - 4)}{7} = x + 2$
$$7\left[\frac{2(5x - 6) - 3(x - 4)}{7}\right] = 7(x + 2)$$
$$2(5x - 6) - 3(x - 4) = 7(x + 2)$$
$$10x - 12 - 3x + 12 = 7x + 14$$
$$7x = 7x + 14$$
$$0 = 14$$
No solution

79. $3(x - 4) = x$
$$3x - 12 = x$$
$$2x = 12$$
$$\frac{2x}{2} = \frac{12}{2}$$
$$x = 6$$

Projects or Group Activities 6.3

81.   Answers will vary

Check Your Progress: Chapter 6

1. $y - 5 = -8$
$$y = -3$$
The solution is −3.

2. $6 + n = -9$
$$n = -15$$
The solution is −15.

3. $-7 = c + 4$
$$-11 = c$$
The solution is −11.

4. $1 + m - 3 = 2$
$$m - 2 = 2$$
$$m = 4$$
The solution is 4.

5. $5d + 6 - 2d = -12$
   $3d + 6 = -12$
   $3d = -18$
   $\dfrac{3d}{3} = \dfrac{-18}{3}$
   $d = -6$
   The solution is −6.

6. $x - \dfrac{5}{8} = \dfrac{1}{8}$

   $x = \dfrac{6}{8}$

   $x = \dfrac{3}{4}$

   The solution is $\dfrac{3}{4}$.

7. $9b = -81$
   $\dfrac{9b}{b} = \dfrac{-81}{9}$
   $b = -9$
   The solution is − 9.

8. $-16t = -64$
   $\dfrac{-16t}{-16} = \dfrac{-64}{-16}$
   $t = 4$
   The solution is 4.

9. $-\dfrac{3a}{8} = 12$

   $-\dfrac{8}{3}\left(-\dfrac{3a}{8}\right) = -\dfrac{8}{3}(12)$

   $a = -32$
   The solution is −32.

10. $y - 5y = 20$
    $-4y = 20$
    $\dfrac{-4y}{-4} = \dfrac{20}{-4}$
    $y = -5$
    The solution is −5.

11. $3z - 8 = 4$
    $3z = 12$
    $\dfrac{3z}{3} = \dfrac{12}{3}$
    $z = 4$
    The solution is 4.

12. $-7r + 6 = -15$
    $-7r = -21$
    $\dfrac{-7r}{-7} = \dfrac{-21}{-7}$
    $r = 3$
    The solution is 3.

13. $10 - p = 19$
    $-p = 9$
    $\dfrac{-p}{-1} = \dfrac{9}{-1}$
    $r = -9$
    The solution is −9.

14. $5 - \dfrac{3}{2}w = 6$

    $-\dfrac{3}{2}w = 1$

    $-\dfrac{2}{3}\left(-\dfrac{3}{2}w\right) = -\dfrac{2}{3}(1)$

    $w = -\dfrac{2}{3}$

    The solution is $-\dfrac{2}{3}$.

15. $8n + 4 - 2n = 28$
    $6n + 4 = 28$
    $6n = 24$
    $\dfrac{6n}{6} = \dfrac{24}{6}$
    $n = 4$
    The solution is 4.

16. $5x - 4 = 3x - 10$
    $2x - 4 = -10$
    $2x = -6$
    $\dfrac{2x}{2} = \dfrac{-6}{2}$
    $n = -3$
    The solution is −3.

17. $3x - 7 = 5x - 7$
    $-7 = 2x - 7$
    $0 = 2x$
    $\dfrac{0}{2} = \dfrac{2x}{2}$
    $0 = x$
    The solution is 0.

18. $-7 - 3x = 9 - 2x$
$-7 = 9 + x$
$-16 = x$
The solution is -16.

19. $8x + 3 - 4x = 5 + x$
$4x + 3 = 5 + x$
$3x + 3 = 5$
$3x = 2$
$\dfrac{3x}{3} = \dfrac{2}{3}$
$x = \dfrac{2}{3}$

The solution is $\dfrac{2}{3}$.

20. $5x + 3(x - 2) = 10$
$5x + 3x - 6 = 10$
$8x - 6 = 10$
$8x = 16$
$\dfrac{8x}{8} = \dfrac{16}{8}$
$x = 2$

The solution is 2.

21. $5 - 4(x + 6) = 21$
$5 - 4x - 24 = 21$
$-4x - 19 = 21$
$-4x = 40$
$\dfrac{-4x}{-4} = \dfrac{40}{-4}$
$x = -10$
The solution is -10.

22. $9x - 3(2x + 5) = 4(5x + 2) - 6$
$9x - 6x - 15 = 20x + 8 - 6$
$3x - 15 = 20x + 2$
$-15 = 17x + 2$
$-17 = 17x$
$\dfrac{-17}{17} = \dfrac{17x}{17}$
$-1 = x$
The solution is -1.

23. $5[6 - 2(5x + 1)] = 8x - 9$
$5[6 - 10x - 2] = 8x - 9$
$5[4 - 10x] = 8x - 9$
$20 - 50x = 8x - 9$
$20 = 58x - 9$
$29 = 58x$
$\dfrac{29}{58} = \dfrac{58x}{58}$
$\dfrac{1}{2} = x$

The solution is $\dfrac{1}{2}$.

24. **Strategy**   To find the initial velocity, substitute 174 for s and 3 for t.

Solution $s = 16t^2 + vt$
$174 = 16(3)^2 + v(3)$
$174 = 16(9) + 3v$
$174 = 144 + 3v$
$30 = 3v$
$\dfrac{30}{3} = \dfrac{3v}{3}$
$10 = v$
The initial velocity is 10 ft/s.

25. **Strategy**   To find the number of miles, substitute 12.35 for F in the given equation

Solution $F = 2.75 + 2.40(m - 1)$
$12.35 = 2.75 + 2.40(m - 1)$
$12.35 = 2.75 + 2.40m - 2.40$
$12.35 = 2.40m + 0.35$
$12.00 = 2.40\,m$
$\dfrac{12.00}{2.40} = \dfrac{2.40m}{2.40}$
$5 = m$
The passenger traveled 5 miles.

Section 6.4

Concept Check

1. $3x = -30$

3. $x + 12 = 3$

5. **a.** 6; 5

   b. 472

Objective A Exercises

7. The unknown number: x

Six less than a number	is	five

   $x - 6 = 5$
   $x - 6 + 6 = 5 + 6$
   $x = 11$
   The number is 11.

9. The unknown number: x

The product of a number and eight	is	Negative forty

   $8x = -40$
   $\dfrac{8x}{8} = \dfrac{-40}{8}$
   $x = -5$
   The number is -5.

11. The unknown number: x

The difference between nine and a number	is	seven

    $9 - x = 7$
    $9 - 9 - x = 7 - 9$
    $-x = -2$
    $\dfrac{-1x}{-1} = \dfrac{-2}{-1}$
    $x = 2$
    The number is 2.

13. The unknown number: x

The quotient of a number and six	is	twelve

    $\dfrac{x}{6} = 12$

    $\dfrac{6}{1}\left(\dfrac{1}{6}x\right) = \dfrac{6}{1}(12)$

    $x = 72$
    The number is 72.

15. The unknown number: x

| The sum of twice a number and five | is | fifteen |

$2x + 5 = 15$

$2x + 5 - 5 = 15 - 5$

$2x = 10$

$\dfrac{2x}{2} = \dfrac{10}{2}$

$x = 5$

The number is 5.

17. The unknown number: x

| six less than four times a number | is | twenty-two |

$4x - 6 = 22$

$4x - 6 + 6 = 22 + 6$

$4x = 28$

$\dfrac{4x}{4} = \dfrac{28}{4}$

$x = 7$

The number is 7.

19. The unknown number: x

| Eight less than the product of eleven and a number | is | negative nineteen |

$11x - 8 = -19$

$11x - 8 + 8 = -19 + 8$

$11x = -11$

$\dfrac{11x}{11} = \dfrac{-11}{11}$

$x = -1$

The number is -1.

21. The unknown number: x

| twenty-three | is | the difference between eight and the product of five and a number |

$23 = 8 - 5x$

$23 - 8 = 8 - 8 - 5x$

$15 = -5x$

$\dfrac{15}{-5} = \dfrac{-5x}{-5}$

$x = -3$

The number is -3.

23. The unknown number: x

| four times a number | is | three times the difference between thirty-five and the number |

$4x = 3(35 - x)$

$4x = 105 - 3x$

$4x + 3x = 105 - 3x + 3x$

$7x = 105$

$\dfrac{7x}{7} = \dfrac{105}{7}$

$x = 15$

The number is 15.

25. The smaller number: x

The larger number: $15 - x$

| one less than three times the smaller number | is | the larger number |

$3x - 1 = 15 - x$

$3x + x - 1 = 15 - x + x$

$4x - 1 + 1 = 15 + 1$

$4x = 16$

$\dfrac{4x}{4} = \dfrac{16}{4}$

$x = 4$

$15 - x = 15 - 4 = 11$

The smaller number is 4.

The larger number is 11.

27. The smaller number: x

The larger number: $30 - x$

| three times the smaller number | is | twice the larger number |

$3x = 2(30 - x)$

$3x = 60 - 2x$

$3x + 2x = 60 - 2x + 2x$

$5x = 60$

$\dfrac{5x}{5} = \dfrac{60}{5}$

$x = 12$

$30 - x = 30 - 12 = 18$

The smaller number is 12.

The larger number is 18.

Objective B Exercises

29. $42 = 2x$

$\dfrac{42}{2} = \dfrac{2x}{2}$

$x = 21$

There were 21 million passengers.

31. Strategy To find the yearly costs for a robot, write and solve an equation using x to represent the salary and other benefits of a human worker.

Solution | 64,000 | is | 103 times yearly costs for a robot |

$64{,}000 = 103x$

$\dfrac{64{,}000}{103} = \dfrac{103x}{103}$

$621.35 \approx x$

The yearly costs for a robot are $600.

33. Strategy   To find the length of each side, let x represent the length of the third side. Then each of the equal sides would be one 3 m more than three times the length of the third side, or $3x + 2$. These three sides added together would give the perimeter, which is 46 m.

Solution | The sum of the lengths of the sides | is | 46 m |

$(3x + 2) + (3x + 2) + x = 46$

$7x + 4 = 46$

$7x = 42$

$\dfrac{7x}{7} = \dfrac{42}{7}$

$x = 6$

$3x + 2 = 3(6) + 2 = 20$

The length of each equal side is 20 m.

The length of the third side is 6 m.

35. Strategy   To find the number of minutes, write and solve an equation using n to represent the number of minutes used.

Solution | $34.50 | is | $18 plus $1.50 per minute |

$34.50 = 18 + 1.50n$

$16.50 = 1.50n$

$\dfrac{16.50}{1.50} = \dfrac{1.50n}{1.50}$

$11 = n$

The customer used the service for 11 minutes.

37. $.15

39. Strategy   To find the number of minutes, write and solve an equation using x to represent the number of minutes used.

Solution | $99.80 | is | $35 per month plus $.40 per minute used |

$99.80 = 35 + .40x$

$64.80 = .40x$

$\dfrac{64.80}{.40} = \dfrac{.40x}{.40}$

$162 = x$

The executive used the phone for 162 minutes.

41. Strategy   To find the length of each piece, write and solve an equation using x to represent the shorter piece and $12 - x$ to represent the longer piece.

Solution | Twice the shorter piece | is | three feet less than the longer piece |

$2x = (12 - x) - 3$

$2x = 9 - x$

$2x + x = 9 - x + x$

$3x = 9$

$\dfrac{3x}{3} = \dfrac{9}{3}$

$x = 3$

$12 - x = 12 - 3 = 9$

The shorter piece is 3 ft.

The longer piece is 9 ft.

43. Strategy   To find the amount of each scholarship, write and solve an equation using x to represent the smaller scholarship and $7{,}000 - x$ to represent the larger scholarship.

Solution

twice the smaller scholarship	is	1,000 less than the larger scholarship

$2x = (7{,}000 - x) - 1{,}000$
$2x = 7{,}000 - x - 1{,}000$
$2x + x = 6{,}000 - x + x$
$3x = 6{,}000$
$\dfrac{3x}{3} = \dfrac{6{,}000}{3}$
$x = 2{,}000$
$7{,}000 - x = 7{,}000 - 2{,}000 = 5{,}000$
The larger scholarship is \$5,000.

45. Strategy   To find the number of pounds of each coffee in the mixture, write and solve an equation using x to represent the amount of Colombian, $x + 1$ to represent the amount of French Roast, and $(x + 1) + 2 = x + 3$ to represent the amount of Java.

Solution

the total amount of coffee	is	10 lb

$x + (x + 1) + (x + 3) = 10$
$3x + 4 = 10$
$3x = 6$
$\dfrac{3x}{3} = \dfrac{6}{3}$
$x = 2$
$x + 1 = 2 + 1 = 3$
$x + 3 = 2 + 3 = 5$
The coffee mixture contains 2 lb of Colombian, 3 lb of French Roast, and 5 lb of Java.
Critical Thinking 6.4

47.   The problem states that a 4-quart mixture of fruit juice is made from apple juice and cranberry juice. There are 6 more quarts of apple juice than of cranberry juice. If we let x = the number of quarts of cranberry juice, then $x + 6$ = the number of quarts of apple juice. The total number of quarts is 4. Therefore, we can write the equation
$x + (x + 6) = 4$.
$\begin{aligned} x + (x + 6) &= 4 \\ 2x + 6 &= 4 \\ 2x &= -2 \\ x &= -1 \end{aligned}$

Since x = the number of quarts of cranberry juice, there are -1 quarts of cranberry juice in the mixture. We cannot add -1 qt to a mixture. The solution is not reasonable. We can see from the original problem that the answer will not be reasonable. If the total number of quarts in the mixture is 4, we cannot have more than 6 qt of apple juice in the mixture.

Section 6.5

Concept Check

1. right; down

3. Student explanations should include the idea of starting at the origin, then moving 4 units to the left and then moving 3 units up.

5. 20; 300

Objective A Exercises

7. $(5, 4)$ is in quadrant I.

9. $(-8, 1)$ is in quadrant II.

11. y-axis

13. **a.** The abscissa is positive, and the ordinate is positive.

    b. The abscissa is negative, and the ordinate is negative.

15.

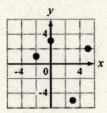

17.

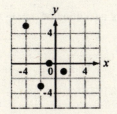

19.

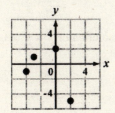

21.

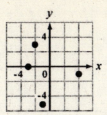

23.

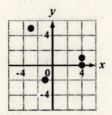

25. $A(0, 2)$
   $B(-4, -1)$
   $C(2, 0)$
   $D(1, -3)$

27. $A(0, 4)$
   $B(-4, 3)$
   $C(-2, 0)$
   $D(2, -3)$

29. $A(3, 5)$
   $B(1, -4)$
   $C(-3, -5)$
   $D(-5, 0)$

31. $A(1, -4)$
   $B(-3, -6)$
   $C(-2, 0)$
   $D(3, 5)$

33. **a.** Abscissa of point A: 2
    Abscissa of point C: -4

    b. Ordinate of point B: 1
    Ordinate of point D: -3

35. **a.** Abscissa of point A: 4
    Abscissa of point C: -3

    b. Ordinate of point B: -2
    Ordinate of point D: 2

37. II

Objective B Exercises

39.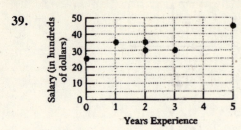

41.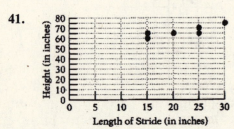

43. The record time for the 800-meter race was 200 s.

45. The point will be graphed with an *x*-coordinate of 1200 and a *y*-coordinate equal to the record time for the 1200-meter race. The graph will have an additional point.

47. **a.** For the car that has 11 mpg in the city has 15 mpg on the highway.

   b. For the car that has 18 mpg on the highway has 13 mpg in the city.

Critical Thinking 6.5

49. 4 units

51. 2 units

53. 5 units

Section 6.6

Concept Check

1. No, this equation is not linear.

3. Yes

5. No, this equation is not in two variables.

7. Yes

9. No, this equation is not linear

11. x; y

13. **a.** -1; -11; $(-1, -11)$

   b. 0; -5; $(0, -5)$

   c. 1; 1; $(1, 1)$

Objective A Exercises

15. $y = -x + 7$

$$4 \,\bigg|\, \begin{array}{l} -(3) + 7 \\ -3 + 7 \\ 4 \end{array}$$

$4 = 4$

Yes, $(3, 4)$ is a solution of $y - x + 7$.

17. $y = \frac{1}{2}x - 1$

$$2 \,\bigg|\, \begin{array}{l} \frac{1}{2}(-1) - 1 \\ -\frac{1}{2} - 1 \\ -\frac{3}{2} \end{array}$$

$2 \neq -\frac{3}{2}$

No, $(-1, 2)$ is not a solution of $y = \frac{1}{2}x - 1$.

19. $y = \frac{1}{4}x + 1$

$$1 \,\bigg|\, \begin{array}{l} \frac{1}{4}(4) + 1 \\ 1 + 1 \\ 2 \end{array}$$

$1 \neq 2$

No, $(4, 1)$ is not a solution of $y = \frac{1}{4}x + 1$.

21. $y = \frac{3}{4}x + 4$

$$4 \,\bigg|\, \begin{array}{l} \frac{3}{4}(0) + 4 \\ 0 + 4 \\ 4 \end{array}$$

$4 = 4$

Yes, $(0, 4)$ is a solution of $y = \frac{3}{4}x + 4$.

23. $y = 3x + 2$

$$0 \,\bigg|\, \begin{array}{l} 3(0) + 2 \\ 0 + 2 \\ 2 \end{array}$$

$0 \neq 2$

No, $(0, 0)$ is not a solution of $y = 3x + 2$.

25. $y = 3x - 2$
$= 3(3) - 2$
$= 9 - 2$
$= 7$
The ordered-pair solution is $(3, 7)$.

27. $y = \frac{2}{3}x - 1$
$= \frac{2}{3}(6) - 1$
$= 4 - 1$
$= 3$
The ordered-pair solution is $(6, 3)$.

29. $y = -3x + 1$
$= -3(0) + 1$
$= 0 + 1$
$= 1$
The ordered-pair solution is $(0, 1)$.

31. $y = \frac{2}{5}x + 2$
$= \frac{2}{5}(-5) + 2$
$= -2 + 2$
$= 0$
The ordered-pair solution is $(-5, 0)$.

33. iv

Objective B Exercises

35.

x	y
0	-4
2	0
4	4

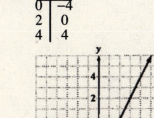

37.

x	y
-2	4
2	0
4	-2

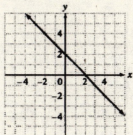

39.

x	y
-1	-4
2	-1
4	1

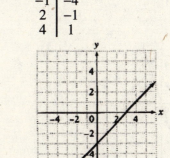

41.

x	y
0	3
1	1
3	-3

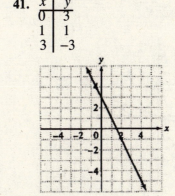

43.

x	y
0	4
1	1
2	-2

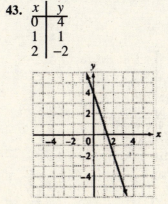

45.

x	y
−1	−3
0	−1
2	3

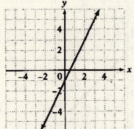

47.

x	y
−1	−3
0	0
1	3

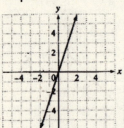

49.

x	y
3	1
0	0
−3	−1

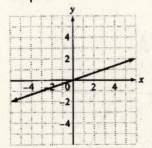

51.

x	y
−3	4
0	0
3	−4

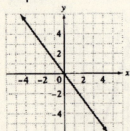

53.

x	y
−2	−4
0	−1
2	2

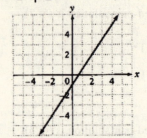

55.

x	y
−5	−3
0	−1
5	1

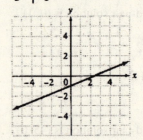

57.

x	y
−3	3
0	1
3	−1

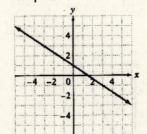

59.

x	y
−3	3
0	−2
1	$-\dfrac{11}{3}$

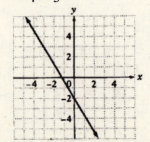

61.

x	y
2	4
0	−1
−2	−6

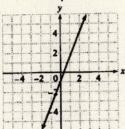

63.

x	y
1	1
0	0
−1	−1

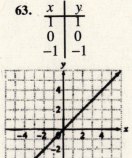

65. When $x = 3, y = 4$.

67. When $x = 4, y = -1$.

69. When $x = 3, y = 2$.

71. When $y = -2, x = 2$.

73. When $y = -2, x = 1$.

75. When $y = -1, x = -3$.

77. same sign

Critical Thinking 6.6

79. To find the coordinates of the point at which the graph of $y = 2x + 1$ crosses the y-axis, substitute 0 for x and solve for y.

$y = 2x + 1$
$y = 2(0) + 1$
$y = 1$

The graph of $y = 2x + 1$ will cross the y-axis at the point $(0,1)$.

Projects or Group Activities 6.6

81. a. Answers will vary
   b. Answers will vary
   c. Answers will vary

Chapter Review Exercises

1. $z + 5 = 2$
$z + 5 - 5 = 2 - 5$
$z = -3$
The solution is -3.

2. $-8x + 4x = -12$
$-4x = -12$
$\dfrac{-4x}{-4} = \dfrac{-12}{-4}$
$x = 3$
The solution is 3.

3. $7 = 8a - 5$
$7 + 5 = 8a - 5 + 5$
$12 = 8a$
$\dfrac{12}{8} = \dfrac{8a}{8}$
$\dfrac{3}{2} = a$
The solution is $\dfrac{3}{2}$.

4. $7 + a = 0$
$7 - 7 + a = 0 - 7$
$a = -7$
The solution is -7.

5. $40 = -\dfrac{5}{3}y$
$-\dfrac{3}{5}(40) = \left(-\dfrac{3}{5}\right)\left(-\dfrac{5}{3}y\right)$
$-24 = y$
The solution is -24.

6. $-\dfrac{3}{8} = \dfrac{4}{5}z$
$\dfrac{5}{4}\left(-\dfrac{3}{8}\right) = \dfrac{5}{4}\left(\dfrac{4}{5}z\right)$
$-\dfrac{15}{32} = z$
The solution is $-\dfrac{15}{32}$.

7. $9 - 5y = -1$

$9 - 9 - 5y = -1 - 9$

$-5y = -10$

$\dfrac{-5y}{-5} = \dfrac{-10}{-5}$

$y = 2$

The solution is 2.

8. $-4(2 - x) = x + 9$

$-8 + 4x = x + 9$

$-8 + 4x - x = x - x + 9$

$-8 + 3x = 9$

$-8 + 8 + 3x = 9 + 8$

$3x = 17$

$\dfrac{3x}{3} = \dfrac{17}{3}$

$x = \dfrac{17}{3}$

The solution is $\dfrac{17}{3}$.

9. $3a + 8 = 12 - 5a$

$3a + 5a + 8 = 12 - 5a + 5a$

$8a + 8 = 12$

$8a + 8 - 8 = 12 - 8$

$8a = 4$

$\dfrac{8a}{8} = \dfrac{4}{8}$

$a = \dfrac{1}{2}$

The solution is $\dfrac{1}{2}$.

10. $12p - 7 = 5p - 21$

$12p - 5p - 7 = 5p - 5p - 21$

$7p - 7 = -21$

$7p - 7 + 7 = -21 + 7$

$7p = -14$

$\dfrac{7p}{7} = \dfrac{-14}{7}$

$p = -2$

The solution is -2.

11. $3(2n - 3) = 2n + 3$

$6n - 9 = 2n + 3$

$6n - 2n - 9 = 2n - 2n + 3$

$4n - 9 = 3$

$4n - 9 + 9 = 3 + 9$

$4n = 12$

$\dfrac{4n}{4} = \dfrac{12}{4}$

$n = 3$

The solution is 3.

12. $3m = -12$

$\dfrac{3m}{3} = \dfrac{-12}{3}$

$m = -4$

The solution is -4.

13. $4 - 3(2p + 1) = 3p + 11$

$4 - 6p - 3 = 3p + 11$

$-6p + 1 = 3p + 11$

$-6p - 3p + 1 = 3p - 3p + 11$

$-9p + 1 = 11$

$-9p + 1 - 1 = 11 - 1$

$-9p = 10$

$\dfrac{-9p}{-9} = \dfrac{10}{-9}$

$p = -\dfrac{10}{9}$

The solution is $-\dfrac{10}{9}$.

14. $1 + 4(2c - 3) = 3(3c - 5)$

$1 + 8c - 12 = 9c - 15$

$8c - 11 = 9c - 15$

$8c - 9c - 11 = 9c - 9c - 15$

$-c - 11 = -15$

$-c - 11 + 11 = -15 + 11$

$-c = -4$

$\dfrac{-1c}{-1} = \dfrac{-4}{-1}$

$c = 4$

The solution is 4.

15. $\dfrac{3x}{4} + 10 = 7$

$\dfrac{3x}{4} + 10 - 10 = 7 - 10$

$\dfrac{3x}{4} = -3$

$\left(\dfrac{4}{3}\right)\left(\dfrac{3}{4}x\right) = \dfrac{4}{3}(-3)$

$x = -4$

The solution is -4.

16. $y = \dfrac{1}{5}x + 2$

$\begin{array}{c|c} 0 & \dfrac{1}{5}(-10) + 2 \\ 0 & -2 + 2 \end{array}$

$0 = 0$

Yes, $(-10, 0)$ is a solution of the equation.

17.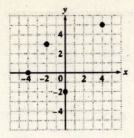

18.

x	y
0	−5
1	−2
2	1

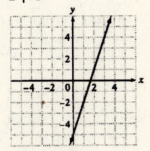

19.

x	y
−2	4
0	3
2	2

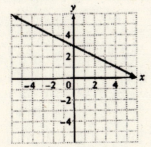

20. $y = 4x - 9$
$y = 4(2) - 9$
$y = 8 - 9$
$y = -1$
The ordered-pair solution is $(2, -1)$.

21. The unknown number: x

the difference between seven and the product of five and a number	is	thirty-seven

$7 - 5x = 37$
$7 - 7 - 5x = 37 - 7$
$-5x = 30$
$\dfrac{-5x}{-5} = \dfrac{30}{-5} = -6$
$x = -6$
The number is −6.

22. Strategy To find the length of each piece, write and solve an equation using x to represent the shorter piece and $24 - x$ to represent the longer piece.

Solution

Twice the length of the shorter piece	equals	the length of the longer piece

$2x = 24 - x$

$2x + x = 24 - x + x$

$3x = 24$

$\dfrac{3x}{3} = \dfrac{24}{3}$

$x = 8$

$24 - x = 24 - 8 = 16$

The longer piece is 16 in.

23. Strategy To find the number of hours of consultation, write and solve an equation using n to represent the number of hours.

Solution

$250 plus $150 per hour	is	$1,300

$250 + 150 \cdot n = 1,300$

$250 - 250 + 150n = 1,300 - 250$

$150n = 1,050$

$\dfrac{150n}{150} = \dfrac{1,050}{150}$

$n = 7$

The consulting fee was for 7 h of consultation.

24. Strategy To find the height of the leaning tower of Pisa, write and solve an equation using h to represent the height of the tower.

Solution

302 m	is	28 m less than six times the height

$302 = 6h - 28$

$302 + 28 = 6h - 28 + 28$

$330 = 6h$

$\dfrac{330}{6} = \dfrac{6h}{6}$

$55 = h$

The leaning tower of Pisa is 55 m high.

25.

26. Strategy To find the force, substitute 18 for d, 6 for x, and 25 for F_1 in the given equation and solve for F_2.

Solution
$$F_1 x = F_2 (d - x)$$
$$25(6) = F_2 (18 - 6)$$
$$150 = 12 F_2$$
$$\frac{150}{12} = \frac{12 F_2}{12}$$
$$12.5 = F_2$$

The force needed to balance the system is 12.5 lb.

27. Strategy To find the number of amplifiers, substitute 38,669 for T, 127 for U, and 20,000 for F in the given equation and solve for N.

Solution
$$T = U \cdot N + F$$
$$38{,}669 = 127 \cdot N + 20{,}000$$
$$38{,}669 - 20{,}000 = 127N$$
$$18{,}669 = 127N$$
$$\frac{18{,}699}{127} = \frac{127N}{127}$$
$$147 = N$$

147 amplifiers were produced during the month.

Chapter Test

1. $7 + x = 2$
   $7 - 7 + x = 2 - 7$
   $a = -5$
   The solution is –5.

2. $-\dfrac{3}{5}y = 6$

   $\left(-\dfrac{5}{3}\right)\left(-\dfrac{3}{5}y\right) = 6\left(-\dfrac{5}{3}\right)$

   $y = -10$
   The solution is –10.

3. $2d - 7 = -13$
   $2d - 7 + 7 = -13 + 7$
   $2d = -6$
   $\dfrac{2d}{2} = \dfrac{-6}{2}$
   $d = -3$
   The solution is –3.

4. $4 - 5c = -11$
   $4 - 4 - 5c = -11 - 4$
   $-5c = -15$
   $\dfrac{-5c}{-5} = \dfrac{-15}{-5}$
   $c = 3$
   The solution is 3.

5. $3x + 4 = 24 - 2x$
   $3x + 2x + 4 = 24 - 2x + 2x$
   $5x + 4 = 24$
   $5x + 4 - 4 = 24 - 4$
   $5x = 20$
   $\dfrac{5x}{5} = \dfrac{20}{5}$
   $x = 4$
   The solution is 4.

6. $7 - 5y = 6y - 26$
   $7 - 5y + 5y = 6y + 5y - 26$
   $7 = 11y - 26$
   $7 + 26 = 11y - 26 + 26$
   $33 = 11y$
   $\dfrac{33}{11} = \dfrac{11y}{11}$
   $y = 3$
   The solution is 3.

7. $2t - 3(4 - t) = t - 8$
   $2t - 12 + 3t = t - 8$
   $5t - 12 = t - 8$
   $5t - t - 12 = t - t - 8$
   $4t - 12 = -8$
   $4t - 12 + 12 = -8 + 12$
   $4t = 4$
   $\dfrac{4t}{4} = \dfrac{4}{4}$
   $t = 1$
   The solution is 1.

8. $12 - 3(n - 5) = 5n - 3$
   $12 - 3n + 15 = 5n - 3$
   $-3n + 27 = 5n - 3$
   $-3n - 5n + 27 = 5n - 5n - 3$
   $-8n + 27 = -3$
   $-8n + 27 - 27 = -3 - 27$
   $-8n = -30$
   $\dfrac{-8n}{-8} = \dfrac{-30}{-8}$
   $n = \dfrac{15}{4}$
   The solution is $\dfrac{15}{4}$.

9. $\dfrac{3}{8} - n = \dfrac{2}{3}$

   $\dfrac{3}{8} - \dfrac{3}{8} - n = \dfrac{2}{3} - \dfrac{3}{8}$

   $-n = \dfrac{7}{24}$

   $\dfrac{-n}{-1} = \left(\dfrac{7}{24}\right)\left(\dfrac{1}{-1}\right)$

   $n = \dfrac{-7}{24}$

   The solution is $-\dfrac{7}{24}$.

10. $3p - 2 + 5p = 2p + 12$
    $8p - 2 = 2p + 12$
    $8p - 2p - 2 = 2p - 2p + 12$
    $6p - 2 = 12$
    $6p - 2 + 2 = 12 + 2$
    $6p = 14$
    $\dfrac{6p}{6} = \dfrac{14}{6}$
    $p = \dfrac{7}{3}$
    The solution is $\dfrac{7}{3}$.

11. $A(-3, 1)$

12.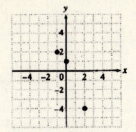

13.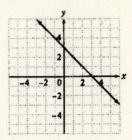

14.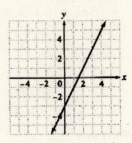

15.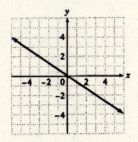

16.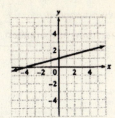

17. $2(4b - 14) = b - 7$
$8b - 28 = b - 7$
$8b - b - 28 = b - b - 7$
$7b - 28 = -7$
$7b - 28 + 28 = -7 + 28$
$7b = 21$
$\dfrac{7b}{7} = \dfrac{21}{7}$
$b = 3$
The solution is 3.

18. $\dfrac{5y}{3} + 12 = 2$
$\dfrac{5y}{3} + 12 - 12 = 2 - 12$
$\dfrac{5y}{3} = -10$
$\left(\dfrac{3}{5}\right)\left(\dfrac{5y}{3}\right) = -10\left(\dfrac{3}{5}\right)$
$y = -6$
The solution is –6.

19. $y = \dfrac{1}{3}x - 4$
$y = \dfrac{1}{3}(6) - 4$
$y = 2 - 4$
$y = -2$
The ordered-pair solution is (6, –2).

20. The unknown number: n

| Four plus one third of a number | is | nine |

$$4 + \frac{1}{3}n = 9$$

$$4 - 4 + \frac{1}{3}n = 9 - 4$$

$$\frac{1}{3}n = 5$$

$$\left(\frac{3}{1}\right)\left(\frac{1}{3}n\right) = 5\left(\frac{3}{1}\right)$$

$$n = 15$$

The number is 15.

21. The unknown number: x

| Sum of eight and the product of two and a number | is | negative 4 |

$$8 + 2x = -4$$

$$8 - 8 + 2x = -4 - 8$$

$$2x = -12$$

$$\frac{2x}{2} = \frac{-12}{2}$$

$$x = -6$$

The number is –6.

22. The smaller number: x

The larger number: $17 - x$

| Four times the smaller number and two times the larger number | is | Forty four |

$$4x + 2(17 - x) = 44$$

$$4x + 34 - 2x = 44$$

$$2x + 34 = 44$$

$$2x + 34 - 34 = 44 - 34$$

$$2x = 10$$

$$\frac{2x}{2} = \frac{10}{2}$$

$$x = 5$$

$$17 - x = 17 - 5 = 12$$

The smaller number is 5.

The larger number is 12.

23.

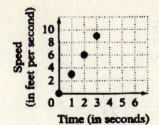

24. Strategy To find the number of hours of labor, write and solve an equation using n to represent the number of hours.

Solution

| $165 plus $58 per hour | is | $455 |

$165 + 58 \cdot n = 455$
$165 - 165 + 58n = 455 - 165$
$58n = 290$
$$\frac{58n}{58} = \frac{290}{58}$$
$n = 5$
The job required 5 h of labor.

25. Strategy To find the depth, substitute 65 for P in the given equation and solve for D.

Solution $P = 15 + \dfrac{1}{2}D$

$65 = 15 + \dfrac{1}{2}D$

$65 - 15 = 15 - 15 + \dfrac{1}{2}D$

$50 = \dfrac{1}{2}D$

$\left(\dfrac{2}{1}\right)50 = \left(\dfrac{2}{1}\right)\left(\dfrac{1}{2}D\right)$

$100 = D$
The depth is 100 ft.

Cumulative Review Exercises

1. $-3ab$
$-3(-2)(3) = 6(3)$
$= 18$

2. $-3(4p - 7) = -12p + 21$

3. $\left(\dfrac{2}{3}\right)\left(-\dfrac{9}{8}\right) + \dfrac{3}{4} = -\dfrac{2 \cdot 9}{3 \cdot 8} + \dfrac{3}{4}$
$= -\dfrac{3}{4} + \dfrac{3}{4}$
$= 0$

4. $-\dfrac{2}{3}y = 12$

$\left(-\dfrac{3}{2}\right)\left(-\dfrac{2}{3}y\right) = -\dfrac{3}{2}(12)$
$y = -18$
The solution is -18.

5. $(-b)^3$
$\left[-(-2)\right]^3 = 2^3$
$= 8$

6. $4xy^2 - 2xy$
$4(-2)(3^2) - 2(-2)(3)$
$= 4(-2)(9) - 2(-2)(3)$
$= -8(9) - 2(-2)(3)$
$= -72 - 2(-2)(3)$
$= -72 - (-4)(3)$
$= -72 - (-12)$
$= -72 + 12$
$= -60$

7. $\sqrt{121} = 11$

8. $\sqrt{48} = \sqrt{16 \cdot 3}$
$= \sqrt{16} \cdot \sqrt{3}$
$= 4\sqrt{3}$

9. $4(3v - 2) - 5(2v - 3)$
$= 12v - 8 - 10v + 15$
$= (12v - 10v) + (-8 + 15)$
$= 2v + 7$

10. $-4(-3m) = [(-4)(-3)]m$
$= 12m$

11. $\dfrac{-5d = -45}{-5(-9) \mid -45}$
$\qquad 45 \neq -45$
No, -9 is not a solution of the equation.

12. $5 - 7a = 3 - 5a$
$5 - 7a + 5a = 3 - 5a + 5a$
$5 - 2a = 3$
$5 - 5 - 2a = 3 - 5$
$-2a = -2$
$\dfrac{-2a}{-2} = \dfrac{-2}{-2}$
$a = 1$
The solution is 1.

13. $6 - 2(7z - 3) + 4z = 6 - 14z + 6 + 4z$
   $= (-14z + 4z) + (6 + 6)$
   $= -10z + 12$

14. $\dfrac{a^2 + b^2}{2ab}$

   $\dfrac{(-2)^2 + (-1)^2}{2(-2)(-1)} = \dfrac{4 + 1}{4}$

   $= \dfrac{5}{4}$

15. $8z - 9 = 3$
   $8z - 9 + 9 = 3 + 9$
   $8z = 12$
   $\dfrac{8z}{8} = \dfrac{12}{8}$
   $z = \dfrac{3}{2}$

   The solution is $\dfrac{3}{2}$.

16. $\left(2m^2 n^5\right)^5 = 2^{1 \cdot 5} m^{2 \cdot 5} n^{5 \cdot 5}$
   $= 32 m^{10} n^{25}$

17. $-3a^3 \left(2a^2 + 3ab - 4b^2\right)$
   $= -3a^3 \left(2a^2\right) + \left(-3a^3\right)(3ab) - \left(-3a^3\right)\left(4b^2\right)$
   $= -6a^5 - 9a^4 b + 12a^3 b^2$

18. $(2x - 3)(3x + 1) = 6x^2 + 2x - 9x - 3$
   $= 6x^2 - 7x - 3$

19. $2^{-4} = \dfrac{1}{2^4} = \dfrac{1}{16}$

20. $\dfrac{x^8}{x^2} = x^{8-2} = x^6$

21. $\left(-5x^3 y\right)\left(-3x^5 y^2\right)$

   $= \left[(-5)(-3)\right]\left(x^3 x^5\right)\left(y y^2\right)$

   $= 15 x^8 y^3$

22. $5 - 3(2x - 8) = -2(1 - x)$
   $5 - 6x + 24 = -2 + 2x$
   $-6x + 29 = -2 + 2x$
   $-6x - 2x + 29 = -2 + 2x - 2x$
   $-8x + 29 = -2$
   $-8x + 29 - 29 = -2 - 29$
   $-8x = -31$
   $\dfrac{-8x}{-8} = \dfrac{-31}{-8}$
   $x = \dfrac{31}{8}$

   The solution is $\dfrac{31}{8}$.

23.

x	y
3	6
0	1
-3	-4

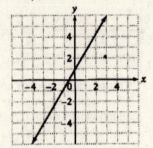

24.

x	y
5	-2
0	0
-5	2

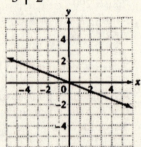

25. $3.5 \times 10^{-8} = 0.000000035$

26. The <u>product</u> of five and the <u>sum</u> of a number and two
   $5(n + 2)$
   $5n + 10$

27. Strategy   To find the time, replace v by 98 and v_0 by 50 in the given formula and solve for t.

   Solution $v = v_0 + 32t$
   $98 = 50 + 32t$
   $98 - 50 = 50 - 50 + 32t$
   $48 = 32t$
   $\dfrac{48}{32} = \dfrac{32t}{32}$
   $1.5 = t$
   The time is 1.5 s.

28. The number of wolves in the world: w
   The number of dogs in the world: $1000w$

29. Strategy   To find the total box office gross, add the totals for the four films.

   Solution $461.0 + 290.2 + 260.0 + 181.3 = 1,192.5$
   The total box office gross for all four films is $1,192.5 million.

30. Strategy   To find the average student debt, write and solve an equation using d to represent the average debt from the previous year.

   Solution

$22,900	is	$1700 up from previous year

   $22,900 = d + 1700$
   $21,200 = d$
   The average student debt of the class of 2010 was $21,200.

31. Strategy   To find the number of stories, divide the depth of the Aleutian Trench (8,100) by the height of one story (4.2).

   Solution $8,100 \div 4.2 \approx 1,929$
   A skyscraper with 1,929 stories is as tall as the Aleutian Trench.

32. Strategy   To find the amount donated to each charity, write and solve an equation using x to represent the donation to one charity and $2x$ to represent the donation to the other charity.

   Solution $x + 2x = 12,000$
   $3x = 12,000$
   $$\frac{3x}{3} = \frac{12,000}{3}$$
   $x = 4,000$
   $2x = 2(4,000) = 8,000$
   One charity received $4,000 and the other charity received $8,000.

Chapter 7: Measurement and Proportion

Prep Test

1. $\dfrac{8}{10} = \dfrac{4}{5}$

2. $\dfrac{372}{15} = 24.8$

3. $\dfrac{5}{3} \times 6 = \dfrac{5 \cdot 2 \cdot \cancel{3}}{\cancel{3}} = 10$

4. $5\dfrac{3}{4} \times 8 = \dfrac{23}{4} \times 8$

   $= \dfrac{23 \cdot 2 \cdot \cancel{2} \cdot \cancel{2}}{\cancel{2} \cdot \cancel{2}} = 46$

5.
   $\begin{array}{r}
   238 \\
   3\overline{)714} \\
   \underline{6} \\
   11 \\
   \underline{9} \\
   24 \\
   \underline{24} \\
   0
   \end{array}$

6. 37,320

7. 0.04107

8. 5.125

9. 0.13

10. $35 \times \dfrac{1.61}{1} = 35 \times 1.61 = 56.35$

11. $1.67 \times \dfrac{1}{3.34} = 1.67 \div 3.34$

    $= 0.5$

12. $315 \div 84 = 3.75$

Section 7.1

Concept Check

1. In the metric system, the basic unit of length is the meter. The basic unit of liquid measure is the liter. The basic unit of weight is the gram.

Objective A Exercises

3. **a.** giga-: 10^9
   mega-: 1 000 000
   kilo-: k, 10^3
   hecto-: 10^2
   deca-: 10
   deci-: 0.1
   centi-: c, 0.01
   milli-: m, $\dfrac{1}{10^3}$
   micro-: 0.000 001
   nano-: 0.000 000 001
   pico-: $\dfrac{1}{10^{12}}$

   b. The exponent 10 indicates the number of places to move the decimal point. For prefixes tera-, giga-, mega-, kilo-, hecto-, and deca-, move the decimal point to the right. For other prefixes shown, move the decimal point to the left.

5. kilogram

7. milliliter

9. kiloliter

11. centimeter

13. kilogram

15. liter

17. gram

19. meter or centimeter

21. kilogram

23. milliliter

25. kilogram

27. kilometer

29. 91 cm = 910 mm

31. 1 856 g = 1.856 kg

33. 7 285 ml = 7.285 L

35. 8 m = 8 000 mm

37. 34 mg = 0.034 g

39. 0.0297 L = 29.7 ml

41. 7 530 m = 7.530 km

43. 9.2 kg = 9 200 g

45. 0.036 kl = 36 L

47. 2.35 km = 2 350 m

49. 0.083 g = 83 mg

51. 71.6 cm = 0.716 m

53. 3 206 L = 3.206 kl

55. 0.99 m = 99 cm

57. 605 mm = 60.5 cm

59. 0.05 m = 5 cm

61. 28 300 mg = 2 830 cg

63. 0.7 km = 70 000 cm

65. Strategy To find the number of tablets:
→Convert 500 mg to grams.
→Divide 2 by the amount
of calcium in each tablet.

Solution 500 mg = 0.5 g
2 ÷ 0.5 = 4
The patient should take 4
tablets per day.

67. Strategy To find the number of liters:
→Convert 12 ml to liters.
→Multiply the total number of
students by the amount of
acid for each student.

Solution 12 ml = 0.012 L
$90 \cdot 4 \cdot 0.012 = 4.32$
The assistant should
order 5 L.

69. Strategy To find the number of
kilometers:
→Convert 1,400 m, 1,200 m,
and 1,800 m to km.
→Add up the three distances.

Solution 1,400 m = 1.4 km
1,200 m = 1.2 km
1,800 m = 1.8 km
1.4 + 1.2 + 1.8 = 4.4
The walk is 4.4 km long.

71. Strategy To find the number of
kilograms:
→Multiply 9 by the eight of
each automobile.

Solution $9 \cdot 1,405 = 12,645$
The total weight of the
automobiles is 12,645 kg.

73. Strategy To find the number of patients:
→Convert 3 cm³ to L
→Divide 12 by the dosage per
patient.

Solution 3 cm³ = .003 L
12 ÷ .003 = 4,000
4000 patients can be
immunized.

75. Strategy To find the number of meters:
→Convert 8,502 km to m.
→Divide this by 3772.

Solution 8,502 km = 8,502,000 m
8,502,000 ÷ 3772 ≈ 2253.97
Each group has adopted an
average of 2254 m.

77. Strategy To find the profit:
→Find the number of jars needed by converting 5 L to milliliters, divide the amount by 250 ml.
→To find the total cost, multiply the number of jars by $0.55 and add to $195.
→To find revenue, multiply number of jars by price per jar.
→To find profit, subtract cost from revenue.

Solution 5 L = 5,000 ml
5,000 ÷ 250 = 20 jars
20 · 0.55 + 195 = 206
20 · 23.89 = 477.8
477.8 − 206 = 271.8
The profit is $271.80.

Critical Thinking 7.1

79. Strategy To find remaining amount:
→Convert 3 L to milliliters
→Subtract 280 ml serving from amount in bottle.
→Convert answer from milliliters to liters.

Solution 3 L = 3 000 ml
3,000 − 280 = 2,720
2 720 ml = 2.720 L
The water that remains is 2,720 ml or 2.720 L.

Section 7.2

Concept Check

1. are not; are

Objective A Exercises

3. $\dfrac{3 \text{ pints}}{15 \text{ pints}} = \dfrac{3}{15} = \dfrac{1}{5}$
3 pints: 15 pints = 1:5
3 pints to 15 pints = 1 to 5

5. $\dfrac{\$40}{\$20} = \dfrac{40}{20} = \dfrac{2}{1}$
$40: $20 = 2:1
$40 to $20 = 2 to 1

7. $\dfrac{3 \text{ miles}}{8 \text{ miles}} = \dfrac{3}{8}$
3 miles: 8 miles = 3:8
3 miles to 8 miles = 3 to 8

9. $\dfrac{6 \text{ min}}{6 \text{ min}} = \dfrac{6}{6} = \dfrac{1}{1}$
6 min: 6 min = 1:1
6 min to 6 min = 1 to 1

11. $\dfrac{35 \text{ cents}}{50 \text{ cents}} = \dfrac{35}{50} = \dfrac{7}{10}$
35 cents: 50 cents = 7:10
35 cents to 50 cents = 7 to 10

13. $\dfrac{160 \text{ miles}}{6 \text{ hours}} = \dfrac{80 \text{ miles}}{3 \text{ hours}}$

15. 1

17. 1

19. $\dfrac{\$51,000}{12 \text{ months}} = \$4250 / \text{month}$

21. $\dfrac{3750 \text{ words}}{15 \text{ pages}} = 250 \text{ words} / \text{page}$

23. $\dfrac{628.8 \text{ miles}}{12 \text{ hours}} = 52.4 \text{ miles} / \text{hour}$

25. $\dfrac{\$11.05}{3.4 \text{ pounds}} = \$3.25 / \text{pound}$

27. Strategy To find the ratio, divide the number of rookies (50) by the number of seniors (2,800).

Solution $\dfrac{50}{2,800} = \dfrac{1}{56}$
The ratio of rookies to seniors is $\dfrac{1}{56}$.

29. Strategy   To find the number of messages per day, divide each year's messages per day (billions) by the number of users (millions).

Solution $\dfrac{156{,}000}{1050} \approx 148.571$

$\dfrac{220{,}000}{1200} \approx 183.333$

$\dfrac{320{,}000}{1330} \approx 240.601$

2009: 148.6 messages
2010: 183.3 messages
2011: 240.6 messages

31. Television advertising rates are given as dollars per second, so the numerator is dollars and the denominator is seconds.

Section 7.3

Concept Check

1. denominator; numerator

3. $\dfrac{1\,\text{m}}{1.09\,\text{yd}}$

Objective A Exercises

5. $64\,\text{in.} = \dfrac{64\,\text{in.}}{1} \cdot \dfrac{1\,\text{ft}}{12\,\text{in.}} = 5\dfrac{1}{3}\,\text{ft}$

7. $42\,\text{oz} = \dfrac{42\,\text{oz}}{1} \cdot \dfrac{1\,\text{lb}}{16\,\text{oz}} = 2\dfrac{5}{8}\,\text{lb}$

9. $7{,}920\,\text{ft} = \dfrac{7{,}920\,\text{ft}}{1} \cdot \dfrac{1\,\text{mi}}{5{,}280\,\text{ft}} = 1\dfrac{1}{2}\,\text{mi}$

11. $500\,\text{lb} = \dfrac{500\,\text{lb}}{1} \cdot \dfrac{1\,\text{ton}}{2{,}000\,\text{lb}} = \dfrac{1}{4}\,\text{ton}$

13. $10\,\text{qt} = \dfrac{10\,\text{qt}}{1} \cdot \dfrac{1\,\text{gal}}{4\,\text{qt}} = 2\dfrac{1}{2}\,\text{gal}$

15. $2\dfrac{1}{2}\,\text{c} = \dfrac{5\,\text{c}}{2} \cdot \dfrac{8\,\text{fl oz}}{1\,\text{c}} = 20\,\text{fl oz}$

17. $2\dfrac{1}{4}\,\text{mi} = \dfrac{9\,\text{mi}}{4} \cdot \dfrac{5{,}280\,\text{ft}}{1\,\text{mi}} = 11{,}880\,\text{ft}$

19. $7\dfrac{1}{2}\,\text{in} = \dfrac{15\,\text{in}}{2} \cdot \dfrac{1\,\text{ft}}{12\,\text{in}} = \dfrac{5}{8}\,\text{ft}$

21. $60\,\text{fl oz} = \dfrac{60\,\text{fl oz}}{1} \cdot \dfrac{1\,\text{c}}{8\,\text{fl oz}} = 7\dfrac{1}{2}\,\text{c}$

23. $7\dfrac{1}{2}\,\text{pt} = \dfrac{15\,\text{pt}}{2} \cdot \dfrac{2\,\text{c}}{1\,\text{pt}} \cdot \dfrac{1\,\text{qt}}{4\,\text{c}} = 3\dfrac{3}{4}\,\text{qt}$

25. $\dfrac{24\,\text{hr}}{1\,\text{day}} \cdot \dfrac{60\,\text{min}}{1\,\text{hr}} \cdot \dfrac{60\,\text{s}}{1\,\text{min}} = \dfrac{86{,}400\,\text{s}}{1\,\text{day}}$

Objective B Exercises

27. $35\,\text{yr} = \dfrac{35\,\text{yr}}{1} \cdot \dfrac{365\,\text{day}}{1\,\text{yr}} \cdot \dfrac{24\,\text{hr}}{1\,\text{day}} \cdot \dfrac{60\,\text{min}}{1\,\text{hr}} \cdot \dfrac{60\,\text{s}}{1\,\text{min}}$
$= 1{,}103{,}760{,}000\,\text{s}$

29. Strategy   To find the amount of punch:
→Multiply the number of guests (200) by the amount of punch for each guest (1 c) to find the number of cups of punch to be prepared.
→Convert cups to gallons.

Solution $200 \cdot 1 = 200$ cups
200 cups
$= \dfrac{200\,\text{c}}{1} \cdot \dfrac{1\,\text{pt}}{2\,\text{c}} \cdot \dfrac{1\,\text{qt}}{2\,\text{pt}} \cdot \dfrac{1\,\text{gal}}{4\,\text{qt}}$
$= 12\dfrac{1}{2}\,ga$

The guest list would require $12\dfrac{1}{2}$ gal of punch.

31. Strategy   To find the number of gallons:
→Convert from quarts to gal.
→Multiply the amount by the number of days and students.

Solution $2\,\text{qt} = \dfrac{2\,\text{qt}}{1} \cdot \dfrac{1\,\text{gal}}{4\,\text{qt}} = \dfrac{1}{2}\,\text{gal}$

$\dfrac{1}{2} \cdot 3 \cdot 5 = 7\dfrac{1}{2}$

They need $7\dfrac{1}{2}$ gallons of water.

33. Strategy   To find the amount of oil:
→Convert from qt to gallons.
→Multiply the number of oil changes by the amount of oil required for each change.

Solution $5\,\text{qt} = \dfrac{5\,\text{qt}}{1} \cdot \dfrac{1\,\text{gal}}{4\,\text{qt}} = \dfrac{5}{4}\,\text{gal}$

$7 \cdot \dfrac{5}{4} = \dfrac{35}{4} = 8\dfrac{3}{4}$

The farmer used $8\dfrac{3}{4}$ gallons of oil.

35. Strategy   To find the cost of the lot:
→Separate the lot into two rectangles.
→Use the formula $A = LW$ to find each area.
→Add the areas together.
→Convert from square feet to acres.
→Multiply the acre by $20,000 to find the cost.

Solution $A = LW = 170 \cdot 270 = 45{,}900$
$A = LW = 350(350-170) = 63{,}000$
$45{,}900 + 63{,}000 = 108{,}900\,\text{ft}^2$

$\dfrac{108{,}900\,\text{ft}^2}{1} \cdot \dfrac{1\,\text{acre}}{43{,}560\,\text{ft}^2} = 2.5\,\text{acre}$
$2.5 \cdot 20{,}000 = 45{,}000$
The cost of the lot is $50,000.

37. Strategy   To find the cost:
→Convert from acres to ft.
→Multiply lot size by price.

Solution $\dfrac{1\,\text{acre}}{2} \cdot \dfrac{43{,}560\,\text{ft}^2}{1\,\text{acre}} = 21{,}780\,\text{ft}^2$
$21{,}780 \cdot 3 = 65{,}340$
The cost of the lot is $65,340.

39. Strategy   To find how many quarts:
→Convert from cups to quarts.
→Multiply by the number of cartons sold.

Solution $\dfrac{1\,\text{cup}}{1} \cdot \dfrac{1\,\text{quart}}{4\,\text{cups}} = \dfrac{1}{4}\,\text{quarts}$

$124 \cdot \dfrac{1}{4} = 31$

The cafeteria sold 31 quarts of milk.

Objective C Exercises

41. $145\,\text{lb} = \dfrac{145\,\text{lb}}{1} \cdot \dfrac{1\,\text{kg}}{2.2\,\text{lb}} \approx 65.91\,\text{kg}$

43. $2\,\text{L} = \dfrac{2\,\text{L}}{1} \cdot \dfrac{1.06\,\text{qt}}{1\,\text{L}} \cdot \dfrac{4\,\text{c}}{1\,\text{qt}} = 8.48\,\text{c}$

45. $14.3\,\text{gal} = \dfrac{14.3\,\text{gal}}{1} \cdot \dfrac{3.97\,\text{L}}{1\,\text{gal}} \approx 54.20\,\text{L}$

47. $86\,\text{kg} = \dfrac{86\,\text{kg}}{1} \cdot \dfrac{2.2\,\text{lb}}{1\,\text{kg}} = 189.2\,\text{lb}$

49. $35\,\text{mm} = \dfrac{35\,\text{mm}}{1} \cdot \dfrac{1\,\text{cm}}{10\,\text{mm}} \cdot \dfrac{1\,\text{in.}}{2.54\,\text{cm}} \approx 1.38\,\text{in.}$

51. $24\,\text{L} = \dfrac{24\,\text{L}}{1} \cdot \dfrac{1\,\text{gal}}{3.79\,\text{L}} \approx 6.33\,\text{gal}$

53. $60\,\text{ft/s} = \dfrac{60\,\text{ft}}{1\,\text{s}} \cdot \dfrac{1\,\text{m}}{3.28\,\text{ft}} \approx 18.29\,\text{m/s}$

55. $\$0.99/\text{lb} = \dfrac{\$0.99}{1\,\text{lb}} \cdot \dfrac{2.2\,\text{lb}}{1\,\text{kg}} \approx \$2.18/\text{kg}$

57. $80\,\text{km/h} = \dfrac{80\,\text{km}}{1\,\text{h}} \cdot \dfrac{1\,\text{mi}}{1.61\,\text{km}} \approx 49.69\,\text{mph}$

59. $\$0.785/\text{L} = \dfrac{\$0.785}{1\,\text{L}} \cdot \dfrac{3.79\,\text{L}}{1\,\text{gal}} \approx \$2.98/\text{gal}$

61. Strategy   To find the distance in kilometers, convert the distance from miles to km.

Solution $93{,}000{,}000\,\text{mi}$
$= \dfrac{93{,}000{,}000\,\text{mi}}{1} \cdot \dfrac{1.61\,\text{km}}{1\,\text{mi}}$
$= 149\,730\,000\,\text{km}$
The distance is $149\,730\,000$ km.

63. $\dfrac{3.28\,\text{ft}}{1\,\text{m}} \cdot \dfrac{12\,\text{in.}}{1\,\text{ft}} = \dfrac{39.36\,\text{in.}}{1\,\text{m}}$

Critical Thinking 7.3

65. a. False. 3.79 L ≈ 1 gal.

   b. False. 1 m ≈ 1.09 yd.

   c. True.

$$30 \text{mph} = \frac{30 \text{ mi}}{1 \text{ h}} \cdot \frac{1.61 \text{ km}}{1 \text{ mi}} = 48.3 \text{ km/h}$$

   d. True. 1 kg ≈ 2.2 lb.

   e. False. 1 oz ≈ 28.35 g.

67. Strategy Convert inches to miles, multiply by the number of adults and compare result to the distance of the equator.

   Solution $19 \text{ in} = \dfrac{19 \text{ in}}{1} \cdot \dfrac{1 \text{ ft}}{12 \text{ in}} \cdot \dfrac{1 \text{ mi}}{5280 \text{ ft}}$

   ≈ 0.0002998 mi

   200,000,000 · 0.0002998

   ≈ 59,975

   59,975 > 25,000

   Yes, they could reach around the equator.

Projects or Group Activities 7.3

69. Answers will vary.

Check Your Progress: Chapter 7

1. 64 cm = 640 mm

2. 3,275 ml = 3.275 L

3. 3.6 kg = 3,600 g

4. 3.45 km = 3450 m

5. 2,760 L = 2.760 kl

6. 0.05 g = 50 mg

7. $\dfrac{30 \text{ min}}{60 \text{ min}} = \dfrac{30}{60} = \dfrac{1}{2}$

   30 min: 60 min = 1:2

   30 min to 60 min = 1 to 2

8. $\dfrac{12 \text{ qt}}{18 \text{ qt}} = \dfrac{12}{18} = \dfrac{12}{3}$

   12 qt: 18 qt = 2:3

   12 qt to 18 qt = 2 to 3

9. $\dfrac{\$39}{6 \text{ T-shirts}} = \dfrac{\$13}{2 \text{ T-shirts}}$

10. $\dfrac{\$38,700}{12 \text{ months}} = \$3225 / \text{month}$

11. $\dfrac{\$19.08}{4.5 \text{ lb}} = \$4.24 / \text{lb}$

12. $20 \text{ qt} = \dfrac{20 \text{ qt}}{1} \cdot \dfrac{1 \text{ gal}}{4 \text{ qt}} = 5 \text{ gal}$

13. $8 \text{ lb} = \dfrac{8 \text{ lb}}{1} \cdot \dfrac{16 \text{ oz}}{1 \text{ lb}} = 128 \text{ oz}$

14. $4 \text{ mi} = \dfrac{4 \text{ mi}}{1} \cdot \dfrac{5,280 \text{ ft}}{1 \text{ mi}} = 21,120 \text{ ft}$

15. $400 \text{ m} = \dfrac{400 \text{ m}}{1} \cdot \dfrac{3.28 \text{ ft}}{1 \text{ m}} = 1312 \text{ ft}$

16. $22 \text{ lb} = \dfrac{22 \text{ lb}}{1} \cdot \dfrac{1 \text{ kg}}{2.2 \text{ lb}} = 10 \text{ kg}$

17. $5 \text{ gal} = \dfrac{5 \text{ gal}}{1} \cdot \dfrac{3.97 \text{ L}}{1 \text{ gal}} = 19.85 \text{ L}$

18. Strategy To find the weight:
   →Multiply the number of carats (15) by the number of milligrams in one carat (200).
   →Convert milligrams to grams.

   Solution 15(200) = 3,000
   3 000 mg = 3 g
   A 15-carat precious stone weights 3 g.

19. Strategy To find the number of kilometers, convert meters to kilometers.

   Solution 50,000 m = 50 km
   The entrants walk 50 km.

20. Strategy   To find the number of servings:
→Convert 230 ml to liters.
→Divide 3.78 by the amount of each serving.

Solution   230 ml = 0.23 L
$3.78 \div 0.23 \approx 16.43$
The container of milk will provide 16 servings.

21. $\dfrac{24 \text{ teeth}}{36 \text{ teeth}} = \dfrac{24}{36} = \dfrac{2}{3}$

22. Strategy   To find the rate of travel, substitute 1,155 for d and 2.5 for t in the given equation and solve for r.

Solution $r = \dfrac{d}{t}$

$r = \dfrac{1,155}{2.5}$

$r = 462$
The rate of travel for the plane was 462 mph.

23. $32 \text{ yd} = \dfrac{32 \text{ yd}}{1} \cdot \dfrac{3 \text{ ft}}{1 \text{ yd}} = 96 \text{ ft}$

24. Strategy   To find the number of pounds:
→Convert the tons eaten to pounds.
→Divide by the number of cattle.
→Divide by the number of days.

Solution $15 \text{ tons} = \dfrac{15 \text{ tons}}{1} \cdot \dfrac{2,000 \text{ lb}}{1 \text{ ton}}$

$= 30,000 \text{ lb}$

$30,000 \div 100 = 300$

$300 \div 15 = 20$

Each cow has eaten 15lb/day

25. $8 \text{ gal} = \dfrac{8 \text{ gal}}{1} \cdot \dfrac{4 \text{ qt}}{1 \text{ gal}} \cdot \dfrac{4 \text{ c}}{1 \text{ qt}} = 128 \text{ c}$

Section 7.4

Concept Check

1. Proportion; a; d; b; c; equal

3. $\dfrac{n \text{ mi}}{30 \text{ min}}$

Objective A Exercises

5. $\dfrac{27}{8} \diagdown \dfrac{9}{4} \longrightarrow \begin{matrix} 72 \\ 108 \end{matrix}$
The product of the extremes does not equal the product of the means.
The proportion is not true.

7. $\dfrac{45}{135} \diagdown \dfrac{3}{9} \longrightarrow \begin{matrix} 405 \\ 405 \end{matrix}$
The product of the extremes equals the product of the means.
The proportion is true.

9. $\dfrac{6 \text{ min}}{5 \text{ cents}} \diagdown \dfrac{30 \text{ min}}{25 \text{ cents}} \longrightarrow \begin{matrix} 150 \\ 150 \end{matrix}$
The product of the extremes equals the product of the means.
The proportion is true.

11. $\dfrac{300 \text{ ft}}{4 \text{ rolls}} \diagdown \dfrac{450 \text{ ft}}{7 \text{ rolls}} \longrightarrow \begin{matrix} 1,800 \\ 2,100 \end{matrix}$
The product of the extremes does not equal the product of the means.
The proportion is not true.

13. $\dfrac{2}{3} = \dfrac{n}{15}$
$2 \cdot 15 = 3 \cdot n$
$30 = 3n$
$\dfrac{30}{3} = \dfrac{3n}{3}$
$10 = n$

15. $\dfrac{n}{5} = \dfrac{12}{25}$
$n \cdot 25 = 5 \cdot 12$
$25n = 60$
$\dfrac{25n}{25} = \dfrac{60}{25}$
$n = 2.4$

17. $\dfrac{3}{8} = \dfrac{n}{12}$
$3 \cdot 12 = 8 \cdot n$
$36 = 8n$
$\dfrac{36}{8} = \dfrac{8n}{8}$
$4.5 = n$

19. $\dfrac{3}{n} = \dfrac{7}{40}$
$3 \cdot 40 = n \cdot 7$
$120 = 7n$
$\dfrac{120}{7} = \dfrac{7n}{7}$
$17.14 \approx n$

21. $\dfrac{16}{n} = \dfrac{25}{40}$

$16 \cdot 40 = n \cdot 25$

$640 = 25n$

$\dfrac{640}{25} = \dfrac{25n}{25}$

$25.6 = n$

23. $\dfrac{120}{n} = \dfrac{144}{25}$

$120 \cdot 25 = n \cdot 144$

$3,000 = 144n$

$\dfrac{3,000}{144} = \dfrac{144n}{144}$

$20.83 \approx n$

25. $\dfrac{0.5}{2.3} = \dfrac{n}{20}$

$0.5 \cdot 20 = 2.3 \cdot n$

$10 = 23n$

$\dfrac{10}{2.3} = \dfrac{2.3n}{2.3}$

$n \approx 4.35$

27. $\dfrac{0.7}{1.2} = \dfrac{6.4}{n}$

$0.7 \cdot n = 1.2 \cdot 6.4$

$0.7n = 7.68$

$\dfrac{0.7n}{0.7} = \dfrac{7.68}{0.7}$

$n \approx 10.97$

29. $\dfrac{x}{6.25} = \dfrac{16}{87}$

$x \cdot 87 = 6.25 \cdot 16$

$87x = 100$

$\dfrac{87x}{87} = \dfrac{100}{87}$

$x \approx 1.15$

31. $\dfrac{1.2}{0.44} = \dfrac{y}{14.2}$

$1.2 \cdot 14.2 = 0.44 \cdot y$

$17.04 = 0.44y$

$\dfrac{17.04}{0.44} = \dfrac{0.44y}{0.44}$

$38.73 \approx y$

33. $\dfrac{n+2}{5} = \dfrac{1}{2}$

$(n+2)2 = 5 \cdot 1$

$2n + 4 = 5$

$2n = 1$

$\dfrac{2n}{2} = \dfrac{1}{2}$

$n = \dfrac{1}{2} = 0.5$

35. $\dfrac{4}{3} = \dfrac{n-2}{6}$

$4 \cdot 6 = 3(n-2)$

$24 = 3n - 6$

$30 = 3n$

$\dfrac{30}{3} = \dfrac{3n}{3}$

$10 = n$

37. $\dfrac{2}{n+3} = \dfrac{7}{12}$

$2 \cdot 12 = (n+3)(7)$

$24 = 7n + 21$

$3 = 7n$

$\dfrac{3}{7} = \dfrac{7n}{7}$

$0.43 \approx n$

39. $\dfrac{7}{10} = \dfrac{3+n}{2}$

$7 \cdot 2 = 10(3+n)$

$14 = 30 + 10n$

$-16 = 10n$

$\dfrac{-16}{10} = \dfrac{10n}{10}$

$-1.6 = n$

41. $\dfrac{x-4}{3} = \dfrac{3}{4}$

$(x-4)4 = 3 \cdot 3$

$4x - 16 = 9$

$4x = 25$

$\dfrac{4x}{4} = \dfrac{25}{4}$

$x = 6.25$

43. $\dfrac{6}{1} = \dfrac{x-2}{5}$

$6 \cdot 5 = 1(x-2)$

$30 = x - 2$

$32 = x$

45. $\dfrac{5}{8} = \dfrac{2}{x-3}$

$5(x-3) = 8 \cdot 2$

$5x - 15 = 16$

$5x = 31$

$\dfrac{5x}{5} = \dfrac{31}{5}$

$x = 6.2$

47. $\dfrac{3}{x-4} = \dfrac{5}{3}$

$3 \cdot 3 = (x-4)5$

$9 = 5x - 20$

$29 = 5x$

$\dfrac{29}{5} = \dfrac{5x}{5}$

$5.8 = x$

49. a. For example, $\dfrac{2}{9} = \dfrac{4}{18}$.

b. For example, $\dfrac{2}{4} = \dfrac{4}{8}$.

Objective B Exercises

51. Strategy To find the weight on the moon, write and solve a proportion using n to represent the weight.

Solution $\dfrac{1 \text{ lb}}{6 \text{ lb}} = \dfrac{n}{174 \text{ lb}}$

$1 \cdot 174 = 6 \cdot n$

$174 = 6n$

$\dfrac{174}{6} = \dfrac{6n}{6}$

$29 = n$

The weight on the moon would be 29 lb.

53. Strategy To find the number of robes, write and solve a proportion using n to represent the number of robes.

Solution $\dfrac{6 \text{ robes}}{6.5 \text{ yd}} = \dfrac{n}{26 \text{ yd}}$

$6 \cdot 26 = 6.5n$

$156 = 6.5n$

$\dfrac{156}{6.5} = \dfrac{6.5n}{6.5}$

$24 = n$

24 robes can be made.

55. Strategy To find the property tax, write and solve a proportion using x to represent the amount of tax.

Solution $\dfrac{\$4,320}{\$180,000} = \dfrac{x}{\$280,000}$

$4,320 \cdot 280,000 = 180,000 \cdot x$

$1,209,600,000 = 180,000x$

$\dfrac{1,209,600,000}{180,000} = \dfrac{180,000x}{180,000}$

$6,720 = x$

The property tax is \$6,720.

57. Strategy To find the distance, write and solve a proportion using n to represent the miles driven.

Solution $\dfrac{84 \text{ mi}}{3 \text{ gal}} = \dfrac{n}{14.5 \text{ gal}}$

$84 \cdot 14.5 = 3 \cdot n$

$1218 = 3n$

$\dfrac{1218}{3} = \dfrac{3n}{3}$

$406 = n$

The car would travel 406 mi on 14.5 gal of gasoline.

59. Strategy To find the number of light fixtures, write and solve a proportion using n to represent the required number of light fixtures.

Solution $\dfrac{5 \text{ lights}}{400 \text{ ft}^2} = \dfrac{n}{35,000 \text{ ft}^2}$

$5 \cdot 35,000 = 400 \cdot n$

$175,000 = 400n$

$\dfrac{175,000}{400} = \dfrac{400n}{400}$

$437.5 = n$

The office will require 438 light fixtures.

61. Strategy   To find the time necessary to lose 36 lb, write and solve a proportion using *n* to represent the time.

Solution $\dfrac{3 \text{ lb}}{5 \text{ weeks}} = \dfrac{36 \text{ lb}}{n}$

$3 \cdot n = 5 \cdot 36$

$3n = 180$

$\dfrac{3n}{3} = \dfrac{180}{3}$

$n = 60$

The dieter will lose 36 lb in 60 weeks.

63. Strategy   To find the number of miles, write and solve a proportion using *n* to represent the number of miles walked.

Solution $\dfrac{5 \text{ mi}}{650 \text{ calories}} = \dfrac{n}{3{,}500 \text{ calories}}$

$5 \cdot 3{,}500 = 650 \cdot n$

$17{,}500 = 650n$

$\dfrac{17{,}500}{650} = \dfrac{650n}{650}$

$26.92 \approx n$

A person would have to walk approximately 26.92 mi to lose one pound.

65. Strategy   To find the number of voters, write and solve a proportion using *n* to represent the number of voters.

Solution $\dfrac{2 \text{ voters}}{3 \text{ voters}} = \dfrac{n}{240{,}000 \text{ voters}}$

$2 \cdot 240{,}000 = 3 \cdot n$

$480{,}000 = 3n$

$\dfrac{480{,}000}{3} = \dfrac{3n}{3}$

$160{,}000 = n$

160,000 people are expected to vote.

67. Strategy   To find the distance between the two points, write and solve a proportion using *n* to represent the distance.

Solution $\dfrac{\frac{1}{2} \text{ in.}}{8 \text{ mi}} = \dfrac{1\frac{1}{4} \text{ in.}}{n}$

$\dfrac{1}{2} \cdot n = 8 \cdot 1\dfrac{1}{4}$

$\dfrac{1}{2} n = 8 \cdot \dfrac{5}{4}$

$\dfrac{1}{2} n = 10$

$\dfrac{2}{1}\left(\dfrac{1}{2} n\right) = \dfrac{2}{1}(10)$

$n = 20$

The distance between the two points is 20 mi.

Critical Thinking 7.4

69. $\dfrac{2}{3}$ cast a vote in favor of the amendment.

$\dfrac{3}{4}$ cast a vote against the amendment.

$\dfrac{2}{5} + \dfrac{3}{4} = \dfrac{8}{20} + \dfrac{15}{20} = \dfrac{23}{20}$

The number 1 represents the total population. The fraction $\dfrac{23}{20}$ indicates that there were more votes than voters. This is not possible.

71. Strategy   To find the first person's share, first add up the amount of money the three people put in ($90) and then write and solve a proportion using *n* to represent the share.

Solution $\dfrac{\$25}{\$90} = \dfrac{n}{\$4{,}500{,}000}$

$25 \cdot 4{,}500{,}000 = 90 \cdot n$

$112{,}500{,}000 = 90n$

$\dfrac{112{,}500{,}000}{90} = \dfrac{90n}{90}$

$1{,}250{,}000 = n$

The first person's share would be $1.25 million.

Projects or Groups Activities 7.4

73. $\dfrac{4 \text{ earned runs}}{21 \text{ innings}} = \dfrac{x}{9 \text{ innings}}$

$4 \cdot 9 = 21 \cdot x$

$36 = 21x$

$\dfrac{36}{21} = \dfrac{21x}{21}$

$1.714 \approx x$

Reardon's ERA for 1979 was 1.71.

75. For 2008,

$\dfrac{76 \text{ earned runs}}{246 \text{ innings}} = \dfrac{x}{9 \text{ innings}}$

$76 \cdot 9 = 246 \cdot x$

$684 = 246x$

$\dfrac{684}{246} = \dfrac{246x}{246}$

$2.78 \approx x$

For 2009,

$\dfrac{74 \text{ earned runs}}{239 \text{ innings}} = \dfrac{x}{9 \text{ innings}}$

$74 \cdot 9 = 239 \cdot x$

$666 = 239x$

$\dfrac{666}{239} = \dfrac{239x}{239}$

$2.79 \approx x$

$2.79 - 2.78 = 0.01$

Halladay's ERA was lower in 2008 by 0.01.

77. Answers will vary

Section 7.5

Concept Check

1. a. The expression $y = kx$ is a direction variation because, on the right side of the equation, the constant of variation, k, is multiplied times x.

   b. The expression $y = \dfrac{k}{x}$ is not a direct variation because, on the right side of the equation, the constant of variation, k, is divided by x, not multiplied times x.

   c. The expression $y = k + x$ is not a direct variation because, on the right side of the equation, the constant of variation, k, is added to x, not multiplied times x.

   d. The expression $y = \dfrac{k}{x^2}$ is not a direct variation because, on the right side of the equation, the constant of variation, k, is divided by x^2, not multiplied times x.

3. a. 1.8; 1.5

   b. divide; 1.5; 1.2

   c. 1.2

   d. 3.5; 3.5; 4.2

   e. 4.2

5. a. 8; 2^2

   b. 4

   c. multiply; 4; 32

   d. 32

Objective A Exercises

7. Strategy To find the constant of variation, substitute 25 for t and 150 or s in the direct variation equation $t = ks$ and solve for k.

Solution
$$t = ks$$
$$25 = k \cdot 150$$
$$\frac{25}{150} = k$$
$$\frac{1}{6} = k$$

The constant of variation is $\frac{1}{6}$.

9. Strategy To find the constant of variation, substitute 54 for z and 3 for y in the direct variation equation $z = ky^2$ and solve for k.

Solution
$$z = ky^2$$
$$54 = k \cdot 3^2$$
$$54 = 9k$$
$$\frac{54}{9} = k$$
$$6 = k$$

The constant of variation is 6.

11. Strategy To find T when $S = 7$:
→Write the basic direct variation equation, replace the variables by the given values, and solve for k.
→Write the direct variation equation, replacing k by its value. Substitute 7 for S and solve for T.

Solution
$$T = kS$$
$$36 = k \cdot 9$$
$$\frac{36}{9} = k$$
$$4 = k$$
$$T = 4S$$
$$T = 4 \cdot 7$$
$$T = 28$$

The value of T is 28.

13. Strategy To find A when $B = 30$:
→Write the basic direct variation equation, replace the variables by the given values, and solve for k.
→Write the direct variation equation, replacing k by its value. Substitute 30 for B and solve for A.

Solution
$$A = kB$$
$$6 = k \cdot 18$$
$$\frac{6}{18} = k$$
$$\frac{1}{3} = k$$
$$A = \frac{1}{3} B$$
$$A = \frac{1}{3} \cdot 30$$
$$A = 10$$

The value of A is 10.

15. Strategy To find W when $V = 9$:
→Write the basic direct variation equation, replacing the variables by the given values, and solve for k.
→Write the direct variation equation, replacing k by its value. Substitute 9 for V and solve for W.

Solution
$$W = kV^2$$
$$50 = k \cdot 5^2$$
$$50 = 25k$$
$$2 = k$$
$$W = 2V^2$$
$$W = 2 \cdot 9^2$$
$$W = 2(81)$$
$$W = 162$$

The value of W is 162.

17. Strategy   To find the force:
→Write the basic direct variation equation, replace the variables by the given values, and solve for k.
→Write the direct variation equation, replacing k by its value. Substitute 5 for d and solve for F.

Solution
$$d = kF$$
$$3 = k \cdot 12$$
$$\frac{3}{12} = k$$
$$\frac{1}{4} = k$$
$$d = \frac{1}{4}F$$
$$5 = \frac{1}{4} \cdot F$$
$$20 = F$$
A force of 20 lb will stretch the spring 5 in.

19. Strategy   To find the number of words typed:
→Write the basic direct variation equation, replace the variables by the given values, and solve for k.
→Write the direct variation equation, replacing k by its value. Substitute 15 for t and solve for w.

Solution
$$w = kt$$
$$260 = k \cdot 4$$
$$\frac{260}{4} = k$$
$$65 = k$$
$$w = 65t$$
$$w = 65 \cdot 15$$
$$w = 975$$
The typist types 975 words in 15 min.

21. Strategy   To find the distance:
→Write the basic direct variation equation, replace the variables by the given values, and solve for k.
→Write the direct variation equation, replacing k by its value. Substitute 5 for t and solve for d.

Solution
$$d = kt^2$$
$$8 = k(0.5)^2$$
$$8 = 0.25k$$
$$\frac{8}{0.25} = k$$
$$32 = k$$
$$d = 32t^2$$
$$d = 32(5)^2$$
$$d = 32 \cdot 25$$
$$d = 800$$
The object will fall 800 ft in 5 s.

23. Strategy   To find the time:
→Convert the time (45 min) to hours.
→Write the basic direct variation equation, replace the variables by the given values, and solve for k.
→Write the direct variation equation, replacing k by its value. Substitute 180 for d and solve for t.

Solution
$$45 \text{ min} = \frac{45 \text{ min}}{1} \cdot \frac{1 \text{ h}}{60 \text{ min}}$$
$$= \frac{45}{60}\text{h} = \frac{3}{4}\text{h}$$
$$d = kt$$
$$50 = k \cdot \frac{3}{4}$$
$$\frac{4}{3}(50) = k$$
$$\frac{200}{3} = k$$
$$d = \frac{200}{3}t$$
$$180 = \frac{200}{3}t$$
$$\frac{3}{200} \cdot 180 = t$$
$$2.7 = t$$
It takes 2.7 h to travel 180 mi.

Objective B Exercises

25. Strategy   To find the constant of proportionality, substitute 0.5 for t and 10 for s in the inverse variation equation $t = \dfrac{k}{s}$ and solve for k.

    Solution $t = \dfrac{k}{s}$

$0.5 = \dfrac{k}{10}$

$0.5 \cdot 10 = k$

$5 = k$

The constant of proportionality is 5.

27. Strategy   To find the constant of variation, substitute 5 for w and 3 for v in the inverse variation equation $w = \dfrac{k}{v^2}$ and solve for k.

    Solution $w = \dfrac{k}{v^2}$

$5 = \dfrac{k}{3^2}$

$5 = \dfrac{k}{9}$

$5 \cdot 9 = k$

$45 = k$

The constant of variation is 45.

29. Strategy   To find L when $W = 2$:

→Write the basic inverse variation equation, replace the variables by the given values, and solve for k.

→Write the inverse variation equation, replacing k by its value. Substitute 2 for W and solve for L.

    Solution $W = \dfrac{k}{L}$

$5 = \dfrac{k}{12}$

$5 \cdot 12 = k$

$60 = k$

$W = \dfrac{60}{L}$

$2 = \dfrac{60}{L}$

$L = \dfrac{60}{2}$

$L = 30$

The value of L is 30.

31. Strategy   To find L when $d = 5$:

→Write the basic inverse variation equation, replace the variables by the given values, and solve for k.

→Write the inverse variation equation, replacing k by its value. Substitute 5 for d and solve for L.

    Solution $L = \dfrac{k}{d^2}$

$25 = \dfrac{k}{2^2}$

$25 = \dfrac{k}{4}$

$25 \cdot 4 = k$

$100 = k$

$L = \dfrac{100}{d^2}$

$L = \dfrac{100}{5^2}$

$L = \dfrac{100}{25}$

$L = 4$

The value of L is 4 when $d = 5$.

33. Strategy To find the average speed of the return trip:
→Write the basic inverse variation equation, replace the variables by the given values, and solve for k.
→Write the inverse variation equation, replacing k by its value. Substitute 5 for the time and solve for v.

Solution
$$t = \frac{k}{v}$$
$$4 = \frac{k}{65}$$
$$4 \cdot 65 = k$$
$$260 = k$$
$$t = \frac{260}{v}$$
$$5 = \frac{260}{v}$$
$$5v = 260$$
$$v = \frac{260}{5}$$
$$v = 52$$
The average speed of the return trip is 52 mph.

35. Strategy To find the volume of the gas:
→Write the basic inverse variation equation, replace the variables by the given values, and solve for k.
→Write the inverse variation equation, replacing k by its value. Substitute 4 for the pressure and solve for the volume.

Solution
$$V = \frac{k}{P}$$
$$12 = \frac{k}{15}$$
$$12 \cdot 15 = k$$
$$180 = k$$
$$V = \frac{180}{P}$$
$$V = \frac{180}{4}$$
$$V = 45$$
The volume of the gas is 45 ft^3.

37. Strategy To find the number of items:
→Write the basic inverse variation equation, replace the variables by the given values, and solve for k.
→Write the inverse variation equation, replacing k by its value. Substitute 6.00 for the cost and solve for the number of items.

Solution
$$N = \frac{k}{C}$$
$$390 = \frac{k}{6.40}$$
$$390 \cdot 6.40 = k$$
$$2496 = k$$
$$N = \frac{2496}{C}$$
$$N = \frac{2496}{6.00}$$
$$N = 416$$
When the cost per item is $6.00, 416 items can be purchased.

39. Strategy To find the intensity of light:
→Write the basic inverse equation, replace the variables by the given values, and solve for k.
→Write the inverse variation equation, replacing k by its value. Substitute 5 for the distance and solve for the intensity.

Solution
$$l = \frac{k}{d^2}$$
$$20 = \frac{k}{8^2}$$
$$20 = \frac{k}{64}$$
$$20 \cdot 64 = k$$
$$1{,}280 = k$$
$$l = \frac{1{,}280}{d^2}$$
$$l = \frac{1{,}280}{5^2}$$
$$l = \frac{1{,}280}{25} = 51.2$$
The intensity of light is 51.2 lumens.

41. Let $y = 250$ and $x = 30$. Then for $y = kx$,
$$250 = k(30)$$
$$k = \frac{25}{3}$$
Next, let $y = 800$ and $x = 96$.
$$800 = k(96)$$
$$k = \frac{800}{96} = \frac{25}{3}$$
It is a direct variation.

Critical Thinking 7.5

43. a. True. Because k is a constant, if x becomes larger, then y becomes larger.

    b. False. Rewrite $y = \dfrac{k}{x}$ as the equivalent expression $xy = k$. If x increases, then y must decrease because the product is a constant.

Projects or Group Activities 7.5

45. If we double x in $y = kx$, we obtain $y = 2kx$. Thus if we double x, y doubles.

Chapter Review Exercises

1. $1.25 \text{ km} = 1\,250 \text{ m}$

2. $0.450 \text{ g} = 450 \text{ mg}$

3. $\dfrac{100 \text{ lb}}{100 \text{ lb}} = \dfrac{1}{1},\ 1:1,\ 1 \text{ to } 1$

4. $\dfrac{18 \text{ roof supports}}{9 \text{ ft}} = \dfrac{2 \text{ roof supports}}{1 \text{ ft}}$

5. $\dfrac{\$628}{40 \text{ h}} = \$15.70/\text{h}$

6. $\dfrac{8 \text{ h}}{15 \text{ h}} = \dfrac{8}{15}$

7. $96 \text{ in.} = \dfrac{96 \text{ in.}}{1} \cdot \dfrac{1 \text{ ft}}{12 \text{ in.}} \cdot \dfrac{1 \text{ yd}}{3 \text{ ft}} = 2\dfrac{2}{3} \text{ yd}$

8. $72 \text{ oz} = \dfrac{72 \text{ oz}}{1} \cdot \dfrac{1 \text{ lb}}{16 \text{ oz}} = 4\dfrac{1}{2} \text{ lb}$

9. $36 \text{ fl oz} = \dfrac{36 \text{ fl oz}}{1} \cdot \dfrac{1 \text{ c}}{8 \text{ fl oz}} = 4\dfrac{1}{2} \text{ c}$

10. $1\dfrac{1}{4} \text{ mi} = \dfrac{5 \text{ mi}}{4} \cdot \dfrac{5{,}280 \text{ ft}}{1 \text{ mi}} = 6{,}600 \text{ ft}$

11. $\dfrac{n}{3} = \dfrac{8}{15}$
$$n \cdot 15 = 3 \cdot 8$$
$$15n = 24$$
$$n = \frac{24}{15}$$
$$n = 1.6$$

12. $\dfrac{15 \text{ lb}}{12 \text{ trees}} = \dfrac{5 \text{ lb}}{4 \text{ trees}}$

13. $\dfrac{171 \text{ mi}}{3 \text{ h}} = 57 \text{ mph}$

14. $\dfrac{2}{3.5} = \dfrac{n}{12}$
$$2 \cdot 12 = 3.5 \cdot n$$
$$24 = 3.5n$$
$$\frac{24}{3.5} = n$$
$$6.86 \approx n$$

15. $1 \text{ qt} = \dfrac{1 \text{ qt}}{1} \cdot \dfrac{1 \text{ L}}{1.06 \text{ qt}} \cdot \dfrac{1000 \text{ ml}}{1 \text{ L}} \approx 943.40 \text{ ml}$

16. $29 \text{ ft} = \dfrac{29 \text{ ft}}{1} \cdot \dfrac{1 \text{ m}}{3.28 \text{ ft}} \approx 8.84 \text{ m}$

17. $100 \text{ m} = \dfrac{100 \text{ m}}{1} \cdot \dfrac{3.28 \text{ ft}}{1 \text{ m}} = 328 \text{ ft}$

18. $2.1 \text{ kg} = \dfrac{2.1 \text{ kg}}{1} \cdot \dfrac{2.2 \text{ lb}}{1 \text{ kg}} = 4.62 \text{ lb}$

19. $30 \text{ mph} = \dfrac{30 \text{ mi}}{1 \text{ h}} \cdot \dfrac{1.61 \text{ km}}{1 \text{ mi}} = 48.3 \text{ km}/\text{h}$

20. $75 \text{ km}/\text{h} = \dfrac{75 \text{ km}}{1 \text{ h}} \cdot \dfrac{1 \text{ mi}}{1.61 \text{ km}} \approx 46.58 \text{ mph}$

21. $\dfrac{18 \text{ c}}{24 \text{ pt}} = \dfrac{3 \text{ c}}{4 \text{ pt}}$

22. Strategy To find the constant of variation, substitute 10 for y and 30 for x in the direct variation equation $y = kx$ and solve for k.

Solution
$$y = kx$$
$$10 = k \cdot 30$$
$$\frac{10}{30} = k$$
$$\frac{1}{3} = k$$

The constant of variation is $\frac{1}{3}$.

23. Strategy To find T when $S = 120$:
→Write the basic direct variation equation, replace the variables by the given values, and solve for k.
→Write the direct variation equation, replacing k by its value. Substitute 120 for S and solve for T.

Solution
$$T = kS^2$$
$$50 = k \cdot 5^2$$
$$50 = 25k$$
$$\frac{50}{25} = k$$
$$2 = k$$
$$T = 2S^2$$
$$T = 2 \cdot 120^2$$
$$T = 2 \cdot 14{,}400 = 28{,}800$$
$$T = 28{,}800$$
The value of T is 28,800.

24. Strategy To find y when $x = 25$:
→Write the inverse variation equation, replace the variables by the given values and solve for k.
→Write the inverse variation equation, replacing k by its value. Substitute 25 for x and solve for y.

Solution
$$y = \frac{k}{x}$$
$$0.2 = \frac{k}{5}$$
$$0.2 \cdot 5 = k$$
$$1 = k$$
$$y = \frac{1}{x}$$
$$y = \frac{1}{25} = 0.04$$
The value of y is 0.04.

25. Strategy To find the number of pieces:
→Convert 1.21 m to cm.
→Add the distances.

Solution
1.21 m = 121 cm
42 + 18 + 121 = 181
The total length of the shaft is 181 cm.

26. Strategy To find the ratio:
→Subtract the present price (75) from the original price (125).
→Form the ratio of the decrease in price to the original price (125).

Solution
125 − 75 = 50
$$\frac{\$50}{\$125} = \frac{2}{5}$$
The ratio of the decrease in price to the original price is $\frac{2}{5}$.

27. Strategy To find the number of ounces:
→Convert 12 lb to ounces.
→Divide the amount by 16.

Solution
$$12\,\text{lb} = \frac{12\,\text{lb}}{1} \cdot \frac{16\,\text{oz}}{1\,\text{lb}} = 192\,\text{oz}$$
$$192 \div 16 = 12$$
Each container has 12 oz of hamburger meat.

28. Strategy To find the amount invested, write and solve for proportion using n to represent the amount of money invested.

Solution
$$\frac{8{,}000}{520} = \frac{n}{780}$$
$$8{,}000 \cdot 780 = 520 \cdot n$$
$$6{,}240{,}000 = 520n$$
$$\frac{6{,}240{,}000}{520} = n$$
$$12{,}000 = n$$
The amount of money invested is $12,000.

29. Strategy   To find the distance:
→Write the basic direct variation equation, replace the variables by the given values, and solve for k.
→Write the direct variation equation, replacing k by its value. Substitute 28 for the weight and solve for the distance.

Solution $d = kw$
$2 = k \cdot 5$
$\dfrac{2}{5} = k$
$d = \dfrac{2}{5}w$
$d = \dfrac{2}{5} \cdot 28$
$d = 11.2$
The weight will stretch the spring 11.2 in.

30. Strategy   To find the amount of plant food, write and solve a proportion using n to represent the amount of plant food.

Solution $\dfrac{0.5}{50} = \dfrac{n}{275}$
$0.5 \cdot 275 = 50 \cdot n$
$137.5 = 50n$
$\dfrac{137.5}{50} = n$
$2.75 = n$
2.75 lb of plant food should be used.

31. Strategy   To convert mph to ft per sec, use the conversion factors
$\dfrac{1 \text{ h}}{3,600 \text{ s}}$ and $\dfrac{5,280 \text{ ft}}{1 \text{ mi}}$.

Solution   87 mph
$= \dfrac{87 \text{ mi}}{1 \text{ h}} \cdot \dfrac{1 \text{ h}}{3,600 \text{ s}} \cdot \dfrac{5,280 \text{ ft}}{1 \text{ mi}}$
$= 127.6 \text{ ft/s}$
The speed is 127.6 ft/s.

32. Strategy   To find the volume:
→Write the basic inverse variation equation, replace the variables by the given values, and solve for k.
→Write the inverse variation equation, replacing k by its value. Substitute 12 for P and solve for V.

Solution $V = \dfrac{k}{P}$
$2.5 = \dfrac{k}{6}$
$2.5 \cdot 6 = k$
$15 = k$
$V = \dfrac{15}{P}$
$V = \dfrac{15}{12}$
$V = 1.25$
The volume of the balloon is 1.25ft^3.

33. Strategy   To find the amount the other attorney receives, write and solve a proportion using n to represent the amount.

Solution $\dfrac{3}{2} = \dfrac{96,000}{n}$
$3 \cdot n = 2 \cdot 96,000$
$3n = 192,000$
$\dfrac{3n}{3} = \dfrac{192,000}{3}$
$n = 64,000$
The attorney receives \$64,000.

Chapter Test

1. 4 650 cm = 46.50 m

2. 4.1 L = 4 100 ml

3. $\dfrac{3 \text{ yd}}{24 \text{ yd}} = \dfrac{1}{8}$, 1 : 8, 1 to 8

4. $\dfrac{16 \text{ oz}}{64 \text{ cookies}} = \dfrac{1 \text{ oz}}{4 \text{ cookies}}$

5. $\dfrac{120 \text{ mi}}{200 \text{ min}} = 0.6 \text{ mi} / \text{min}$

6. $\dfrac{200\,\text{ft}}{100\,\text{ft}} = \dfrac{2}{1}$

7. $2\dfrac{3}{5}\,\text{tons} = \dfrac{13\,\text{tons}}{5} \cdot \dfrac{2{,}000\,\text{lb}}{1\,\text{ton}} = 5{,}200\,\text{lb}$

8. $2\dfrac{1}{2}\,\text{c} = \dfrac{5\,\text{c}}{2} \cdot \dfrac{8\,\text{fl oz}}{1\,\text{c}} = 20\,\text{fl oz}$

9. $3\dfrac{1}{4}\,\text{lb} = \dfrac{13\,\text{lb}}{4} \cdot \dfrac{16\,\text{oz}}{1\,\text{lb}} = 52\,\text{oz}$

10. $8\dfrac{1}{2}\,\text{ft} = \dfrac{17\,\text{ft}}{2} \cdot \dfrac{12\,\text{in.}}{1\,\text{ft}} = 102\,\text{in.}$

11. $\dfrac{n}{5} = \dfrac{3}{20}$

    $n \cdot 20 = 5 \cdot 3$

    $20n = 15$

    $n = \dfrac{15}{20}$

    $n = 0.75$

12. $\dfrac{8\,\text{ft}}{4\,\text{s}} = 2\,\text{ft/s}$

13. $4.3\,\text{c} = \dfrac{4.3\,\text{c}}{1} \cdot \dfrac{8\,\text{fl oz}}{1\,\text{c}} = 34.4\,\text{fl oz}$

14. $42\,\text{yd} = \dfrac{42\,\text{yd}}{1} \cdot \dfrac{3\,\text{ft}}{1\,\text{yd}} = 126\,\text{ft}$

15. $\dfrac{2{,}860\,\text{ft}^2}{6\,\text{h}} = 476.67\,\text{ft}^2/\text{h}$

16. $\dfrac{n}{4} = \dfrac{8}{9}$

    $n \cdot 9 = 4 \cdot 8$

    $9n = 32$

    $\dfrac{9n}{9} = \dfrac{32}{9}$

    $n \approx 3.56$

17. $12\,\text{oz} = \dfrac{12\,\text{oz}}{1} \cdot \dfrac{28.35\,\text{g}}{1\,\text{oz}} = 340.2\,\text{g}$

18. $547\,\text{ft} = \dfrac{547\,\text{ft}}{1} \cdot \dfrac{1\,\text{m}}{3.28\,\text{ft}} \approx 166.77\,\text{m}$

19. $1{,}000\,\text{m} = \dfrac{1{,}000\,\text{m}}{1} \cdot \dfrac{1.09\,\text{yd}}{1\,\text{m}} = 1{,}090\,\text{yd}$

20. $1.9\,\text{kg} = \dfrac{1.9\,\text{kg}}{1} \cdot \dfrac{2.2\,\text{lb}}{1\,\text{kg}} = 4.18\,\text{lb}$

21. $35\,\text{mph} = \dfrac{35\,\text{mi}}{1\,\text{h}} \cdot \dfrac{1.61\,\text{km}}{1\,\text{mi}} = 56.35\,\text{km/h}$

22. $60\,\text{km/h} = \dfrac{60\,\text{km}}{1\,\text{h}} \cdot \dfrac{1\,\text{mi}}{1.61\,\text{km}} \approx 37.27\,\text{mph}$

23. **Strategy**   To find the constant of proportionality, substitute 10 for y and 2 for x in the inverse variation equation $y = \dfrac{k}{x}$ and solve for k.

    Solution $y = \dfrac{k}{x}$

    $10 = \dfrac{k}{2}$

    $10 \cdot 2 = k$

    $20 = k$

    The constant of proportionality is 20.

24. **Strategy**   To find P when $R = 15$:

    →Write the basic direct variation equation, replace the variables by the given values, and solve for k.

    →Write the direct variation equation, replacing k by its value. Substitute 15 for R and solve for P.

    Solution $R = kP$

    $4 = k \cdot 20$

    $\dfrac{4}{20} = k$

    $\dfrac{1}{5} = k$

    $R = \dfrac{1}{5}P$

    $15 = \dfrac{1}{5}P$

    $5 \cdot 15 = P$

    $75 = P$

    The value of P is 75.

25. Strategy To find U when $V = 2$:
→Write the inverse variation equation, replace the variables by the given values, and solve for k.
→Write the inverse variation equation, replacing k by its value. Substitute 2 for V and solve for U.

Solution $U = \dfrac{k}{V^2}$

$20 = \dfrac{k}{4^2}$

$20 = \dfrac{k}{16}$

$20 \cdot 16 = k$

$320 = k$

$U = \dfrac{320}{V^2}$

$U = \dfrac{320}{2^2}$

$U = \dfrac{320}{4} = 80$

The value of U is 80.

26. Strategy To find the ratio, form the ratio of the original weight (165) to the increased weight (190).

Solution $\dfrac{165}{190} = \dfrac{33}{38}$

The ratio of the original weight to the increased weight is $\dfrac{33}{38}$.

27. Strategy To find the sales tax, write and solve a proportion using x to represent the amount of tax.

Solution $\dfrac{\$7.60}{\$95} = \dfrac{x}{\$39,200}$

$7.60 \cdot 39{,}200 = 95 \cdot x$

$297{,}920 = 95x$

$\dfrac{297{,}920}{95} = \dfrac{95x}{95}$

$3{,}136 = x$

The sales tax is $3,136.

28. Strategy To find the number of registered voters that would vote, write and solve a proportion using n to represent the number of people that would vote.

Solution $\dfrac{3}{4} = \dfrac{n}{325{,}000}$

$3 \cdot 325{,}000 = 4n$

$975{,}000 = 4n$

$\dfrac{975{,}000}{4} = n$

$243{,}750 = n$

243,750 of the registered voters would vote.

29. Strategy To find the difference:
→Convert from lb to ounces.
→Divide the amount of cheese by the package size.
→To find the selling price, multiply the number of packages by price (7.50).
→Subtract the purchase price from the selling price.

Solution $24 \,\text{lb} = \dfrac{24\,\text{lb}}{1} \cdot \dfrac{16\,\text{oz}}{1\,\text{lb}} = 384\,\text{oz}$

$\dfrac{384}{12} = 32$

$32 \cdot 10.50 = 336$

$336 - 192 = 144$

The difference is $144.

30. Strategy To find the length of the room, write and solve a proportion using n to represent the length.

Solution $\dfrac{4}{1} = \dfrac{n}{12\frac{1}{2}}$

$4 \cdot \dfrac{25}{2} = 1 \cdot n$

$50 = n$

The length of the room is 50 ft.

31. Strategy   To convert mph to ft per sec, use the conversion factors $\dfrac{1\,h}{3{,}600\,s}$ and $\dfrac{5{,}280\,ft}{1\,mi}$.

Solution   52 mph

$= \dfrac{52\,mi}{1\,h} \cdot \dfrac{1\,h}{3{,}600\,s} \cdot \dfrac{5{,}280\,ft}{1\,mi}$

$= 76.27$ ft/s

The speed is 76.27 ft/s.

32. Strategy   To find the stopping distance:
→Write the basic direct variation equation, replace the variables by the given values, and solve for k.
→Write the direct variation equation, replacing k by its value. Substitute 60 for v and solve for d

Solution $d = kv^2$

$130 = k \cdot 40^2$

$130 = 1{,}600k$

$\dfrac{130}{1{,}600} = k$

$d = \dfrac{13}{160}v^2$

$d = \dfrac{13}{160} \cdot 60^2$

$d = \dfrac{13}{160} \cdot 3{,}600$

$d = 292.5$

The stopping distance of the car is 292.5 ft.

33. Strategy   To find the number of revolutions per minute:
→Write the basic inverse variation equation, replace the variables by the given values, and solve for k.
→Write the inverse variation equation, replacing k by its value. Substitute 40 for the number of teeth and solve for the number of revolutions per minute.

Solution $s = \dfrac{k}{t}$

$160 = \dfrac{k}{25}$

$160 \cdot 25 = k$

$4{,}000 = k$

$s = \dfrac{4{,}000}{t}$

$s = \dfrac{4000}{40} = 100$

The gear will make 100 revolutions per minute.

Cumulative Review Exercises

1. $18 \div \dfrac{6-3}{9} - (-3)$

$= 18 \div \dfrac{3}{9} - (-3)$

$= 18 \div \dfrac{1}{3} - (-3)$

$= 18 \cdot \dfrac{3}{1} - (-3)$

$= 54 - (-3)$

$= 54 + 3$

$= 57$

2. 1.2 gal $= 1.2$ gal $\cdot \dfrac{4\,qt}{1\,gal}$

$= 4.8$ qt

3. $7\dfrac{5}{12} - 3\dfrac{5}{9} = 7\dfrac{15}{36} - 3\dfrac{20}{36}$

$= 6\dfrac{51}{36} - 3\dfrac{20}{36}$

$= 3\dfrac{31}{36}$

4. $\dfrac{4}{5} \div \dfrac{4}{5} + \dfrac{2}{3} = \dfrac{4}{5} \cdot \dfrac{5}{4} + \dfrac{2}{3}$

$\qquad = 1 + \dfrac{2}{3}$

$\qquad = 1\dfrac{2}{3}$

5. $342 \div (-3) = -114$

6. $2a - 3ab$

$2(2) - 3(2)(-3)$

$= 4 - 3(2)(-3)$

$= 4 - 6(-3)$

$= 4 - (-18)$

$= 4 + 18$

$= 22$

7. $5x - 20 = 0$

$5x - 20 + 20 = 0 + 20$

$5x = 20$

$\dfrac{5x}{5} = \dfrac{20}{5}$

$x = 4$

The solution is 4.

8. $3(x - 4) + 2x = 3$

$3x - 12 + 2x = 3$

$5x - 12 = 3$

$5x - 12 + 12 = 3 + 12$

$5x = 15$

$\dfrac{5x}{5} = \dfrac{15}{5}$

$x = 3$

The solution is 3.

9. Draw a solid dot one half unit to the left of -3 on the number line.

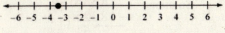

10. Draw a parenthesis at -3. Draw a line to the left of -3. Draw an arrow at the left end of the line.

11. $(-5)^2 - (-8) \div (7 - 5)^2 \cdot 2 - 8$

$= (-5)^2 - (-8) \div 2^2 \cdot 2 - 8$

$= 25 - (-8) \div 4 \cdot 2 - 8$

$= 25 - (-2) \cdot 2 - 8$

$= 25 - (-4) - 8$

$= 25 + 4 - 8$

$= 29 - 8$

$= 21$

12. $\left(-\dfrac{2}{3}\right)\left(-\dfrac{3}{4}\right)^2 = \left(-\dfrac{2}{3}\right)\left(\dfrac{9}{16}\right)$

$= -\dfrac{2}{3} \cdot \dfrac{9}{16}$

$= -\dfrac{2 \cdot 9}{3 \cdot 16}$

$= -\dfrac{2 \cdot 3 \cdot 3}{3 \cdot 2 \cdot 2 \cdot 2 \cdot 2}$

$= -\dfrac{3}{8}$

13. $\sqrt{169} = 13$

14. $5 - 2(1 - 3a) + 2(a - 3)$

$= 5 - 2 + 6a + 2a - 6$

$= 8a - 3$

15. $\left(4a^3 b\right)\left(-5a^2 b^3\right) = \left[4(-5)\right]\left(a^3 a^2\right)\left(b b^3\right)$

$= -20a^5 b^4$

16. $-3y^2 + 3y - y^2 - 6y$

$= \left(-3y^2 - y^2\right) + (3y - 6y)$

$= -4^2 - 3y$

17. $y = 3x - 2$

$y = 3(-1) - 2$

$y = -3 - 2$

$y = -5$

The ordered pair solution is $(-1, -5)$.

18. $\dfrac{30 \text{ cents}}{1 \text{ dollar}} = \dfrac{30 \text{ cents}}{100 \text{ cents}} = \dfrac{3}{10}$

19. $\dfrac{\$19,425}{5 \text{ months}} = \$3,885 / \text{month}$

20. $\$2.97 / \text{gal} = \dfrac{\$2.97}{1 \text{ gal}} \cdot \dfrac{1 \text{ gal}}{3.79 \text{ L}} \approx \$0.78 / \text{L}$

21. $\dfrac{2}{3} = \dfrac{n}{48}$

$2 \cdot 48 = 3 \cdot n$

$96 = 3n$

$\dfrac{96}{3} = n$

$32 = n$

22. $\dfrac{\frac{1}{2}+\frac{3}{4}}{2-\frac{5}{8}}=\dfrac{\frac{2}{4}+\frac{3}{4}}{\frac{16}{8}-\frac{5}{8}}$

$=\dfrac{\frac{5}{4}}{\frac{11}{8}}$

$=\dfrac{5}{4}\div\dfrac{11}{8}$

$=\dfrac{5}{4}\cdot\dfrac{8}{11}$

$=\dfrac{5\cdot 8}{4\cdot 11}$

$=\dfrac{5\cdot 2\cdot 2\cdot 2}{2\cdot 2\cdot 11}=\dfrac{10}{11}$

23. $-2\sqrt{x^2-3y}$

$-2\sqrt{4^2-3(-3)}$

$=-2\sqrt{16-(-9)}$

$=-2\sqrt{16+9}$

$=-2\sqrt{25}$

$=-2\cdot 5=-10$

24. $3x+3(x+4)=4(x+2)$

$3x+3x+12=4x+8$

$6x+12=4x+8$

$6x-4x+12=4x-4x+8$

$2x+12=8$

$2x+12-12=8-12$

$2x=-4$

$\dfrac{2x}{2}=\dfrac{-4}{2}$

$x=-2$

The solution is –2.

25. Strategy To find the monthly difference:

→Find the difference in annual expenses between the northeast (587) and south (243).

→Convert the difference from annual to monthly by dividing by 12.

Solution $587-243=344$

$344\div 12\approx 28.67$

The monthly difference is $28.67.

26. The unknown number: x

five less than two thirds of a number	is	three

$\dfrac{2}{3}x-5=3$

$\dfrac{2}{3}x-5+5=3+5$

$\dfrac{2}{3}x=8$

$\dfrac{3}{2}\left(\dfrac{2}{3}x\right)=\dfrac{3}{2}\cdot 8$

$x=12$

The number is 12.

27. Let the number be x.

The <u>difference</u> between four <u>times</u> a number and three <u>times</u> the sum of the number and two.

$4x-3(x+2)$

$4x-3x-6$

$x-6$

28. Strategy To find the number of miles left to drive:

→Find the number of miles already driven by subtracting the original odometer reading (18,325) from the present odometer reading (18,386).

→Subtract the number of miles already drive from 125.

Solution

$\begin{array}{r}18{,}386\\-18{,}325\\\hline 61\end{array}$

$\begin{array}{r}125\\-61\\\hline 64\end{array}$

You have 64 mi left to drive.

29. Strategy   To find the new checking balance:
→Add the deposit (122.35) to the checking account balance (422.89).
→Subtract the check (279.76) from the new checking account balance.

Solution
$$422.89$$
$$+122.35$$
$$\overline{545.24}$$

$$545.24$$
$$-279.76$$
$$\overline{265.48}$$

The new checking account balance is $265.48.

30. Strategy   To find the part that remains to be completed:
→Add the amount already done $\left(\dfrac{2}{5}+\dfrac{1}{3}\right)$.
→Subtract the amounts already done from the total job (1).

Solution
$$\frac{2}{5}+\frac{1}{3}=\frac{6}{15}+\frac{5}{15}=\frac{11}{15}$$
$$1-\frac{11}{15}=\frac{15}{15}-\frac{11}{15}=\frac{4}{15}$$

$\dfrac{4}{15}$ of the job remains to be done.

31. Strategy   To find the number of votes cast, multiply the number of registered voters (31,281) by the fraction of those who voted $\left(\dfrac{2}{3}\right)$.

Solution
$$31,281\cdot\frac{2}{3}=\frac{31,281\cdot 2}{3}$$
$$=20,854$$
In the city election 20,854 votes were cast.

32. Strategy   To find the number of miles driven per gallon, divide the number of miles driven (402.5) by the number of gallons of gas used (11.5).

Solution
$$\frac{402.5}{11.5}=35$$
The car travels 35 mi on each gallon of gas.

33. Strategy   To find the rpm in third gear, write and solve an equation using n to represent the rpm in third gear.

Solution

$\dfrac{2}{3}$ of the rpm in third gear	is	2,500

$$\frac{2}{3}n=2,500$$
$$\frac{3}{2}\left(\frac{2}{3}n\right)=\frac{3}{2}(2,500)$$
$$n=3,750$$
The rpm of the engine is 3,750 in third gear.

Chapter 8: Percent

Prep Test

1. $\dfrac{19}{100}$

2. 0.23

3. 47

4. 2,850

5.
$$0.015.\overline{)60.000.} \quad \begin{array}{r} 4000. \\ \hline \end{array}$$
$$\begin{array}{r} -60 \\ \hline 00 \\ -0 \\ \hline 00 \\ -0 \\ \hline 0 \end{array}$$

6. $8 \div \dfrac{1}{4} = \dfrac{8}{1} \times \dfrac{4}{1} = 32$

7. $\dfrac{5}{8} \times \dfrac{100}{1} = \dfrac{5 \cdot \overset{1}{\cancel{2}} \cdot \overset{1}{\cancel{2}} \cdot 5 \cdot 5}{\underset{1}{\cancel{2}} \cdot \underset{1}{\cancel{2}} \cdot 2} = \dfrac{125}{2} = 62\dfrac{1}{2} = 62.5$

8. $66\dfrac{2}{3}$

9. $16\overline{)28.00} \quad \begin{array}{r} 1.75 \\ \hline \end{array}$

Section 8.1

Concept Check

1. 29 figures should be circled

3. **a.** Students should explain that to convert a percent to a fraction, drop the percent sign and multiply by $\dfrac{1}{100}$.

   b. Students should explain that to convert a percent to a decimal, drop the percent sign and multiply by 0.01.

5. 0.01; 0.01; 0.53

7. $\dfrac{1}{100}$; $\dfrac{1}{100}$; $\dfrac{4}{5}$

9. 100%; 100%; 46%

11. 100%; 100%; 30%

Objective A Exercises

13. $5\% = 5\left(\dfrac{1}{100}\right) = \dfrac{5}{100} = \dfrac{1}{20}$
    $5\% = 5(0.01) = 0.05$

15. $30\% = 30\left(\dfrac{1}{100}\right) = \dfrac{30}{100} = \dfrac{3}{10}$
    $30\% = 30(0.01) = 0.30$

17. $250\% = 250\left(\dfrac{1}{100}\right) = \dfrac{250}{100} = \dfrac{5}{2}$
    $250\% = 250(0.01) = 2.50$

19. $28\% = 28\left(\dfrac{1}{100}\right) = \dfrac{28}{100} = \dfrac{7}{25}$
    $28\% = 28(0.01) = 0.28$

21. $35\% = 35\left(\dfrac{1}{100}\right) = \dfrac{35}{100} = \dfrac{7}{20}$
    $35\% = 35(0.01) = 0.35$

23. $29\% = 29\left(\dfrac{1}{100}\right) = \dfrac{29}{100}$
    $29\% = 29(0.01) = 0.29$

25. $11\dfrac{1}{9}\% = 11\dfrac{1}{9}\left(\dfrac{1}{100}\right) = \dfrac{100}{9}\left(\dfrac{1}{100}\right) = \dfrac{1}{9}$

27. $37\dfrac{1}{2}\% = 37\dfrac{1}{2}\left(\dfrac{1}{100}\right) = \dfrac{75}{2}\left(\dfrac{1}{100}\right) = \dfrac{3}{8}$

29. $66\dfrac{2}{3}\% = 66\dfrac{2}{3}\left(\dfrac{1}{100}\right) = \dfrac{200}{3}\left(\dfrac{1}{100}\right) = \dfrac{2}{3}$

31. $6\dfrac{2}{3}\% = 6\dfrac{2}{3}\left(\dfrac{1}{100}\right) = \dfrac{20}{3}\left(\dfrac{1}{100}\right) = \dfrac{1}{15}$

33. $\dfrac{1}{2}\% = \dfrac{1}{2}\left(\dfrac{1}{100}\right) = \dfrac{1}{200}$

35. $6\dfrac{1}{4}\% = 6\dfrac{1}{4}\left(\dfrac{1}{100}\right) = \dfrac{25}{4}\left(\dfrac{1}{100}\right) = \dfrac{1}{16}$

37. $7.3\% = 7.3(0.01) = 0.073$

39. $15.8\% = 15.8(0.01) = 0.158$

41. $0.3\% = 0.3(0.01) = 0.003$

43. $121.2\% = 121.2(0.01) = 1.212$

45. $62.14\% = 62.14(0.01) = 0.6214$

47. $8.25\% = 8.25(0.01) = 0.0825$

49. $65\% = 65\left(\dfrac{1}{100}\right) = \dfrac{65}{100} = \dfrac{13}{20}$

51. greater than

53. $50\% = 50\left(\dfrac{1}{100}\right) = \dfrac{50}{100} = \dfrac{1}{2}$

Objective B Exercises

55. $0.37 = 0.37(100\%) = 37\%$

57. $0.02 = 0.02(100\%) = 2\%$

59. $0.125 = 0.125(100\%) = 12.5\%$

61. $1.36 = 1.36(100\%) = 136\%$

63. $0.96 = 0.96(100\%) = 96\%$

65. $0.07 = 0.07(100\%) = 7\%$

67. $\dfrac{83}{100} = \dfrac{83}{100}(100\%) = 83\%$

69. $\dfrac{1}{3} = \dfrac{1}{3}(100\%) \approx 33.3\%$

71. $\dfrac{4}{9} = \dfrac{4}{9}(100\%) \approx 44.4\%$

73. $\dfrac{9}{20} = \dfrac{9}{20}(100\%) = 45\%$

75. $2\dfrac{1}{2} = \dfrac{5}{2} = \dfrac{5}{2}(100\%) = 250\%$

77. $\dfrac{1}{6} = \dfrac{1}{6}(100\%) \approx 16.7\%$

79. $\dfrac{17}{25} = \dfrac{17}{25}(100\%) = 68\%$

81. $\dfrac{9}{16} = \dfrac{9}{16}(100\%) = 56\dfrac{1}{4}\%$

83. $2\dfrac{5}{8} = \dfrac{21}{8} = \dfrac{21}{8}(100\%) = 262\dfrac{1}{2}\%$

85. $2\dfrac{5}{6} = \dfrac{17}{6} = \dfrac{17}{6}(100\%) = 283\dfrac{1}{3}\%$

87. $\dfrac{7}{30} = \dfrac{7}{30}(100\%) = 23\dfrac{1}{3}\%$

89. $\dfrac{2}{9} = \dfrac{2}{9}(100\%) = 22\dfrac{2}{9}\%$

91. $\dfrac{3}{4}$; 75%

93. 0.375; 37.5%

95. $\dfrac{9}{16}$; 0.5625

97. $\dfrac{13}{25}$; 52%

99. 0.18; 18%

101. $\dfrac{3}{4} = \dfrac{3}{4}(100\%) = \dfrac{300}{4}\% = 75\%$

103. Since $\dfrac{4}{3}$ has a numerator greater than the denominator, the value of the fraction is *greater than* 100%.

Critical Thinking 8.1

105. Shaded: $\dfrac{1}{4}$; 0.25; 25%

Not shaded: $\dfrac{3}{4}$; 0.75; 75%

107. a. False

b. Answers will vary. For example, $200\% \times 4 = 2 \times 4 = 8$

Projects and Group Activities 8.1

109. Employee B's salary is the highest after the raise. Explanations will vary. One possible explanation is: Because each employee start with the same original salary, the largest resulting salary will go to the employee with the largest percent raise. Since Employee B is given the largest percent raise, that employee will have the highest salary after the raises.

Section 8.2

Concept Check

1. 0.12; n; 68

3. 0.08; 450; n

5. 36; n; 25

Objective A Exercises

7. $70\% = 0.70$
$0.70 \cdot 30 = 21$
21 of the 30 squares should be shaded.

9. $62.5\% = 0.625$
$0.625 \cdot 40 = 25$
25 of the 40 squares should be shaded.

11. $16\frac{2}{3}\% = 16\frac{2}{3}\left(\frac{1}{100}\right) = \frac{50}{3}\left(\frac{1}{100}\right) = \frac{50}{300} = \frac{1}{6}$

$\frac{1}{6} \cdot 30 = 5$

5 of the 30 squares should be shaded.

13. $56.5\% = 0.565$
$0.565 \cdot 48 = 27$
27 of the 48 people should be circled.

15. Strategy To find the amount, solve the basic percent equation.
Percent = $8\% = 0.08$,
base = 100, amount = n

Solution Percent · base = amount
$0.08 \cdot 100 = n$
$8 = n$
8% of 100 is 8.

17. Strategy To find the amount, solve the basic percent equation.
Percent = $0.05\% = 0.0005$,
base = 150, amount = n.

Solution Percent · base = amount
$0.0005 \cdot 150 = n$
$0.075 = n$
0.05% of 150 is 0.075.

19. Strategy To find the amount, solve the basic percent equation.
Percent = n, base = 90,
amount = 15

Solution Percent · base = amount
$n \cdot 90 = 15$

$n = \frac{15}{90} = 0.16\frac{2}{3}$

$n = 16\frac{2}{3}\%$

15 is $16\frac{2}{3}\%$ of 90.

21. Strategy To find the percent, solve the basic percent equation.
Percent = n, base = 16,
amount = 6

Solution Percent · base = amount
$n \cdot 16 = 6$
$n = \frac{6}{16} = 0.375$
$n = 37.5\%$

23. Strategy To find the base, solve the basic percent equation.
Percent = $10\% = 0.10$,
base = n, amount = 10

Solution Percent · base = amount
$0.10 \cdot n = 10$
$n = \frac{10}{0.10}$
$n = 100$
10 is 10% of 100.

25. Strategy To find the base, solve the basic percent equation.
Percent = $2.5\% = 0.025$,
base = n, amount = 30

Solution Percent · base = amount
$0.025 \cdot n = 30$
$n = \frac{30}{0.025}$
$n = 1,200$
2.5% of 1,200 is 30.

27. Strategy To find the amount, solve the basic percent equation.
Percent = $10.7\% = 0.107$, base = 485, amount = n

Solution Percent · base = amount
$0.107 \cdot 485 = n$
$51.895 = n$
10.7% of 485 is 51.895.

29. Strategy To find the amount, solve the basic percent equation.
Percent = $80\% = 0.80$,
base = 16.25, amount = n

Solution Percent · base = amount
$0.80 \cdot 16.25 = n$
$13 = n$
80% of 16.25 is 13.

31. Strategy   To find the percent, solve the basic percent equation. Percent = n, base = 2,000, amount = 54

Solution   Percent · base = amount
$n \cdot 2{,}000 = 54$
$n = \dfrac{54}{2{,}000} = 0.027$
$n = 2.7\%$
54 is 2.7% of 2,000.

33. Strategy   To find the percent, solve the basic percent equation. Percent = n, base = 4.1, amount = 16.4

Solution   Percent · base = amount
$n \cdot 4.1 = 16.4$
$n = \dfrac{16.4}{4.1} = 4$
$n = 400\%$
16.4 is 400% of 4.1.

35. Strategy   To find the percent, solve the basic percent equation. Percent = 240% = 2.40, base = n, amount = 18

Solution   Percent · base = amount
$2.40 \cdot n = 18$
$n = \dfrac{18}{2.40}$
$n = 7.5$
18 is 240% of 7.5.

37. Strategy   29.8% responded yes, so 100% − 29.8% = 70.2% did not respond yes. To find the percent, solve the basic percent equation. Percent = 70.2% = 0.702, base = 8878, amount = n

Solution   Percent · base = amount
$0.702 \cdot 8878 = n$
$6232.356 = n$
$6232 \approx n$
6232 respondents did not answer yes.

39. Strategy   To find the percent, solve the basic percent equation. Percent = 11% = 0.11, base = n, amount = 1.738

Solution   Percent · base = amount
$0.11 \cdot n = 1.738$
$n = \dfrac{1.738}{0.11} = 15.5$
About 15.8 million travelers allowed their children to miss school to go on a trip.

41. $x < y$

Objective B Exercises

43. Percent = 18, base = 150, amount = n
$\dfrac{\text{percent}}{100} = \dfrac{\text{amount}}{\text{base}}$
$\dfrac{18}{100} = \dfrac{n}{150}$
$18 \cdot 150 = 100 \cdot n$
$2{,}700 = 100n$
$27 = n$
18% of 150 is 27.

45. Percent = n, base = 150, amount = 33
$\dfrac{\text{percent}}{100} = \dfrac{\text{amount}}{\text{base}}$
$\dfrac{n}{100} = \dfrac{33}{150}$
$n \cdot 150 = 3{,}300$
$150n = 3{,}300$
$n = \dfrac{3{,}300}{150}$
$n = 22$
33 is 22% of 150.

47. Percent = 84, base = n, amount = 126
$\dfrac{\text{percent}}{100} = \dfrac{\text{amount}}{\text{base}}$
$\dfrac{84}{100} = \dfrac{126}{n}$
$84 \cdot n = 100 \cdot 126$
$84n = 12{,}600$
$n = \dfrac{12{,}600}{84}$
$n = 150$
126 is 84% of 150.

49. Percent = n, base = 50, amount = 750
$$\frac{\text{percent}}{100} = \frac{\text{amount}}{\text{base}}$$
$$\frac{n}{100} = \frac{750}{50}$$
$$n \cdot 50 = 100 \cdot 750$$
$$50n = 75{,}000$$
$$n = \frac{75{,}000}{50}$$
$$n = 1{,}500$$
750 is 1,500% of 50.

51. Percent = 2.4, base = n, amount = 21
$$\frac{\text{percent}}{100} = \frac{\text{amount}}{\text{base}}$$
$$\frac{2.4}{100} = \frac{21}{n}$$
$$2.4 \cdot n = 100 \cdot 21$$
$$2.4n = 2{,}100$$
$$n = \frac{2{,}100}{2.4}$$
$$n = 875$$
21 is 2.4% of 875.

53. Percent = 96, base = 75, amount = n
$$\frac{\text{percent}}{100} = \frac{\text{amount}}{\text{base}}$$
$$\frac{96}{100} = \frac{n}{75}$$
$$96 \cdot 75 = 100 \cdot n$$
$$7{,}200 = 100n$$
$$72 = n$$
72 is 96% of 75.

55. Percent = n, base = 1,500, amount = 693
$$\frac{\text{percent}}{100} = \frac{\text{amount}}{\text{base}}$$
$$\frac{n}{100} = \frac{693}{1{,}500}$$
$$n \cdot 1{,}500 = 100 \cdot 693$$
$$1{,}500n = 69{,}300$$
$$n = \frac{69{,}300}{1{,}500} = 46.2$$
693 is 46.2% of 1,500.

57. Percent = 87.4, base = 225, amount = n
$$\frac{\text{percent}}{100} = \frac{\text{amount}}{\text{base}}$$
$$\frac{87.4}{100} = \frac{n}{225}$$
$$87.4 \cdot 225 = 100 \cdot n$$
$$19{,}665 = 100n$$
$$196.65 = n$$
196.65 is 87.4% of 225.

59. Percent = 175, base = n, amount = 14
$$\frac{\text{percent}}{100} = \frac{\text{amount}}{\text{base}}$$
$$\frac{175}{100} = \frac{14}{n}$$
$$175 \cdot n = 100 \cdot 14$$
$$175n = 1{,}400$$
$$n = \frac{1{,}400}{175}$$
$$n = 8$$
175% of 8 is 14.

61. Percent = 325, base = 4.4, amount = n
$$\frac{\text{percent}}{100} = \frac{\text{amount}}{\text{base}}$$
$$\frac{325}{100} = \frac{n}{4.4}$$
$$325 \cdot 4.4 = 100 \cdot n$$
$$1{,}430 = 100n$$
$$14.3 = n$$
325% of 4.4 is 14.3.

63. Percent = n, base = 38, amount = 95
$$\frac{\text{percent}}{100} = \frac{\text{amount}}{\text{base}}$$
$$\frac{n}{100} = \frac{95}{38}$$
$$n \cdot 38 = 100 \cdot 95$$
$$38n = 9{,}500$$
$$n = \frac{9{,}500}{38}$$
$$n = 250$$
95 is 250% of 38.

65. Strategy   To find the percent, solve the basic percent equation.
Percent = n, base = 4450, amount = 17.8

Solution   Percent · base = amount
$$n \cdot 4450 = 17.8$$
$$n = \frac{17.8}{4450} = 0.004$$
$$n = 0.4\%$$
0.4% of the total energy production was generated by wind machines.

67. $\dfrac{\frac{1}{4}}{100} = \dfrac{1}{4} \cdot \dfrac{1}{100} = \dfrac{1}{400}$
The answer is (ii).

69. True. Multiplication is associative and commutative.

Objective C Exercises

71. Strategy   To find the percent of the number of the false alarms, use the basic percent equation.
Percent = n, base = 200, amount = 24

Solution   Percent · base = amount
$n \cdot 200 = 24$
$n = \dfrac{24}{200}$
$n = 0.12 = 12\%$
The percent of the false alarms was 12%.

73. Strategy   To find the percent of the individual giving, use the basic percent equation.
Percent = n, base = 260.28, amount = 199

Solution   Percent · base = amount
$n \cdot 260.28 = 199$
$n = \dfrac{199}{260.28}$
$n \approx 0.765 = 76.5\%$
The percent of the individual giving was 76.5%, which is more than three-quarters.

75. Strategy   To find the number of workers 65 and older, use the basic percent equation.
Percent = 25% = 0.25, base = 24.6, amount = n

Solution   Percent · base = amount
$0.25 \cdot 24.6 = n$
$6.15 = n$
The were 6.15 million workers 65 and older in 2006.

77. Strategy   To find number of couples, use the basic percent equation.
Percent = 4.4% = 0.044, base = 114, amount = n

Solution   Percent · base = amount
$0.044 \cdot 114 = n$
$5.016 = n$
There are 5 million couples.

79. Strategy   To find the cash generated, use the basic percent equation.
a.
Percent = 25% = 0.25, base = 700, amount = n
b.
Percent = 9% = 0.9, base = 700, amount = n

Solution   **a.**
Percent · base = amount
$0.25 \cdot 700 = n$
$175 = n$
The cash generated by the sale of Thin Mints is $175 million.
b.
Percent · base = amount
$0.09 \cdot 700 = n$
$63 = n$
The cash generated by the sale of Trefoil shortbread is $63 million.

81. Strategy   To find the percent, use the basic percent equation.
Percent = n, base = 572, amount = 291.72

Solution   Percent · base = amount
$n \cdot 572 = 291.72$
$n = \dfrac{291.72}{572}$
$n = 0.51 = 51\%$
Wisconsin produced 51% of the cranberry crop.

83. Strategy   To find the total revenue, use the basic percent equation.
Percent = 16% = 0.16, base = n, amount = 2,240,000

Solution   Percent · base = amount
$0.16 \cdot n = 2,240,000$
$n = \dfrac{2,240,000}{0.16}$
$n = 14,000,000$
14,000,000 oz of gold were minded.

85. Strategy   To find the percent, use the basic percent equation.
Add all the costs to find the base.
Percent = n, base = (total),
amount = 4,020

Solution $1,400 + 1,070 + 1,220 + 2,960 + 3,930 + 4,020 = 14,600$
Percent · base = amount
$n \cdot 14,600 = 4,020$
$n = \dfrac{4,020}{14,600}$
$n \approx 0.275 = 27.5\%$
Of the total cost, 27.5% is spent on food.

87. Strategy   To find number of people, use the basic percent equation.
Percent = 30% = 0.30,
base = 44, amount = n

Solution   Percent · base = amount
$0.30 \cdot 44 = n$
$13.2 = n$
About 13.2 million people aged 18 to 24 do not have health insurance.

89. Strategy   To find the percent,
→Add the number of people who have been diagnosed and who have not been diagnosed to find the base $(14.6 + 6.2 = 20.8)$.
→Use the basic percent equation.
Percent = n, base = 20.8,
amount = 6.2

Solution   Percent · base = amount
$n \cdot 20.8 = 6.2$
$n = \dfrac{6.2}{20.8}$
$n \approx 0.29807 \approx 29.8\%$
$n = 12,150$
Around 29.8% of Americans with diabetes have not been diagnosed.

91. Strategy   To find how many fewer faculty members described their views as conservative than middle of the road,
→Subtract the percent of Conservative from the percent of Middle of the road
→Use the basic percent equation.
Percent = difference, base = 32,840, amount = n

Solution   Percent · base = amount
$34.3 - 17.7 = 16.6\% = 0.166$
$0.166 \cdot 32,840 = n$
$n = 5,451$
There were 5,451 fewer faculty members who described their views as conservative than middle of the road.

Critical Thinking 8.2

93. Increase 100 by 10% $100 + 0.10(100) = 100 + 10 = 110$

Now decrease 110 by 10% $110 - 0.10(110) = 110 - 11 = 99$

No, the new number is not the original number. The 10% increase applied to the number 100, while the decrease in 10% applied to 110, thus the decrease was greater than the increase.

Projects or Group Activities 8.2

95. a. $\dfrac{100\ \text{lb}}{100\ \text{squares}} = 1\ \text{lb per square}$

b. Percent · base = amount
$0.01 \cdot 100 = n$
$1 = n$
1 lb is 1% of 100 lb.

c. $50 \cdot 1\ \text{lb per square} = 50\ \text{lb}$

d. Percent · base = amount
$0.50 \cdot 100 = n$
$50 = n$
50 lb is 50% of 100 lb.

e. $25 \cdot 1\ \text{lb per square} = 25\ \text{lb}$

f. Percent · base = amount
$0.25 \cdot 100 = n$
$25 = n$
25 lb is 25% of 100 lb.

g. 70 squares should be shaded as each square represents 1 lb

h. Percent · base = amount
$n \cdot 100 = 70$
$n = \dfrac{70}{100} = 0.70 = 70\%$
70 lb is 70% of 100 lb.

97. a. $\dfrac{20\ \text{lb}}{100\ \text{squares}} = 0.2\ \text{lb per square}$

b. Percent · base = amount
$0.01 \cdot 20 = n$
$0.22 = n$
0.2 lb is 1% of 20 lb.

c. $50 \cdot 0.2\ \text{lb per square} = 10\ \text{lb}$

d. Percent · base = amount
$0.50 \cdot 20 = n$
$10 = n$
10 lb is 50% of 20 lb.

e. $25 \cdot 0.2\ \text{lb per square} = 5\ \text{lb}$

f. Percent · base = amount
$0.25 \cdot 20 = n$
$5 = n$
5 lb is 25% of 20 lb.

g. 40 squares should be shaded as each square represents 0.2 lb

h. Percent · base = amount
$n \cdot 20 = 8$
$n = \dfrac{8}{20} = 0.40 = 40\%$
8 lb is 40% of 20 lb.

Section 8.3

Concept Check

1. **a.** original; new; 18; 15; 3

    b. 15; 3

Objective A Exercises

3. Strategy   To find percent increase:
                →Find the increase in
                the consumption of bison.
                →Use the basic percent
                equation. Percent = n,
                base = 17,674,
                amount = amount of
                increase

    Solution $50,000 - 17,674 = 32,326$
                 Percent · base = amount
                 $n \cdot 17{,}674 = 32{,}326$
                 $n = \dfrac{32{,}326}{17{,}674}$
                 $n \approx 1.82901 \approx 182.9\%$
                 The percent increase in the
                 consumption of bison is
                 182.9%.

5. Strategy   To find the percent increase:
                →Find the increase in the
                 number of events.
                →Use the basic percent
                 equation. Percent = n,
                 base = 14,
                 amount = amount of
                 increase

    Solution $86 - 14 = 72$
                 Percent · base = amount
                 $n \cdot 14 = 72$
                 $n = \dfrac{72}{14}$
                 $n \approx 5.14285 \approx 514.3\%$
                 The percent increase in the
                 number of events is 514.3%.

7. Strategy   To find percent increase:
                →Find the increase in
                 the number of Americans
                 living in poverty.
                →Use the basic percent
                 equation. Percent = n,
                 base = 32.9,
                 amount = amount of
                 increase

    Solution $43.6 - 32.9 = 10.7$
                 Percent · base = amount
                 $n \cdot 32.9 = 10.7$
                 $n = \dfrac{10.7}{32.9}$
                 $n \approx 0.32522 = 32.5\%$
                 The percent increase in the
                 number of Americans living in
                 poverty is 32.5%.

9. Strategy   To find the percent increase:
                →Find the increase in the
                 cost for public college.
                →Use the basic percent
                 equation. Percent = n,
                 base = 5,074,
                 amount = amount of
                 increase

    Solution $7{,}020 - 5{,}074 = 1{,}946$
                 Percent · base = amount
                 $n \cdot 5{,}074 = 1{,}946$
                 $n = \dfrac{1{,}946}{5{,}074}$
                 $n \approx 0.38352 = 38.4\%$
                 The percent increase in the cost
                 to attend public college is
                 38.4%.

11. Strategy   To find the percent increase:
→Find the increase in the applicants for Teach for America.
→Use the basic percent equation. Percent = n, base = 35,120, amount = amount of increase

Solution $46,359 - 35,120 = 11,239$
Percent · base = amount
$n \cdot 35,120 = 11,239$
$n = \dfrac{11,239}{35,120}$
$n \approx 0.32001 \approx 32.0\%$
The percent increase in Teach for America applicants is 32.0%

13. Yes. The process described would yield
$\$12 \cdot 0.10 + \$12 = \$1.20 + \$12 = \$13.20$.
Likewise, $\$12 * 1.10 = 13.20$.

Objective B Exercises

15. Strategy   To find percent decrease:
→Find the decrease in the passports.
→Use the basic percent equation. Percent = n, base = 18.3, amount = amount of increase

Solution $18.3 - 13.8 = 4.5$
Percent · base = amount
$n \cdot 18.3 = 4.5$
$n = \dfrac{4.5}{18.3}$
$n \approx 0.2459 \approx 24.6\%$
The percent increase in the passports is 24.6%.

17. Strategy   To find the percent decrease in size, use the basic percent equation.
Percent = n, base = 3.3, amount = the decrease in cleaning time

Solution $3.3 - 2.6 = 0.7$
Percent · base = amount
$n \cdot 3.3 = 0.7$
$n = \dfrac{0.7}{3.3}$
$n \approx 0.21212 \approx 21.2\%$
The size of the average U.S. household decreased by 21.2%.

19. Strategy   To find the percent decrease, use the basic percent equation.
Percent = n, base = 148,014, amount = the decrease in cleaning time

Solution $148,014 - 137,444 = 10,570$
Percent · base = amount
$n \cdot 148,014 = 10,570$
$n = \dfrac{10,570}{148,014}$
$n \approx 0.07141 \approx 7.1\%$
The number of people who took the LSAT in the past three years decreased by 7.1%.

21. Strategy   To find the value after 1 year:
→Find the decrease in value by using the basic percent equation.
Percent = 30% = 0.30, base = 21,900, amount = n
→Subtract the decrease in value from 21,900.

Solution   Percent · base = amount
$0.30 \cdot 21,900 = n$
$6,570 = n$
$21,900 - 6,570 = 15,330$
The value of the car after 1 year is $15,330.

23. Strategy   To find the branch in which the percent decrease was the greatest, find the percent decrease for all military branches and compare. Note that the Marines increased, so a percent decrease cannot be computed.

   Solution $732 - 553 = 179$
   Percent · base = amount
   $n \cdot 732 = 179$
   $n = \dfrac{197}{732}$
   $n \approx 0.2445 = 24.5\%$ Army

   $579 - 330 = 249$
   Percent · base = amount
   $n \cdot 579 = 249$
   $n = \dfrac{249}{579}$
   $n \approx 0.43005 = 43.0\%$ Navy

   $535 - 335 = 200$
   Percent · base = amount
   $n \cdot 535 = 200$
   $n = \dfrac{200}{535}$
   $n \approx 0.3738 = 37.4\%$ Air Force

   $43.0\% > 37.4\% > 24.5\%$
   The Navy had the greatest decrease of personnel at 43.0%.

Critical Thinking 8.3

25. Answers will vary

27. Answers will vary

29. Answers will vary

Projects and Group Activities 8.3

31. The order in which the coupons are applied to the purchase will not affect the purchase price. By multiplying the percents times the purchase price, and by the Commutative and Associative Properties of Multiplication, we can multiply factors in any order and the product will be the same.

Check Your Progress: Chapter 8

1. $38.7\% = 38.7(0.01) = 0.387$

2. $44\% = 44\left(\dfrac{1}{100}\right) = \dfrac{44}{100} = \dfrac{11}{25}$

3. $5\dfrac{1}{3}\% = 5\dfrac{1}{3}\left(\dfrac{1}{100}\right) = \dfrac{16}{3}\left(\dfrac{1}{100}\right) = \dfrac{16}{300} = \dfrac{4}{75}$

4. $0.725 = 0.725(100\%) = 72.5\%$

5. $\dfrac{7}{12} = \dfrac{7}{12}(100\%) = \dfrac{700}{12}\% = \dfrac{175}{3}\% = 58\dfrac{1}{3}\%$

6. $\dfrac{9}{16} = 0.5625(100\%) = 56.25\% \approx 56.3\%$

7. **Strategy**   To find the amount, solve the basic percent equation.
Percent $= 21\% = 0.21$,
base $= 50$, amount $= n$

    Solution   Percent $\cdot$ base $=$ amount
$0.21 \cdot 50 = n$
$10.5 = n$
21% of 50 is 10.5.

8. **Strategy**   To find the percent, solve the basic percent equation.
Percent $= n$, base $= 80$,
amount $= 50$

    Solution   Percent $\cdot$ base $=$ amount
$n \cdot 80 = 50$
$n = \dfrac{50}{80} = 0.625$
$n = 62.5\%$
50 is 62.5% of 80.

9. **Strategy**   To find the base, solve the basic percent equation.
Percent $= 15\% = 0.15$,
base $= n$, amount $= 30$

    Solution   Percent $\cdot$ base $=$ amount
$0.15 \cdot n = 30$
$n = \dfrac{30}{0.15}$
$n = 200$
15% of 200 is 30.

10. Percent $= 18$, base $= 425$, amount $= n$
$\dfrac{\text{percent}}{100} = \dfrac{\text{amount}}{\text{base}}$
$\dfrac{18}{100} = \dfrac{n}{425}$
$18 \cdot 425 = 100 \cdot n$
$7{,}650 = 100n$
$76.5 = n$
18% of 425 is 76.5.

11. Percent $= n$, base $= 95$, amount $= 38$
$\dfrac{\text{percent}}{100} = \dfrac{\text{amount}}{\text{base}}$
$\dfrac{n}{100} = \dfrac{38}{95}$
$95 \cdot n = 100 \cdot 38$
$95n = 3800$

    $n = \dfrac{3800}{95}$
$n = 40$
38 is 40% of 95.

12. Percent $= 140$, base $= n$, amount $= 56$
$\dfrac{\text{percent}}{100} = \dfrac{\text{amount}}{\text{base}}$
$\dfrac{140}{100} = \dfrac{56}{n}$
$140 \cdot n = 100 \cdot 56$
$140n = 5600$
$n = \dfrac{5600}{140}$
$n = 40$
56 is 140% of 40.

13. **Strategy**   To find the percent, use the basic percent equation.
Percent $= n$, base $= 4500$, amount $= 180$

    Solution   Percent $\cdot$ base $=$ amount
$n \cdot 4500 = 180$
$n = \dfrac{180}{4500}$
$n = 0.04 = 4\%$
The sales tax is 4% of the purchase price.

14. Strategy   To find percent decrease:
→ Find the decrease in area.
→ Use the basic percent equation. Percent = n, base = 2.7, amount = amount of increase

Solution $2.7 - 1.9 = 0.8$
Percent · base = amount
$n \cdot 2.7 = 0.8$
$n = \dfrac{0.8}{2.7}$
$n \approx 0.2962 = 29.6\%$
The percent decrease in the area of the Artic ice cap is 29.6%.

15. Strategy   To find the percent, use the basic percent equation.
Percent = 54% = 0.54, base = 800,000, amount = n

Solution   Percent · base = amount
$0.54 \cdot 800{,}000 = n$
$432{,}000 = n$
$432,000 was reinvested in research and development.

16. Strategy   To find percent increase:
→ Find the increase in the consumption of bison.
→ Use the basic percent equation. Percent = n, base = 0.506 billion, amount = amount of increase

Solution $1.67 - 0.506 = 1.164$
Percent · base = amount
$n \cdot 0.506 = 1.164$
$n = \dfrac{1.164}{0.506}$
$n \approx 2.3003 \approx 230.0\%$
The percent increase in Netflix's revenue was 230.0%.

17. Strategy   To find the purchase price, use the basic percent equation.
Percent = 15% = 0.15, base = n, amount = 4110

Solution   Percent · base = amount
$0.15 \cdot n = 4110$
$n = \dfrac{4110}{0.15}$
$n = 27{,}400$
The purchase price of the car is $27,400.

Section 8.4

Concept Check

1. pays for a product

3. selling price; cost; cost

5. an amount of money; a percent

Objective A Exercises

7. a. $M = P - C$

b. $M = P - C$; $M = r \cdot C$

c. $M = r \cdot C$; $P = C + M$

9. Strategy   To find the markup, solve the formula $M = r \cdot C$ for M.
$r = 30\%$, $C = 315$

Solution $M = r \cdot C$
$M = 0.30 \cdot 315$
$M = 94.50$
The markup on the cost is $94.50.

11. Strategy   To find the markup rate:
→ Solve the formula $M = S - C$ for M.
$S = 630$, $C = 360$
→ Solve the formula $M = r \cdot C$ for r.

Solution $M = S - C$
$M = 630 - 360$
$M = 270$
$M = r \cdot C$
$270 = r \cdot 360$
$\dfrac{270}{360} = r$
$0.75 = r$
The markup rate on the cost is 75%.

13. Strategy   To find the markup rate:
→Solve the formula
$M = S - C$ for M.
$S = 479, C = 320$
→Solve the formula
$M = r \cdot C$ for r.

Solution
$M = S - C$
$M = 479 - 320$
$M = 159$
$M = r \cdot C$
$159 = r \cdot 320$
$\dfrac{159}{320} = r$
$0.4969 \approx r$
The markup rate on the cost is approximately 49.7%.

15. Strategy   To find the selling price, solve the formula
$S = (1 + r)$ for S.
$r = 125\%, C = 7.60$

Solution
$S = (1 + r)C$
$S = (1 + 1.25) \cdot 7.60$
$S = 2.25 \cdot 7.60$
$S = 17.10$
The selling price of the strawberries is \$17.10.

17. Strategy   To find the selling price, solve the formula
$S = (1 + r)$ for S.
$r = 58\%, C = 225$

Solution
$S = (1 + r)C$
$S = (1 + 0.58) \cdot 225$
$S = 1.58 \cdot 225$
$S = 355.50$
The selling price of the leather jacket is \$355.50.

Objective B Exercises

19. Strategy   To find the markdown, solve the formula
$D = R - S$ for D.
$R = 460, S = 350$

Solution
$D = R - S$
$D = 460 - 350$
$D = 110$
The markdown is \$110.

21. Strategy   To find the markdown rate:
→Solve the formula
$D = R - S$ for D.
$R = 1,295, S = 995$
→Solve the formula
$D = r \cdot R$ for r.

Solution
$D = R - S$
$D = 1,295 - 995$
$D = 300$
$D = r \cdot R$
$300 = r \cdot 1,295$
$\dfrac{300}{1,295} = r$
$0.2317 \approx r$
The discount rate is approximately 23.2%.

23. Strategy   To find the discount rate:
→Solve the formula
$D = R - S$ for D.
$R = 162.50, S = 94.25$
→Solve the formula
$D = r \cdot R$ for r.

Solution
$D = R - S$
$D = 162.50 - 94.25$
$D = 68.25$
$D = r \cdot R$
$68.25 = r \cdot 162.50$
$\dfrac{68.25}{162.50} = r$
$0.42 = r$
The discount rate is 42%.

25. Strategy   To find the sale price, solve the formula
$S = (1 - r)R$ for S.
$r = 30\%, R = 1,995$

Solution
$S = (1 - r)R$
$S = (1 - 0.30) \cdot 1,995$
$S = 0.70 \cdot 1,995$
$S = 1,396.50$
The sale price is \$1,396.50

27. Strategy   To find the sale price, solve the formula
$S = (1 - r)R$ for S.
$r = 40\%, R = 42$

Solution
$S = (1 - r)R$
$S = (1 - 0.40) \cdot 42$
$S = 0.60 \cdot 42$
$S = 25.20$
The sale price is \$25.20.

29. Strategy   To find the regular price, solve the formula
$S(1-r)R$ for R.
$r = 40\%, S = 180$

Solution $S = (1-r)R$
$180 = (1 - 0.40) \cdot R$
$180 = 0.60R$
$\dfrac{180}{0.60} = R$
$300 = R$
The regular price is $300.

31. Strategy   To find the regular price, solve the formula
$S(1-r)R$ for R.
$r = 35\%, S = 80$

Solution $S = (1-r)R$
$80 = (1 - 0.35) \cdot R$
$80 = 0.65R$
$\dfrac{80}{0.65} = R$
$123.08 \approx R$
The regular price is approximately $123.08.

Critical Thinking 8.4

33. Suppose the regular price is $100. Then the sale price is $75 [$100 - 0.25(100) = 100 - 25 = 75$]. The promotional sale offers 25% off the sale price of $75. So the price is lowered to $56.25
[$75 - 0.25(75) = 75 - 18.75 = 56.25$].
A sale that offers 50% off the regular price of $100 offers the product for
$50 [$100 - 0.50(100) = 100 - 50 = 50$].
Since $50 < $56.25, the better price is the one that is 50% off the regular price.

Projects and Group Activities 8.4

35. Answers will vary

Section 8.5

Concept Check

1. Answers will vary

Objective A Exercises

3. $I = Prt$

1 month: $I = 5,000 \cdot 0.06 \cdot \dfrac{1}{12} = \$25$

2 month: $I = 5,000 \cdot 0.06 \cdot \dfrac{2}{12} = \$50$

3 month: $I = 5,000 \cdot 0.06 \cdot \dfrac{3}{12} = \$75$

4 month: $I = 5,000 \cdot 0.06 \cdot \dfrac{4}{12} = \$100$

5 month: $I = 5,000 \cdot 0.06 \cdot \dfrac{5}{12} = \$125$

5. To calculate the interest for a 7-month loan, multiply the interest due on a one month loan by 7.

7. Strategy   To find the simple interest, solve the simple interest formula $I = Prt$ for I.
$P = 7,500, \; t = \dfrac{75}{365},$
$r = 0.045$

Solution $I = Prt$
$I = (7,500)(0.045)\left(\dfrac{75}{365}\right)$
$I \approx 69.349$
The interest on the loan is $69.35.

9. Strategy   To find the simple interest, solve the simple interest formula $I = Prt$ for I.
$P = 50,000, \; t = \dfrac{8}{12},$
$r = 0.035$

Solution $I = Prt$
$I = (50,000)(0.035)\left(\dfrac{8}{12}\right)$
$I \approx 1,166.666$
The interest on the loan is $1,166.67.

11. Strategy   To find the simple interest, solve the simple interest formula $I = Prt$ for I.
$P = 800$,
$rt = 0.02$ (rate per month)

Solution $I = Prt$
$I = (800)(0.02)$
$I = 16$
The interest owed to the credit union is $16.

13. Strategy   To find the simple interest, solve the interest formula $I = Prt$ for I.
$P = 1{,}500$, $t = 1\dfrac{1}{2}$,
$r = 0.075$

Solution $I = Prt$
$I = (1{,}500)(0.075)\left(\dfrac{3}{2}\right)$
$I = 168.75$
The interest on the loan is $168.75.

15. Strategy   To find the maturity value:
→Solve the formula $I = Prt$ for I.
$P = 25{,}000$, $r = 0.082$, $t = 1$
→Use the formula for the maturity value of a simple interest loan, $M = P + I$.

Solution $I = Prt$
$I = 25{,}000(0.082)(1)$
$I = 2{,}050$
$M = P + I$
$M = 25{,}000 + 2{,}050$
$M = 27{,}050$
The maturity value is $27,050.

17. Strategy   To find the maturity value:
→Solve the formula $I = Prt$ for I.
$P = 5{,}000$, $r = 0.069$, $t = \dfrac{18}{12}$.
→Use the formula for the maturity value of a simple interest loan, $M = P + I$.

Solution $I = Prt$
$I = (5{,}000)(0.069)\left(\dfrac{18}{12}\right)$
$I = 517.50$
$M = P + I$
$M = 5{,}000 + 517.50$
$M = 5{,}517.50$
The maturity value is $5,517.50

19. Strategy   To find the interest rate, solve the formula $I = Prt$ for r.
$I = 168.75$, $P = 3{,}000$, $t = \dfrac{9}{12}$

Solution $I = Prt$
$168.75 = (3{,}000)(r)\left(\dfrac{9}{12}\right)$
$168.75 = 2{,}250r$
$\dfrac{168.75}{2{,}250} = r$
$0.075 = r$
The simple interest rate is 7.5%.

21. Strategy   To find the rate, solve the formula $I = Prt$ for r.
$I = 604.80$, $P = 18{,}000$, $t = \dfrac{210}{365}$

Solution $I = Prt$
$604.80 = 18{,}000(r)\left(\dfrac{210}{365}\right)$
$604.80 = \dfrac{756{,}000}{73}r$
$\dfrac{73 \cdot 604.80}{756{,}000} = r$
$0.0584 = r$
The interest rate is 5.84%.

Critical Thinking 8.5

23. a. At 8.75%,
$I = Prt$

$I = (10{,}000)(0.0875)\left(\dfrac{6}{12}\right)$

$I = 437.5$
$M = P + I$
$M = 10{,}000 + 437.5$
$M = 10{,}437.5$
The maturity value is $10,437.50

At 6.25%,
$I = Prt$

$I = (10{,}000)(0.0625)\left(\dfrac{6}{12}\right)$

$I = 312.5$
$M = P + I$
$M = 10{,}000 + 312.5$
$M = 10{,}312.50$
The maturity value is $10,312.50

The difference between the maturity values
is $10,437.50 - $10,312.50 = $125.

b. The difference in the monthly payments
would be the difference in the maturity
values divided by the number of months:
$\dfrac{125}{6} \approx 20.8333$

The difference between the monthly
payments would be $20.83.

Chapter Review Exercises

1. $32\% = 32\left(\dfrac{1}{100}\right) = \dfrac{32}{100} = \dfrac{8}{25}$

2. $22\% = 22(0.01) = 0.22$

3. $25\% = 25\left(\dfrac{1}{100}\right) = \dfrac{25}{100} = \dfrac{1}{4}$
$25\% = 25(0.01) = 0.25$

4. $3\dfrac{2}{5}\% = 3\dfrac{2}{5}\left(\dfrac{1}{100}\right) = \dfrac{17}{5}\left(\dfrac{1}{100}\right) = \dfrac{17}{500}$

5. $\dfrac{7}{40} = \dfrac{7}{40}(100\%) = 17.5\%$

6. $1\dfrac{2}{7} = 1\dfrac{2}{7}(100\%) = \dfrac{9}{7}(100\%) = \dfrac{900}{7}\%$
$\approx 128.6\%$

7. $2.8 = 2.8(100\%) = 280\%$

8. Strategy   To find the amount, solve the
basic percent equation.
Percent = 42% = 0.42,
base = 50, amount = n

Solution   Percent · base = amount
$0.42 \cdot 50 = n$
$21 = n$
42% of 50 is 21.

9. Strategy   To find the percent, solve the
basic percent equation.
Percent = n, base = 3,
amount = 15

Solution   Percent · base = amount
$n \cdot 3 = 15$
$n = \dfrac{15}{3}$
$n = 5 = 500\%$
500% of 3 is 15.

10. Strategy   To find the percent, solve the
basic percent equation.
Percent = n, base = 18,
amount = 12

Solution   Percent · base = amount
$n \cdot 18 = 12$
$n = \dfrac{12}{18}$
$n \approx 0.667 = 66.7\%$
12 is approximately 66.7% of
18.

11. Strategy   To find the amount, solve the
basic percent equation.
Percent = 150% = 1.50,
base = 20, amount = n

Solution   Percent · base = amount
$1.50 \cdot 20 = n$
$30 = n$
150% of 20 is 30.

12. Strategy — To find the amount, solve the basic percent equation.
Percent = 18% = 0.18, base = 85, amount = n

Solution — Percent · base = amount
$0.18 \cdot 85 = n$
$15.3 = n$
18% of 85 is 15.3.

13. Strategy — To find the base, solve the basic percent equation.
Percent = 32% = 0.32, base = n, amount = 180

Solution — Percent · base = amount
$0.32 \cdot n = 180$
$n = \dfrac{180}{0.32}$
$n = 562.5$
32% of 562.5 is 180.

14. Strategy — To find the percent, solve the basic percent equation.
Percent = n, base = 80, amount = 4.5

Solution — Percent · base = amount
$n \cdot 80 = 4.5$
$n = \dfrac{4.5}{80}$
$n = 0.05625 = 5.625\%$
4.5 is 5.625% of 80.

15. Strategy — To find the amount, solve the basic percent equation.
Percent = 0.58% = 0.0058, base = 2.54, amount = n

Solution — Percent · base = amount
$0.0058 \cdot 2.54 = n$
$0.014732 = n$
0.58% of 2.54 is 0.014732.

16. Strategy — To find the base, solve the basic percent equation.
Percent = 0.05% = 0.0005, base = n, amount = 0.0048

Solution — Percent · base = amount
$0.0005 \cdot n = 0.0048$
$n = \dfrac{0.0048}{0.0005}$
$n = 9.6$
0.0048 is 0.05% of 9.6.

17. Strategy — To find the percent visiting in China:
→Add to find the total amount of tourists visiting the four countries.
→Use the basic percent equation to find the percent.
Percent = n, base = the total amount of tourists visiting the four countries, amount = 137 million

Solution — $137 + 93 + 71 + 102 = 403$
Percent · base = amount
$n \cdot 403 = 137$
$n = \dfrac{137}{403}$
$n \approx 0.340 = 34.0\%$
About 34.0% of the tourists will be visiting China.

18. Strategy — To find the amount of the budget spent for advertising, use the basic percent equation.
Percent = 7% = 0.07, base = 120,000, amount = n

Solution — Percent · base = amount
$0.07 \cdot 120{,}000 = n$
$8{,}400 = n$
$8,400 of the budget was spent for advertising.

19. Strategy   To find the number of phones that were not defective:
→Find the number of defective phones by using the basic percent equation. Percent = 1.2% = 0.012, base = 4,000, amount = n
→Subtract the number of defective phones from 4,000.

Solution   Percent · base = amount
$0.012 \cdot 4,000 = n$
$48 = n$
$4,000 - 48 = 3,952$
3,952 of the telephones were not defective.

20. Strategy   To find the percent of the week, use the basic percent equation.
Percent = n, base = 168, amount = 61.35.

Solution   Percent · base = amount
$n \cdot 168 = 61.35$
$n = \dfrac{61.35}{168}$
$n \approx 0.365 = 36.5\%$
The percent is approximately 36.5% of the week.

21. Strategy   To find the expected profit, use the basic percent equation.
Percent = 22% = 0.22, base = 750,000, amount = n

Solution   Percent · base = amount
$0.22 \cdot 750,000 = n$
$165,000 = n$
The expected profit is $165,000.

22. Strategy   To find the number of seats added, use the basic percent equation.
Percent = 18% = 0.18, base = 9,000, amount = n

Solution   Percent · base = amount
$0.18 \cdot 9,000 = n$
$1,620 = n$
1,620 seats were added to the auditorium.

23. Strategy   To find the number of tickets sold:
→Use the basic percent equation to find the number of seats overbooked. Percent = 12% = 0.12, base = 175, amount = n
→Add the number of seats overbooked to 175.

Solution   Percent · base = amount
$0.12 \cdot 175 = n$
$21 = n$
$21 + 175 = 196$
The airline would sell 196 tickets.

24. Strategy   To find the percent of registered voters that voted, use the basic percent equation.
Percent = n, based = 112,000, amount = 25,400

Solution   Percent · base = amount
$n \cdot 112,000 = 25,400$
$n = \dfrac{25,400}{112,000}$
$n \approx 0.227 = 22.7\%$
Approximately 22.7% of the registered voters voted in the city election.

25. Strategy   To find the clerk's new hourly wage:
→Find the increase by using the basic percent equation. Percent = 8% = 0.08, base = 10.50, amount = n
→Add the increase to 10.50.

Solution   Percent · base = amount
$0.08 \cdot 10.50 = n$
$0.84 = n$
$0.84 + 10.50 = 11.34$
The clerk's new wage is $11.34 per hour.

26. Strategy To find the percent decrease in cost:
→Subtract to find the dollar decrease in the cost of the computer.
→Find the percent decrease by using the basic percent equation.
Percent = n, base = 2,400, amount = decrease in price

Solution 2,400 − 1,800 = 600
Percent · base = amount
$n \cdot 2{,}400 = 600$
$n = \dfrac{600}{2{,}400}$
$n = 0.25 = 25\%$
The computer decreased 25% in cost.

27. Strategy To find the selling price of the car:
→Use the basic percent equation to find the markup.
Percent = 6% = 0.06, base = 18,500, amount = n
→Add the markup to 18,500.

Solution Percent · base = amount
$0.06 \cdot 18{,}500 = n$
$1{,}110 = n$
1,110 + 18,500 = 19,610
The selling price of the car is $19,610.

28. Strategy To find the markup rate:
→Solve the formula
$M = S − C$ for M.
$S = 181.50, C = 110$
→Solve the formula
$M = r \cdot C$ for r.

Solution $M = S − C$
$M = 181.50 − 110$
$M = 71.50$
$M = r \cdot C$
$71.50 = r \cdot 110$
$\dfrac{71.50}{110} = r$
$0.65 = r$
The markup rate of the parka is 65%.

29. Strategy To find the sale price, solve the formula
$S = (1 − r)R$ for S.
$r = 0.30, R = 80$

Solution $S = (1 − r)R$
$S = (1 − 0.30) \cdot 80$
$S = 0.70 \cdot 80$
$S = 56$
The sale price of the tennis racket is $56.

30. Strategy To find the sale price, solve the formula
$S = (1 − r)R$ for S.
$r = 0.40, R = 650$

Solution $S = (1 − r)R$
$S = (1 − 0.40) \cdot 650$
$S = 0.60 \cdot 650$
$S = 390$
The sale price of the ticket is $390.

31. Strategy To find the simple interest, solve the formula
$I = Prt$ for I.
$P = 3{,}000, r = 0.086, t = \dfrac{45}{365}$

Solution $I = Prt$
$I = 3{,}000(0.086)\left(\dfrac{45}{365}\right)$
$I = 31.81$
The interest on the loan is $31.81.

32. Strategy To find the rate, solve the formula
$I = Prt$ for r.
$I = 7{,}397.26, P = 500{,}000,$
$t = \dfrac{60}{365}$

Solution $I = Prt$
$7{,}397.26 = 500{,}000(r)\left(\dfrac{60}{365}\right)$
$7{,}397.26 = \dfrac{6{,}000{,}000}{73}r$
$\dfrac{73 \cdot 7{,}397.26}{6{,}000{,}000} = r$
$0.09 = r$
The interest rate is 9.00%.

33. Strategy   To find the maturity value:
→Solve the formula $I = Prt$ for I.
$P = 10,000, r = 0.084$,
$t = \dfrac{9}{12}$.
→Use the formula for the maturity value of a simple interest loan, $M = P + I$.

Solution $I = Prt$

$I = (10,000)(0.084)\left(\dfrac{9}{12}\right)$

$I = 640$
$M = P + I$
$M = 10,000 + 630$
$M = 10,630$
The maturity value is \$10,630.

Chapter Test

1. $86.4\% = 86.4(0.01) = 0.864$

2. $0.4 = 0.4(100\%) = 40\%$

3. $\dfrac{5}{4} = \dfrac{5}{4}(100\%) = 125\%$

4. $83\dfrac{1}{3}\% = 83\dfrac{1}{3}\left(\dfrac{1}{100}\right) = \dfrac{250}{3}\left(\dfrac{1}{100}\right) = \dfrac{5}{6}$

5. $32\% = 32\left(\dfrac{1}{100}\right) = \dfrac{32}{100} = \dfrac{8}{25}$

6. $1.18 = 1.18(100\%) = 118\%$

7. Strategy   To find the base, solve the basic percent equation.
Percent $= 20\% = 0.20$,
base $= n$, amount $= 18$

Solution   Percent $\cdot$ base $=$ amount
$0.20 \cdot n = 18$
$n = \dfrac{18}{0.20} = 90$
20% of 90 is 18.

8. Strategy   To find the amount, solve the basic percent equation.
Percent $= 68\% = 0.68$,
base $= 73$, amount $= n$

Solution   Percent $\cdot$ base $=$ amount
$0.68 \cdot 73 = n$
$49.64 = n$
68% of 73 is 49.64.

9. Strategy   To find the percent, solve the basic percent equation.
Percent $= n$, base $= 320$,
amount $= 180$

Solution   Percent $\cdot$ base $=$ amount
$n \cdot 320 = 180$
$n = \dfrac{180}{320}$
$n = 0.5625 = 56.25\%$
56.25% of 320 is 180.

10. Strategy   To find the base, solve the basic percent equation.
Percent $= 14\% = 0.14$,
base $= n$, amount $= 28$

Solution   Percent $\cdot$ base $=$ amount
$0.14 \cdot n = 28$
$n = \dfrac{28}{0.14} = 200$
14% of 200 is 28.

11. Strategy   To find the amount of expected accidents, use the basic percent equation.
Percent $= 2.2\% = 0.022$,
base $= 1,500$, amount $= n$

Solution   Percent $\cdot$ base $=$ amount
$0.022 \cdot 1,500 = n$
$33 = n$
33 accidents are expected.

12. Strategy   To find the percent answered correctly, use the basic percent equation.
Percent $= n$, base $= 90$,
amount $= 90 - 16 = 74$.

Solution   Percent $\cdot$ base $=$ amount
$n \cdot 90 = 74$
$n = \dfrac{74}{90}$
$n \approx 0.822 = 82.2\%$
The percent is approximately 82.2% correct.

13. Strategy   To find the dollar increase:
→Find the increase by using the basic percent equation.
Percent = 120% = 1.20, base = n, amount = 480
→Subtract the increase from 480.

Solution   Percent · base = amount
$1.20 \cdot n = 480$

$n = \dfrac{480}{1.20}$

$n = 400$
$480 - 400 = 80$
The dollar increase is $80.

14. Strategy   To find the percent decrease in cost:
→Subtract to find the dollar increase in the cost for public tuition.
→Find the percent increase by using the basic percent equation.
Percent = n, base = 12,127, amount = increase in price

Solution $29,026 - 12,127 = 16,899$
Percent · base = amount
$n \cdot 12,127 = 16,899$

$n = \dfrac{16,899}{12,127}$

$n \approx 1.394 = 139.4\%$
The tuition increased 139.4%.

15. Strategy   To find the percent decrease in cost:
→Subtract to find the increase in trainees.
→Find the percent increase by using the basic percent equation.
Percent = n, base = 36, amount = increase in trainees

Solution $42 - 36 = 6$
Percent · base = amount
$n \cdot 36 = 6$

$n = \dfrac{6}{36}$

$n \approx 0.1667 = 16\dfrac{2}{3}\%$

The number of trainees increased by approximately $16\dfrac{2}{3}\%$.

16. Strategy

a. To find the percent decrease in fat:
→Subtract to find the decrease in fat.
→Find the percent decrease by using the basic percent equation.
Percent = n, base = 24, amount = decrease in fat.

b. To find the percent decrease in cholesterol:
→Subtract to find the decrease in cholesterol.
→Find the percent decrease by using the basic percent equation.
Percent = n, base = 75, amount = decrease in cholesterol.

c. To find the percent decrease in calories:
→Subtract to find the decrease in calories.
→Find the percent decrease by using the basic percent equation.
Percent = n, base = 280, amount = decrease in calories.

Solution

a. $24 - 4 = 20$
Percent · base = amount
$n \cdot 24 = 20$
$n = \dfrac{20}{24}$
$n \approx 0.8333 = 83\dfrac{1}{3}\%$

The fat content decreased by $83\dfrac{1}{3}\%$.

b. $75 - 0 = 75$
Percent · base = amount
$n \cdot 75 = 75$
$n = \dfrac{75}{75}$
$n = 1 = 100\%$
The cholesterol percent decreased by 100%.

c. $280 - 140 = 140$
Percent · base = amount
$n \cdot 280 = 140$
$n = \dfrac{140}{280}$
$n = 0.5 = 50\%$
The calorie percent decreased by 50%.

17. Strategy To find the decrease in personnel:
→Find the decrease in travel expenses.
→Use the basic percent equation to find the percent decrease in expenses. Percent = n, base = 25,000, amount = decrease in expenses.

Solution $25{,}000 - 23{,}000 = 2{,}000$
Percent · base = amount
$n \cdot 25{,}000 = 2{,}000$
$n = \dfrac{2{,}000}{25{,}000}$
$n = 0.08 = 8\%$
The amount of travel expenses decreased by 8%.

18. Strategy To find the dollar increase:
→Find the price from last year by using the basic percent equation.
Percent = 125% = 1.25, base = n, amount = 1,500
→Subtract last year's price from 1,500.

Solution Percent · base = amount
$1.25n = 1{,}500$
$n = \dfrac{1{,}500}{1.25} = 1{,}200$
$1{,}500 - 1{,}200 = 300$
The dollar increase is $300.

19. Strategy To find the markup, solve the formula $M = r \cdot C$ for M.
$r = 60\%, C = 21$

Solution $M = r \cdot C$
$M = 0.60 \cdot 21$
$M = 12.60$
The markup on the cost is $12.60.

20. Strategy To find the markup rate:
→Solve the formula
$M = S - C$ for M.
$S = 349, C = 225$
→Solve the formula
$M = r \cdot C$ for r.

Solution $M = S - C$
$M = 349 - 225$
$M = 124$
$M = r \cdot C$
$124 = r \cdot 225$
$\dfrac{124}{225} = r$
$0.551 \approx r$
The markup rate on the cost is approximately 55.1%.

21. Strategy To find the regular price, solve the formula
$S(1 - r)R$ for R.
$r = 40\%, S = 180$

Solution $S = (1 - r)R$
$180 = (1 - 0.40) \cdot R$
$180 = 0.6R$
$\dfrac{180}{0.6} = R$
$300 = R$
The regular price is $300.

22. Strategy To find the discount rate:
→Solve the formula
$D = r \cdot R$ for r, $D = 51.80$,
$R = 370$.

Solution $D = r \cdot R$
$51.80 = r \cdot 370$
$\dfrac{51.80}{370} = r$
$0.14 = r$
The discount rate is 14%.

23. Strategy To find the simple interest, solve the interest formula
$I = Prt$ for I.
$P = 5,000$, $t = \dfrac{9}{12}$,
$r = 0.054$

Solution $I = Prt$
$I = (5,000)(0.054)\left(\dfrac{9}{12}\right)$
$I = 202.5$
The interest on the loan is $202.50.

24. Strategy To find the maturity value:
→Solve the formula $I = Prt$ for I.
$P = 40,000, r = 0.0925$,
$t = \dfrac{150}{365}$.
→Use the formula for the maturity value of a simple interest loan, $M = P + I$.

Solution $I = Prt$
$I = (40,000)(0.0925)\left(\dfrac{150}{365}\right)$
$I = 1,520.55$
$M = P + I$
$M = 40,000 + 1,520.55$
$M = 41,520.55$
The maturity value is $41,520.55.

25. Strategy To find the rate, solve the formula $I = Prt$ for r.
$I = 672, P = 12,000$, $t = \dfrac{8}{12}$

Solution $I = Prt$
$672 = 12,000(r)\left(\dfrac{8}{12}\right)$
$672 = 8,000r$
$\dfrac{672}{8,000} = r$
$0.084 = r$
The interest rate is 8.4%.

Cumulative Review Exercises

1. $a - b$
$102.5 - 77.546 = 24.954$

2. $5^4 = 5 \cdot 5 \cdot 5 \cdot 5$
$= 625$

3. $(4.67)(3.007) = 14.04269$

4. $(2x - 3)(2x - 5) = 4x^2 - 10x - 6x + 15$
$= 4x^2 - 16x + 15$

5. $3\dfrac{5}{8} \div 2\dfrac{7}{12} = \dfrac{29}{8} \div \dfrac{31}{12}$

$\phantom{3\dfrac{5}{8} \div 2\dfrac{7}{12}} = \dfrac{29}{8} \cdot \dfrac{12}{31}$

$\phantom{3\dfrac{5}{8} \div 2\dfrac{7}{12}} = \dfrac{29 \cdot 12}{8 \cdot 31}$

$\phantom{3\dfrac{5}{8} \div 2\dfrac{7}{12}} = \dfrac{29 \cdot 2 \cdot 2 \cdot 3}{2 \cdot 2 \cdot 2 \cdot 31}$

$\phantom{3\dfrac{5}{8} \div 2\dfrac{7}{12}} = \dfrac{87}{62} = 1\dfrac{25}{62}$

6. $-2a^2b\left(-3ab^2 + 4a^2b^3 - ab^3\right)$

$ = 6a^3b^3 - 8a^4b^4 + 2a^3b^4$

7. Strategy   To find the amount, use the basic percent equation.
Percent = 120% = 1.20, base = 35, amount = n

   Solution   Percent · base = amount
   $1.20 \cdot 35 = n$
   $42 = n$
   120% of 35 is 42.

8. $x - 2 = -5$
$x - 2 + 2 = -5 + 2$
$x = -3$
The solution is -3.

9. $1.005 \times 10^5 = 100,500$

10. $-\dfrac{5}{8} - \left(-\dfrac{3}{4}\right) + \dfrac{5}{6} = -\dfrac{5}{8} + \dfrac{3}{4} + \dfrac{5}{6}$

$\phantom{-\dfrac{5}{8} - \left(-\dfrac{3}{4}\right)} = \dfrac{-15}{24} + \dfrac{18}{24} + \dfrac{20}{24}$

$\phantom{-\dfrac{5}{8} - \left(-\dfrac{3}{4}\right)} = \dfrac{-15 + 18 + 20}{24}$

$\phantom{-\dfrac{5}{8} - \left(-\dfrac{3}{4}\right)} = \dfrac{23}{24}$

11. $\dfrac{3 - \dfrac{7}{8}}{\dfrac{11}{12} + \dfrac{1}{4}} = \dfrac{\dfrac{24}{8} - \dfrac{7}{8}}{\dfrac{11}{12} + \dfrac{3}{12}} = \dfrac{\dfrac{17}{8}}{\dfrac{14}{12}} = \dfrac{\dfrac{17}{8}}{\dfrac{7}{6}}$

$\phantom{\dfrac{3 - \dfrac{7}{8}}{\dfrac{11}{12} + \dfrac{1}{4}}} = \dfrac{17}{8} \div \dfrac{7}{6}$

$\phantom{\dfrac{3 - \dfrac{7}{8}}{\dfrac{11}{12} + \dfrac{1}{4}}} = \dfrac{17}{8} \cdot \dfrac{6}{7}$

$\phantom{\dfrac{3 - \dfrac{7}{8}}{\dfrac{11}{12} + \dfrac{1}{4}}} = \dfrac{17 \cdot 6}{8 \cdot 7} = \dfrac{51}{28}$

$\phantom{\dfrac{3 - \dfrac{7}{8}}{\dfrac{11}{12} + \dfrac{1}{4}}} = 1\dfrac{23}{28}$

12. $\left(-3a^2b\right)\left(4a^5b^4\right) = (-3 \cdot 4)\left(a^2 a^5\right)\left(b b^4\right)$

$ = -12a^7b^5$

13.

x	y
2	1
1	3
0	5

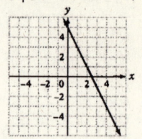

14.

x	y
3	3
0	-2
-3	-7

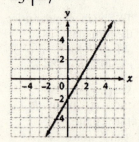

15. $\dfrac{7}{8} \div \dfrac{5}{16} = \dfrac{7}{8} \cdot \dfrac{16}{5}$

$\phantom{\dfrac{7}{8} \div \dfrac{5}{16}} = \dfrac{7 \cdot 16}{8 \cdot 5}$

$\phantom{\dfrac{7}{8} \div \dfrac{5}{16}} = \dfrac{7 \cdot 2 \cdot 2 \cdot 2 \cdot 2}{2 \cdot 2 \cdot 2 \cdot 5}$

$\phantom{\dfrac{7}{8} \div \dfrac{5}{16}} = \dfrac{14}{5} = 2\dfrac{4}{5}$

16. $4 - (-3) + 5 - 8 = 4 + 3 + 5 + (-8)$

$ = 7 + 5 + (-8)$

$ = 12 + (-8)$

$ = 4$

17. $\dfrac{3}{4}x = -9$

$\dfrac{4}{3}\left(\dfrac{3}{4}x\right) = \dfrac{4}{3}(-9)$

$x = -12$
The solution is -12.

18. $6x - 9 = -3x + 36$
$6x + 3x - 9 = -3x + 3x + 36$
$9x - 9 = 36$
$9x - 9 + 9 = 36 + 9$
$9x = 45$
$\dfrac{9x}{9} = \dfrac{45}{9}$
$x = 5$
The solution is 5.

19. $\dfrac{322.4 \text{ mi}}{5 \text{ h}} = 64.48 \text{ mph}$

20. $\dfrac{32}{n} = \dfrac{5}{7}$
$32 \cdot 7 = n \cdot 5$
$224 = 5n$
$\dfrac{224}{5} = \dfrac{5n}{5}$
$44.8 = n$

21. Strategy   To find the percent, use the basic percent equation. Percent = n, base = 30, amount = 2.5

Solution   Percent · base = amount
$n \cdot 30 = 2.5$
$n = \dfrac{2.5}{30} \approx 0.0833$
2.5 is approximately 8.3% of 30.

22. Strategy   To find the amount, use the basic percent equation. Percent = 42% = 0.42, base = 160, amount = n

Solution   Percent · base = amount
$0.42 \cdot 160 = n$
$67.2 = n$
42% of 160 is 67.2.

23. $44 - (-6)^2 \div (-3) + 2 = 44 - 36 \div (-3) + 2$
$= 44 - (-12) + 2$
$= 44 + 12 + 2$
$= 56 + 2$
$= 58$

24. $3(x - 2) + 2 = 11$
$3x - 6 + 2 = 11$
$3x - 4 = 11$
$3x - 4 + 4 = 11 + 4$
$3x = 15$
$\dfrac{3x}{3} = \dfrac{15}{3}$
$x = 5$
The solution is 5.

25. Strategy   To find the fraction, convert 10% to a fraction.

Solution $10\% = 10\left(\dfrac{1}{100}\right) = \dfrac{10}{100} = \dfrac{1}{10}$
$\dfrac{1}{10}$ of the population aged 75–84 are affected by Alzheimer's Disease.

26. Strategy   To find the sale price, solve the formula $S = (1 - r)R$ for S. $r = 0.36$, $R = 202.50$

Solution $S = (1 - r)R$
$S = (1 - 0.36) \cdot 202.50$
$S = 0.64 \cdot 202.50$
$S = 129.60$
The sale price is $129.60.

27. Strategy   To find the cost of the calculator, solve the formula $S = (1 + r)C$ for C. $S = 67.20$, $r = 0.60$

Solution $S = (1 + r)C$
$67.20 = (1 + 0.60)C$
$67.20 = 1.60C$
$\dfrac{67.20}{1.60} = C$
$42 = C$
The cost of the calculator is $42.

28. Strategy   To find the number of games the team will win, write and solve a proportion using n to represent the number of games won.

Solution $\dfrac{13}{18} = \dfrac{n}{162}$
$13 \cdot 162 = 18 \cdot n$
$2,106 = 18n$
$\dfrac{2,106}{18} = n$
$117 = n$
The team will win 117 games.

29. Strategy To find the amount of weight to lose:
→Add the amount already lost $\left(3\frac{1}{2}+2\frac{1}{4}\right)$.
→Subtract the amount lost from 8.

Solution $3\frac{1}{2}+2\frac{1}{4}=3\frac{2}{4}+2\frac{1}{4}=5\frac{3}{4}$

$8-5\frac{3}{4}=7\frac{4}{4}-5\frac{3}{4}=2\frac{1}{4}$

The wrestler must lose another $2\frac{1}{4}$ lb.

30. Strategy To find the speed, substitute 81 for d in the given formula and solve for v.

Solution $v=\sqrt{64d}$
$v=\sqrt{64\cdot81}$
$v=\sqrt{64}\cdot\sqrt{81}$
$v=8\cdot9=72$
$v=72$
The speed of the falling object is 72 ft/s.

31. Strategy To convert meters per second to kilometers per hour, use the conversion factors $\frac{1\text{ h}}{3,600\text{ s}}$ and $\frac{1,000\text{ m}}{1\text{ km}}$.

Solution $\frac{130\text{ km}}{1\text{ h}}$
$=\frac{130\text{ km}}{1\text{ h}}\cdot\frac{1\text{ h}}{3,600\text{ s}}\cdot\frac{1,000\text{ m}}{1\text{ km}}$
≈36.11 m/s
The speed was approximately 36.11 m/s.

32. Strategy To find the number of hours worked:
→Subtract the cost of materials (192) from the total cost (1,632).
→Divide the cost of labor by 40.

Solution $1,632-192=1,440$
$\frac{1,440}{40}=36$
The plumber worked for 36 h.

33. Strategy To find the resistance:
→Write the basic inverse variation equations, replace the variables by the given values, and solve for k.
→Write the inverse variation equation, replacing k by its value. Substitute 8 for I and solve for R.

Solution $I=\frac{k}{R}$
$2=\frac{k}{20}$
$2\cdot20=k$
$40=k$
$I=\frac{40}{R}$
$8=\frac{40}{R}$
$8R=40$
$R=\frac{40}{8}=5$
The resistance is 5 ohms.

Chapter 9: Geometry

Prep Test

1. $2(18) + 2(10) = 36 + 20 = 56$

2. abc
$= (2)(3.14)(9)$
$= (6.28)(9)$
$= 56.52$

3. xyz^3
$= \left(\dfrac{4}{3}\right)(3.14)(3)^3$
$= 113.04$

4. $\quad x + 47 = 90$
$x + 47 - 47 = 90 - 47$
$\qquad\qquad x = 43$

5. $\quad 32 + 97 + x = 180$
$\qquad 129 + x = 180$
$129 - 129 + x = 180 - 129$
$\qquad\qquad x = 51$

6. $\dfrac{5}{12} = \dfrac{6}{x}$
$5x = 12 \times 6$
$\dfrac{5x}{5} = \dfrac{36}{5}$
$\quad x = 14.4$

Section 9.1

Concept Check

1. $12 = 5 + x + 4$

3. $x + 160° + 140° = 360°$

5. $a; b$

7. $c; d; 180°$

9. a. $\angle a, \angle b,$ and $\angle c$

b. $\angle y$ and $\angle z$

c. $\angle x$

Objective A Exercises

11. The measure of the given angle is approximately 40°. The measure of the angle is between 0° and 90°. The angle is an acute angle.

13. The measure of the given angle is approximately 115°. The measure of the angle is between 90° and 180°. The angle is an obtuse angle.

15. The measure of the given angle is approximately 90°. The angle is a right angle.

17. Strategy Complementary angle are two angles whose sum is 90°. To find the complement, let x represent the complement of a 62° angle. Write an equation and solve for x.

Solution $x + 62° = 90°$
$x = 28°$
The complement of a 62° angle is a 28° angle.

19. Strategy Supplementary angles are two angles whose sum is 180°. To find the supplement, let x represent the supplement of a 162° angle. Write an equation and solve for x.

Solution $x + 162° = 180°$
$x = 18°$
The supplement of a 162° angle is an 18° angle.

21. $AB + BC + CD = AD$
$12 + BC + 9 = 35$
$21 + BC = 35$
$BC = 14$
$BC = 14$ cm

23. $QR + RS = QS$
$QR + 3(QR) = QS$
$7 + 3 \cdot 7 = QS$
$7 + 21 = QS$
$28 = QS$
$QS = 28$ ft

25. $EF + FG = EG$
$EF + \dfrac{1}{2}(EF) = EG$
$20 + \dfrac{1}{2}(20) = EG$
$20 + 10 = EG$
$30 = EG$
$EG = 30$ m

27. $\angle LOM + \angle MON = \angle LON$
$53° + \angle MON = 139°$
$\angle MON = 139° - 53°$
$\angle MON = 86°$

29. Strategy   To find the measure of ∠x, write an equation using the fact that the sum of the measures of ∠x and ∠2x is 90°. Solve for ∠x.

Solution $x + 2x = 90°$
$3x = 90°$
$x = 30°$
The measure of x is 30°.

31. Strategy   To find the measure of ∠x, write an equation using the fact that the sum of x and $x + 18°$ is 90°. Solve for x.

Solution $x + x + 18° = 90°$
$2x + 18° = 90°$
$2x = 72°$
$x = 36°$
The measure of ∠x is 36°.

33. Strategy   To find the measure of ∠x, write an equation using the fact that the sum of the measure of ∠x and 74° is 145°. Solve for ∠x.

Solution $∠x + 74° = 145°$
$∠x = 71°$
The measure of ∠x is 71°

35. Strategy   To find the measure of ∠a, write an equation using the fact that the sum of the measure of ∠a and 53° is 180°. Solve for ∠a.

Solution $∠a + 53° = 180°$
$∠a = 127°$
The measure of ∠a is 127°.

37. Strategy   The sum of the measures of the three angles shown is 360°. To find ∠a, write an equation and solve for ∠a.

Solution $∠a + 76° + 168° = 360°$
$∠a + 244° = 360°$
$∠a = 116°$
The measure of ∠a is 116°.

39. Strategy   The sum of the measures of the three angles shown is 180°. To find x, write an equation and solve for x.

Solution $3x + 4x + 2x = 180°$
$9x = 180°$
$x = 20°$
The measure of x is 20°.

41. Strategy   The sum of the measures of the three angles shown is 180°. To find x, write an equation and solve for x.

Solution $5x + (x + 20°) + 2x = 180°$
$8x + 20° = 180°$
$8x = 160°$
$x = 20°$
The measure of x is 20°.

43. Strategy   The sum of the measures of the four angles shown is 360°. To find x, write an equation and solve for x.

Solution $3x + 4x + 6x + 5x = 360°$
$18x = 360°$
$x = 20°$
The measure of x is 20°.

45. Strategy

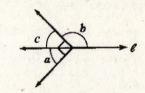

To find the measure of ∠b:
→Use the fact that ∠a and ∠c are complementary angles.
→Find ∠b by using the fact that ∠c and ∠b are supplementary angles.

Solution $∠a + ∠c = 90°$
$51° + ∠c = 90°$
$∠c = 39°$
$∠b + ∠c = 180°$
$∠b + 39° = 180°$
$∠b = 141°$
The measure of ∠b is 141°.

47. $90° - x$

Objective B Exercises

49. Strategy   The angles labeled are adjacent angles of intersecting lines and are, therefore, supplementary angles. To find x, write an equation and solve for x.

Solution $x + 74° = 180°$
$x = 106°$
The measure of x is $106°$.

51. Strategy   The angles labeled are vertical angles and are, therefore, equal. To find x, write an equation and solve for x.

Solution $5x = 3x + 22°$
$2x = 22°$
$x = 11°$
The measure of x is $11°$.

53. Strategy   →To find the measure of $\angle a$, use the fact that corresponding angles of parallel lines are equal.
→To find the measure of $\angle b$, use the fact that adjacent angles of intersecting lines are supplementary.

Solution $\angle a = 38°$
$\angle b + \angle a = 180°$
$\angle b + 38° = 180°$
$\angle b = 142°$
The measure of $\angle a$ is $38°$.
The measure of $\angle b$ is $142°$.

55. Strategy   →To find the measure of $\angle a$, use the fact that alternate interior angles of parallel lines are equal.
→To find the measure of $\angle b$, use the fact that adjacent angles of intersecting lines are supplementary.

Solution $\angle a = 47°$
$\angle a + \angle b = 180°$
$47° + \angle b = 180°$
$\angle b = 133°$
The measure of $\angle a$ is $47°$.
The measure of $\angle b$ is $133°$.

57. Strategy

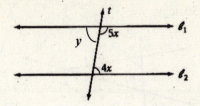

$4x = y$ because alternate interior angles have the same measure. $y + 5x = 180°$ because adjacent angles of intersecting lines are supplementary. Substitute $4x$ for y and solve for x.

Solution $4x + 5x = 180°$
$9x = 180°$
$x = 20°$
The measure of x is $20°$.

59. False

61. True

Objective C Exercises

63. Strategy

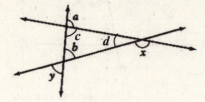

→To find the measure of
∠y, use the fact that
∠b and ∠y are vertical
angles.

→To find the measure of
∠x:
Find the measure of ∠c
by using the fact that the
sum of an interior and
exterior angle is 180°.
Find the measure of ∠d
by using the fact that the
sum of the interior angles
of a triangle is 180°.
Find the measure of ∠x,
by using the fact that the
sum of an interior and
exterior angle is 180°.

Solution $\angle y = \angle b = 70°$
$\angle a + \angle c = 180°$
$95° + \angle c = 180°$
$\angle c = 85°$
$\angle b + \angle c + \angle d = 180°$
$70° + 85° + \angle d = 180°$
$155° + \angle d = 180°$
$\angle d = 25°$
$\angle d + \angle x = 180°$
$25° + \angle x = 180°$
$\angle x = 155°$
The measure of ∠x is 155°.
The measure of ∠y is 70°.

65. Strategy

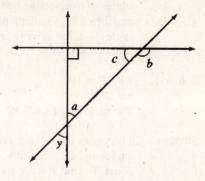

→To find the measure of
∠a, use the fact that
∠a and ∠y are vertical
angles.

→To find the measure of
∠b:
Find the measure of ∠c
by using the fact that the
sum of the interior angles
of a triangle is 180°.
Find the measure of ∠b
by using the fact that the
sum of an interior and
exterior angle is 180°.

Solution $\angle a = \angle y = 45°$
$\angle a + \angle c + 90° = 180°$
$45° + \angle c + 90° = 180°$
$\angle c + 135° = 180°$
$\angle c = 45°$
$\angle c + \angle b = 180°$
$45° + \angle b = 180°$
$\angle b = 135°$
The measure of ∠a is 45°.
The measure of ∠b is 135°.

67. Strategy   To find the measure of the third angle, use the fact that the sum of the measures of the interior angles of a triangle is 180°. Write an equation using x to represent the measure of the third angle. Solve the equation for x.

Solution $x + 90° + 30° = 180°$
$x + 120° = 180°$
$x = 60°$
The measure of the third angle is 60°.

69. Strategy   To find the measure of the third angle, use the fact that the sum of the measures of the interior angles of a triangle is 180°. Write an equation using x to represent the measure of the third angle. Solve the equation for x.

Solution $x + 42° + 103° = 180°$
$x + 145° = 180°$
$x = 35°$
The measure of the third angle is 35°.

71. True

Critical Thinking 9.1

73. Putting the numbers in order, we have −2.5, 2, 3.5 and 5. We can see that 2 is not halfway between −2.5 and 3.5, but 3.5 is halfway between 2 and 5.

75. Let x be the measure of an angle. If the ratio of the angles of a triangle is 2 : 3 : 7, and the sum of the angles of a triangle is 180°, then we need to solve the equation
$2x + 3x + 7x = 180$
$12x = 180$
$x = \dfrac{180}{12} = 15$
So the measure of the largest angle would be $7x = 7(15) = 105°$.

Projects or Group Activities 9.1

77. The sum of the interior angles of a triangle is 180°; therefore, $\angle a + \angle b + \angle c = 180°$. The sum of an interior and exterior angle is 180°; therefore, $\angle c + \angle x = 180°$. Solving this equation for $\angle c$,
$\angle c = 180° - \angle x$.
Substitute $180° - \angle x$ for $/c$ in the equation $\angle a + \angle b + \angle c = 180°$;
$\angle a + \angle b + 180° - \angle x = 180°$.
Add the measure of x to each side of the equation, and subtract 180° from each side of the equation: $\angle a + \angle b = \angle x$.
The measure of an exterior angle of a triangle is equal to the sum of the measures of the two opposite interior angles: $\angle a + \angle c = \angle z$.

79.

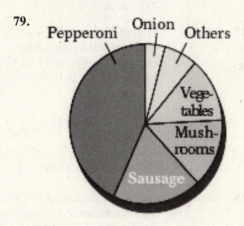

Section 9.2

Concept Check

1. The polygon has 6 sides.
The polygon is a hexagon.

3. The polygon has 5 sides.
The polygon is a pentagon.

5. The triangle has no sides equal.
The triangle is a scalene triangle.

7. The triangle has three sides equal.
The triangle is an equilateral triangle.

9. The triangle has one obtuse angle.
The triangle is an obtuse triangle.

11. The triangle has three acute angles.
The triangle is an acute triangle.

Objective A Exercises

13. Strategy   To find the perimeter, use the formula for the perimeter of a triangle. Substitute 12 for *a*, 20 for *b*, and 24 for *c*. Solve for *P*.

Solution
$$P = a + b + c$$
$$P = 12 + 20 + 24$$
$$P = 56$$
The perimeter is 56 in.

15. Strategy   To find the perimeter, use the formula for the perimeter of a square. Substitute 3.5 for *s* and solve for *P*.

Solution
$$P = 4s$$
$$P = 4 \cdot 3.5$$
$$P = 14$$
The perimeter is 14 ft.

17. Strategy   To find the perimeter, use the formula for the perimeter of a rectangle. Substitute 13 for *L* and 10.5 for *W*. Solve for *P*.

Solution
$$P = 2L + 2W$$
$$P = 2 \cdot 13 + 2 \cdot 10.5$$
$$P = 26 + 21$$
$$P = 47$$
The perimeter is 47 mi.

19. Strategy   To find the circumference, use the circumference formula that involves the radius. For the exact answer, leave the answer in terms of π. For an approximation, use the π key on a calculator.
$r = 4$.

Solution
$$C = 2\pi r$$
$$C = 2\pi (4)$$
$$C = 8\pi$$
$$C \approx 25.13$$
The circumference is 8π cm.
The circumference is approximately 25.13 cm.

21. Strategy   To find the circumference, use the circumference formula that involves the radius. For the exact answer, leave the answer in terms of π. For an approximation use the π key on a calculator.
$r = 5.5$.

Solution
$$C = 2\pi r$$
$$C = 2\pi(5.5)$$
$$C = 11\pi$$
$$C \approx 34.56$$
The circumference is 11π mi.
The circumference is approximately 34.56 mi.

23. Strategy   To find the circumference, use the circumference formula that involves the diameter. For the exact answer leave the answer in terms of π. For an approximation use the π key on a calculator. $d = 17$.

Solution
$$C = \pi d$$
$$C = \pi(17)$$
$$C = 17\pi$$
$$C \approx 53.41$$
The circumference is 17π ft.
The circumference is approximately 53.41 ft.

25. Strategy   To find the perimeter, use the formula for the perimeter of a triangle. Substitute 3.8 for *a*, 5.2 for *b*, and 8.4 for *c*. Solve for *P*.

Solution
$$P = a + b + c$$
$$P = 3.8 + 5.2 + 8.4$$
$$P = 17.4$$
The perimeter is 17.4 cm.

27. Strategy   To find the perimeter, use the formula for the perimeter of a triangle. Substitute $2\frac{1}{2}$ for a and b, and 3 for c. Solve for P.

Solution
$$P = a + b + c$$
$$P = 2\frac{1}{2} + 2\frac{1}{2} + 3$$
$$P = 8$$
The perimeter is 8 cm.

29. Strategy   To find the perimeter, use the formula for the perimeter of a rectangle. Substitute 8.5 for L and 3.5 for W. Solve for P.

Solution
$$P = 2L + 2W$$
$$P = 2(8.5) + 2(3.5)$$
$$P = 17 + 7$$
$$P = 24$$
The perimeter is 24 m.

31. Strategy   To find the perimeter, multiply the measure of one of the equal sides (3.5) by 5.

Solution
$$P = 5(3.5)$$
$$P = 17.5$$
The perimeter is 17.5 in.

33. Strategy   To find the perimeter, use the formula for the perimeter of a square. Substitute 12.2 for s. Solve for P.

Solution
$$P = 4s$$
$$P = 4(12.2)$$
$$P = 48.8$$
The perimeter is 48.8 cm.

35. Strategy   To find the circumference, use the circumference formula that involves the diameter. Leave the answer in terms of π.
$d = 1.5$

Solution
$$C = \pi d$$
$$C = \pi(1.5)$$
$$C = 1.5\pi$$
The circumference is 1.5π in.

37. Strategy   To find the circumference, use the circumference formula that involves the radius. An approximation is asked for; use the π key on a calculator.
$r = 36$.

Solution
$$C = 2\pi r$$
$$C = 2\pi(36)$$
$$C = 72\pi$$
$$C \approx 226.19$$
The circumference is approximately 226.19 cm.

39. Strategy   To find the amount of fencing, use the formula for the perimeter of a rectangle. Substitute 18 for L and 12 for W. Solve for P.

Solution
$$P = 2L + 2W$$
$$P = 2(18) + 2(12)$$
$$P = 36 + 24$$
$$P = 60$$
The perimeter of the garden is 60 ft.

41. Strategy   To find the amount to be nailed down, use the formula for the perimeter of a rectangle. Substitute 12 for L and 10 for W. Solve for P.

Solution
$$P = 2L + 2W$$
$$P = 2(12) + 2(10)$$
$$P = 24 + 20$$
$$P = 44$$
44 ft of carpet must be nailed down.

43. Strategy   To find the length, use the formula for the perimeter of a rectangle. Substitute 440 for P and 100 for W. Solve for L.

Solution
$$P = 2L + 2W$$
$$440 = 2L + 2(100)$$
$$440 = 2L + 200$$
$$240 = 2L$$
$$120 = L$$
The length is 120 ft.

45. Strategy   To find the third side of the
banner, use the formula for the
perimeter of a triangle.
Substitute 46 for P, 18 for a,
and 18 for b. Solve for c.

Solution $P = a + b + c$
$46 = 18 + 18 + c$
$46 = 36 + c$
$10 = c$
The third side of the banner is
10 in.

47. Strategy   To find the length of each side,
use the formula for the
perimeter of a square.
Substitute 48 for P. Solve for s.

Solution $P = 4s$
$48 = 4s$
$12 = s$
The length of each side is
12 in.

49. Strategy   To find the length of the
diameter, use the
circumference formula that
involves the diameter. An
approximation is asked for; use
the π key on a calculator.
$C = 8$.

Solution $C = \pi d$
$8 = \pi d$
$\dfrac{8}{\pi} = d$
$2.55 \approx d$
The diameter is approximately
2.55 cm.

51. Strategy   To find the length of molding,
use the circumference formula
that involves the diameter. An
approximation is asked for; use
the π key on a calculator.
$d = 4.2$.

Solution $C = \pi d$
$C = \pi(4.2)$
$C \approx 13.19$
The length of molding is
approximately 13.19 ft.

53. Strategy   To find the distance:
→Convert the diameter to feet.
→Multiply the circumference
by 8. An approximation is
asked for; use the π key on a
calculator.

Solution $24 \text{ in.} = 24 \text{ in.} \cdot \dfrac{1 \text{ ft}}{12 \text{ in.}} = 2 \text{ ft}$
distance $= 8C$
distance $= 8\pi d$
distance $= 8\pi(2)$
distance $= 16\pi$
distance ≈ 50.27
The bicycle travels
approximately 50.27 ft.

55. Strategy   To find the circumference of
the earth, use the
circumference formula that
involves the radius. An
approximation is asked for; use
the π key on a calculator. $r =$
6,356.

Solution $C = 2\pi r$
$C = 2\pi(6,356)$
$C = 12,712\pi$
$C \approx 39,935.93$
The circumference of the earth
is approximately 39,935.93
km.

57. A square whose side is 1 ft

Objective B Exercises

59. Strategy   To find the area, use the
formula for the area of a
rectangle. Substitute 12 for L
and 5 for W. Solve for A.

Solution $A = LW$
$A = 12(5)$
$A = 60$
The area is 60 ft^2.

61. Strategy   To find the area, use the formula for the area of the square. Substitute 4.5 for s. Solve for A.

Solution
$$A = s^2$$
$$A = (4.5)^2$$
$$A = 20.25$$
The area is 20.25 in^2.

63. Strategy   To find the area, use the formula for the area of a triangle. Substitute 42 for b and 26 for h. Solve for A.

Solution
$$A = \frac{1}{2}bh$$
$$A = \frac{1}{2}(42)(26)$$
$$A = 546$$
The area is 546 ft.^{2}.

65. Strategy   To find the area, use the formula for the area of a circle. Substitute 4 for r. Solve for A. For the exact answer, leave the answer in terms of π. For an approximation, use the π key on a calculator.

Solution
$$A = \pi r^2$$
$$A = \pi(4)^2$$
$$A = 16\pi$$
$$A \approx 50.27$$
The area is 16π cm^2.
The area is approximately 50.27 cm^2.

67. Strategy   To find the area, use the formula for the area of a circle. Substitute 5.5 for r. Solve for A. For the exact answer, leave the answer in terms of π. For an approximation, use the π key on a calculator.

Solution
$$A = \pi r^2$$
$$A = \pi(5.5)^2$$
$$A = 30.25\pi$$
$$A \approx 95.03$$
The area is 30.25π mi^2.
The area is approximately 95.03 mi^2.

69. Strategy   To find the area:
→Find the radius of the circle.
→Use the formula for the area of a circle. For an exact answer, leave the answer in terms of π. For an approximation, use the π key on a calculator.

Solution
$$r = \frac{1}{2}d = \frac{1}{2}(17) = 8.5$$
$$A = \pi r^2$$
$$A = \pi(8.5)^2$$
$$A = 72.25\pi$$
$$A \approx 226.98$$
The area is 72.25π ft^2.
The area is approximately 226.98 ft^2.

71. Strategy   To find the area, use the formula for the area of a square. Substitute 12.5 for s. Solve for A.

Solution
$$A = s^2$$
$$A = (12.5)^2$$
$$A = 156.25$$
The area is 156.25 cm^2.

73. Strategy   To find the area, use the formula for the area of a rectangle. Substitute 38 for L and 15 for W. Solve for A.

Solution
$$A = LW$$
$$A = 38(15)$$
$$A = 570$$
The area is 570 in^2.

75. Strategy   To find the area, use the formula for the area of a parallelogram. Substitute 16 for b and 12 for h. Solve for A.

Solution
$$A = bh$$
$$A = 16(12)$$
$$A = 192$$
The area is 192 in^2.

77. Strategy   To find the area, use the formula for the area of a triangle. Substitute 6 for b and 4.5 for h. Solve for A.

Solution
$$A = \frac{1}{2}bh$$
$$A = \frac{1}{2}(6)(4.5)$$
$$A = 13.5$$
The area is 13.5 ft^2.

79. Strategy   To find the area, use the formula for the area of a trapezoid. Substitute 35 for b_1, 20 for b_2, and 12 for h. Solve for A.

Solution
$$A = \frac{1}{2}h(b_1 + b_2)$$
$$A = \frac{1}{2} \cdot 12(35 + 20)$$
$$A = 330$$
The area is 330 cm^2.

81. Strategy   To find the area, use the formula for the area of a circle. Leave the answer in terms of π. $r = 5$.

Solution
$$A = \pi r^2$$
$$A = \pi(5)^2$$
$$A = 25\pi$$
The area is 25π in^2.

83. Strategy   To approximate the area, use the formula for the area of a rectangle. Substitute 150 for L and 70 for W. Solve for A.

Solution
$$A = LW$$
$$A = 150(70)$$
$$A = 10,500$$
The area of the reserve is approximately 10,500 mi^2.

85. Strategy   To find the area, use the formula for the area of a circle. Leave the answer in terms of π. $r = 50$.

Solution
$$A = \pi r^2$$
$$A = \pi(50)^2$$
$$A = 2,500\pi$$
The area is 2,500π ft^2.

87. Strategy   To find the area, use the formula for the area of a square. Substitute 8.5 for s. Solve for A.

Solution
$$A = s^2$$
$$A = (8.5)^2$$
$$A = 72.25$$
The area of the patio is 72.25 m^2.

89. Strategy   To find the amount of turf, use the formula for the area of a rectangle. Substitute 100 for L and 75 for W. Solve for A.

Solution
$$A = LW$$
$$A = 100(75)$$
$$A = 7,500$$
7,500 yd^2 of artificial turf must be purchased.

91. Strategy   To find the width, use the formula for the area of a rectangle. Substitute 300 for A and 30 for L. Solve for W.

Solution
$$A = LW$$
$$300 = 30W$$
$$10 = W$$
The width of the rectangle is 10 in.

93. Strategy   To find the length of the base, use the formula for the area of a triangle. Substitute 50 for A and 5 for h. Solve for b.

Solution
$$A = \frac{1}{2}bh$$
$$50 = \frac{1}{2}b(5)$$
$$50 = \frac{5}{2}b$$
$$20 = b$$
The base of the triangle is 20 m.

95. Strategy   To find the number of quarts of stain:
→Use the formula for the area of a rectangle to find the area of the deck.
→Divide the area of the deck by the area one quart will cover (50).

Solution
$$A = LW$$
$$A = 10(8)$$
$$A = 80$$
$$80 \div 15 = 1.6$$
Because a portion of a second quart is needed, 2 qt of stain should be purchased.

97. Strategy   To find the cost of the wallpaper:
→Use the formula for the area of a rectangle to find the areas of the two walls.
→Add the areas of the two walls.
→Divide the total area by the area in one roll (40) to find the total number of rolls.
→Multiply the number of rolls by 37

Solution
$$A_1 = LW = 9(8) = 72$$
$$A_2 = LW = 11(8) = 88$$
$$A = A_1 + A_2 = 72 + 88 = 160$$
$$160 \div 40 = 4$$
$$4 \cdot 37 = 148$$
The cost to wallpaper the two walls is $148.

99. Strategy   Compute the area of each storage unit and compare to see which is just over the needed space (175 ft^2).

Solution
Area of $10 \times 5 = 50$ ft^2
Area of $10 \times 10 = 100$ ft^2
Area of $10 \times 15 = 150$ ft^2
Area of $10 \times 20 = 200$ ft^2
Area of $10 \times 25 = 250$ ft^2
Area of $10 \times 30 = 300$ ft^2
You should select the 10×20 unit.

101. Strategy   To find the increase in area:
→Use the formula for the area of a circle to find the area of a circle with $r = 6$.
→Use the formula for the area of a circle to find the area of a circle with $r = 2(6) = 12$.
→Subtract the area of the smaller circle from the area of the larger circle. An approximation is asked for; use the π key on a calculator.

Solution

$$A_1 = \pi r^2$$
$$A_1 = \pi(6)^2 = 36\pi$$
$$A_2 = \pi(12)^2 = 144\pi$$
$$A_2 - A_1 = 144\pi - 36\pi$$
$$= 108\pi \approx 339.29$$

The area is increased by 339.29 cm^2.

103. Strategy   To find the cost of the paint:
→Use the formula for the area of a rectangle to find the area of the two walls that measure 15 by 9 and the two walls that measure 12 by 9.
→Add the area to find the total area.
→Divide the total area by the area that one gallon will cover (400).
→Multiply the number of gallons by 29.98.

Solution

$$A_1 = 2(LW) = 2[15(9)] = 270$$
$$A_1 = 2(LW) = 2[12(9)] = 216$$
$$A_1 + A_2 = 270 + 216 = 486$$
$$486 \div 400 = 1.22$$

Because the portion of a second gallon is need, 2 gal of paint should be purchased.
$$2(29.98) = 59.96$$
The paint will cost $59.96.

105. Strategy   To find the amount of material:
→Use the formula for the area of a rectangle to find the area of the drapes. Substitute $2 \cdot 2 = 4$ for W and $4 + 1 = 5$ for L.
→Multiply the area of one drape by 4.

Solution $A = LW = 5(4) = 20$
$$4(20) = 80$$
the drapes will required 80 ft^2 of material.

Critical Thinking 9.2

107. 4 times

Projects or Group Activities

109. 5 units by 5 units

111. There are a great number of quilt patterns that incorporate regular polygons, for example, Nine Patch Block, Grandmother's Flower Garden, Hour Glasses, Sunshine and Shadow, Field of Diamonds, and Trip Around the World.

Check Your Progress: Chapter 9

1. $AB + BC = AC$

$$\frac{1}{3}(BC) + BC = AC$$
$$\frac{1}{3}(15) + 15 = AC$$
$$5 + 15 = AC$$
$$20 = AC$$
$$AC = 20 \text{ ft}$$

2. Strategy   Supplementary angles are two angles whose sum is 180°. To find the supplement, let x represent the supplement of a 12° angle. Write an equation and solve for x.

Solution $x + 12° = 180°$
$$x = 168°$$
The supplement of a 12° angle is an 168° angle.

3. $\angle a + \angle b = 180°$

$42° + \angle b = 180°$

$\angle b = 138°$

$\angle c = \angle a = 40°$

$\angle c = 40°$

$\angle d = \angle b = 138°$

$\angle d = 138°$

4. Strategy   To find the measure of the third angle, use the fact that the sum of the measures of the interior angles of a triangle is $180°$. Write an equation using x to represent the measure of the third angle. Solve the equation for x.

Solution $x + 23° + 90° = 180°$

$x + 113° = 180°$

$x = 67°$

The measure of the third angle is $67°$.

5. Strategy   To find the area, use the formula for the area of a square. Substitute 40 for s. Solve for A.

Solution $A = s^2$

$A = (40)^2$

$A = 1600$

The area of the patio is 1600 in^2.

6. Strategy   The sum of the measures of the three angles shown is $180°$. To find x, write an equation and solve for x.

Solution $(4x - 10) + 3x + (2x + 10)$

$= 180°$

$9x = 180°$

$x = 20°$

The measure of x is $20°$.

7. Strategy   To find the circumference, use the circumference formula that involves the diameter. For an approximation, use the π key on a calculator. $d = 12$.

Solution $C = \pi d$

$C = \pi(12)$

$C = 12\pi$

$C \approx 37.699$

The circumference is approximately 37.70 cm.

8. Strategy   →To find the measure of $\angle a$, use the fact that alternate interior angles of parallel lines are equal.

→To find the measure of $\angle b$, use the fact that adjacent angles of intersecting lines are supplementary.

Solution $\angle a = 135°$

$\angle b + \angle a = 180°$

$\angle b + 135° = 180°$

$\angle b = 45°$

The measure of $\angle a$ is $135°$.

The measure of $\angle b$ is $45°$.

9. Strategy   To find the length of the base, use the formula for the area of a triangle. Substitute 8 for h and 20 for A. Solve for b.

Solution $A = \frac{1}{2}bh$

$20 = \frac{1}{2}b(8)$

$20 = 4b$

$\frac{20}{4} = b$

$5 = b$

The length of the base is 5 m.

10. Strategy   To find the area, use the formula for the area of parallelogram. Substitute 14 for b and 7 for h. Solve for A.

Solution $A = bh$

$A = 14(7)$

$A = 98$

The area is 98 m^2.

11. Strategy   To find the length of each side, use the formula for the perimeter of a square. Substitute 38 for P. Solve for s.

Solution $P = 4s$
$38 = 4s$
$9.5 = s$
The length of each side is 9.5 in.

12. Strategy

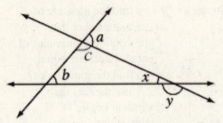

→To find the measure of $\angle x$, first find the measure of $\angle c$ by using use the fact that sum of an interior angle and exterior angle is 180° and then use the fact that the sum of the interior angles of a triangle is 180° to find $\angle x$.
→Find the measure of $\angle y$ by using the fact that the sum of an interior and exterior angle is 180°.

Solution $\angle a + \angle c = 180°$
$72° + \angle c = 180°$
$\angle c = 108°$
$\angle b + \angle c + \angle x = 180°$
$48° + 108° + \angle x = 180°$
$156° + \angle x = 180°$
$\angle x = 24°$
$\angle x + \angle y = 180°$
$24° + \angle y = 180°$
$\angle y = 156°$
The measure of $\angle x$ is 24°.
The measure of $\angle y$ is 156°.

13. Strategy   To find the length, use the formula for the area of a rectangle. Substitute 8 for W and 128 for A. Solve for L.

Solution $A = LW$
$128 = L(8)$
$16 = L$
The length of the rectangle is 16 m.

14. Strategy   To find the perimeter, use the formula for the perimeter of a triangle. Substitute 9 for a, 12 for b, and 15 for c. Solve for P.

Solution $P = a + b + c$
$P = 9 + 12 + 15$
$P = 36$
The perimeter is 36 in.

15. Strategy   To find the area:
→Find the radius of the circle.
→Use the formula for the area of a circle. For an exact answer, leave the answer in terms of π. For an approximation, use the π key on a calculator.

Solution $r = \dfrac{1}{2}d = \dfrac{1}{2}(2.8) = 1.4$
$A = \pi r^2$
$A = \pi (1.4)^2$
$A = 1.96\pi$
$A \approx 6.1575$
The area is 1.96π m^2.
The area is approximately 6.16 m^2.

16. Strategy   To find the area, use the formula for the area of a trapezoid. Substitute 14 for b_1, 10 for b_2, and 6 for h. Solve for A.

Solution $A = \dfrac{1}{2}h(b_1 + b_2)$
$A = \dfrac{1}{2} \cdot 6(14 + 10)$
$A = 72$
The area is 72 cm^2.

17. Strategy   To find the length of molding, use the formula for the perimeter of a rectangle.

Substitute 10 for L and $8\frac{1}{2}$ for W. Solve for P.

Solution $P = 2L + 2W$

$P = 2(10) + 2\left(8\frac{1}{2}\right)$

$P = 20 + 17$
$P = 37$

37 ft of decorative molding would be required.

Section 9.3

Concept Check

1. the hypotenuse; a leg

3. *FD*; *BA*

5. The ratio is $\dfrac{5}{10} = \dfrac{1}{2}$.

7. The ratio is $\dfrac{6}{8} = \dfrac{3}{4}$.

9. $\angle E$

Objective A Exercises

11. Strategy   To find the hypotenuse, use the Pythagorean Theorem.
$a = 3, b = 4$

Solution $c^2 = a^2 + b^2$
$c^2 = 3^2 + 4^2$
$c^2 = 9 + 16$
$c^2 = 25$
$c = \sqrt{25}$
$c = 5$

The length of the hypotenuse is 5 in.

13. Strategy   To find the hypotenuse, use the Pythagorean Theorem.
$a = 5, b = 7$

Solution $c^2 = a^2 + b^2$
$c^2 = 5^2 + 7^2$
$c^2 = 25 + 49$
$c^2 = 74$
$c = \sqrt{74}$
$c \approx 8.6$

The length of the hypotenuse if approximately 8.6 cm.

15. Strategy   To find the measure of the other leg, use the Pythagorean Theorem.
$c = 15, a = 10$

Solution $a^2 + b^2 = c^2$
$10^2 + b^2 = 15^2$
$100 + b^2 = 225$
$b^2 = 125$
$b = \sqrt{125}$
$b \approx 11.2$

The measure of the other leg is approximately 11.2 ft.

17. Strategy   To find the measure of the other leg, use the Pythagorean Theorem. $c = 6, a = 4$

Solution $a^2 + b^2 = c^2$
$4^2 + b^2 = 6^2$
$16 + b^2 = 36$
$b^2 = 20$
$b = \sqrt{20}$
$b \approx 4.5$

The measure of the other leg is approximately 4.5 cm.

19. Strategy   To find the hypotenuse, use the Pythagorean Theorem.
$a = 9, b = 9$

Solution $c^2 = a^2 + b^2$
$c^2 = 9^2 + 9^2$
$c^2 = 81 + 81$
$c^2 = 162$
$c = \sqrt{162}$
$c \approx 12.7$

The length of the hypotenuse is approximately 12.7 yd.

21. iii

23. Strategy   To find the height, use the Pythagorean Theorem to find the other leg. $c = 8, a = 3$

Solution
$$a^2 + b^2 = c^2$$
$$3^2 + b^2 = 8^2$$
$$9 + b^2 = 64$$
$$b^2 = 55$$
$$b = \sqrt{55}$$
$$b \approx 7.4$$
The ladder will reach a height of approximately 7.4 m.

25. Strategy   To find the perimeter:
→Use the Pythagorean Theorem to find the hypotenuse of the triangle. $a = 5, b = 9$
→Use the formula for the perimeter of a triangle to find the perimeter.

Solution
$$c^2 = a^2 + b^2$$
$$c^2 = 5^2 + 9^2$$
$$c^2 = 25 + 81$$
$$c^2 = 106$$
$$c = \sqrt{106}$$
$$c \approx 10.3$$
$$P = a + b + c$$
$$P = 5 + 9 + 10.3$$
$$P = 24.3$$
The perimeter is approximately 24.3 cm.

Objective B Exercises

27. Strategy   To find *DE*, write a proportion using the fact that in similar triangles, the ratios of corresponding sides are equal. Solve the proportion for *DE*.

Solution
$$\frac{AB}{DE} = \frac{AC}{DF}$$
$$\frac{4}{DE} = \frac{5}{9}$$
$$4(9) = (5)DE$$
$$36 = (5)DE$$
$$7.2 = DE$$
The length of *DE* is 7.2 cm.

29. Strategy   To find the height of triangle *DEF*, write a proportion using the fact that, in similar triangles, the ratio of corresponding sides equals the ratio of corresponding sides equals the ratio of corresponding heights. Solve the proportion for the height (*h*).

Solution
$$\frac{AC}{DF} = \frac{2}{h}$$
$$\frac{3}{5} = \frac{2}{h}$$
$$3h = 5(2)$$
$$3h = 10$$
$$h \approx 3.3$$
The height of triangle *DEF* is approximately 3.3 m.

31. Strategy   To find the perimeter:
→Find the side *BC* by writing a proportion using the fact that the ratios of corresponding sides of similar triangles are equal.
→Use the formula for the perimeter of a triangle.

Solution
$$\frac{BC}{EF} = \frac{AC}{DF}$$
$$\frac{BC}{6} = \frac{4}{8}$$
$$(8)BC = 6(4)$$
$$(8)BC = 24$$
$$BC = 3$$
$$P = a + b + c$$
$$P = 3 + 4 + 5$$
$$P = 12$$
The perimeter of triangle *ABC* is 12 m.

33. Strategy   To find the perimeter:
→Find side BC by writing a proportion using the fact that the ratios of corresponding sides of similar triangles are equal.
→Use the formula for the perimeter of a triangle.

Solution $\dfrac{BC}{EF} = \dfrac{AB}{DE}$

$\dfrac{BC}{15} = \dfrac{4}{12}$

$(12)BC = 15(4)$
$(12)BC = 60$
$BC = 5$
$P = a + b + c$
$P = 3 + 4 + 5$
The perimeter of the triangle is 12 in.

35. Strategy   To find the area:
→Find the height (h) of triangle ABC by writing a proportion using the fact that, in similar triangles, the ratio of corresponding sides equals the ratio of corresponding heights.
→Use the formula for the area of a triangle. $b = 15$

Solution $\dfrac{AB}{DE} = \dfrac{h}{20}$

$\dfrac{15}{40} = \dfrac{h}{20}$

$15(20) = 40h$
$300 = 40h$
$7.5 = h$
$A = \dfrac{1}{2}bh$

$A = \dfrac{1}{2}(15)(7.5)$

$A \approx 56.3$
The area of triangle ABC is approximately 56.3 cm^2.

37. Strategy   To find the height, write a proportion using the fact that, in similar triangles, the ratios of corresponding sides are equal.

Solution $\dfrac{24}{8} = \dfrac{\text{height}}{6}$

$24(6) = 8 \cdot \text{height}$
$144 = 8 \cdot \text{height}$
$18 = \text{height}$
The height of the flagpole is 18 ft.

39. Strategy   To find the height, write a proportion using the fact that, in similar triangles, the ratios of corresponding sides are equal.

Solution $\dfrac{\text{height}}{8} = \dfrac{8}{4}$

$\text{height} \cdot 4 = 8(8)$
$4 \cdot \text{height} = 64$
$\text{height} = 16$
The height of the building is 16 m.

Objective C Exercises

41. Strategy   To determine if the triangles are congruent, determine if one of the rules for congruence is satisfied.

Solution $AC = DE$, $AB = EF$, and $\angle A = \angle E$.
Two sides and the included angle of one triangle equal two sides and the included angle of the other triangle.
The triangles are congruent by the SAS rule.

43. Strategy   To determine if the triangles are congruent, determine if one of the rules for congruence is satisfied.

Solution $AB = DE$, $AC = EF$, and $BC = DF$.
Three sides of one triangle equal the three sides of the other triangle. The triangles are congruent by the SSS rule.

45. Strategy   To determine if the triangles are congruent, determine if one of the rules for congruence is satisfied.

Solution $AC = DF$, $\angle A = \angle D$, and $\angle C = \angle F$.
Two angles and the included side of one triangle equal two angles and the included side of the other triangle.
The triangles are congruent by the ASA rule.

47. Strategy   Draw a sketch of the two triangles and determine if one of the rules for congruence is satisfied.

Solution

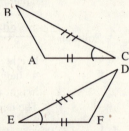

$AC = EF$, $BC = DE$, and $\angle C = \angle E$.
Because two sides and the included angle of one triangle equal two sides and the included angle of the other triangle, the triangles are congruent by the SAS rule.

49. Strategy   Draw a sketch of the two triangles and determine if one of the rules of congruence is satisfied.

Solution

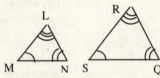

$\angle M = \angle S$, $\angle N = \angle Q$, and $\angle L = \angle R$.
The triangles do not satisfy the SSS rule, the SAS rule, or the ASA rule. The triangles are not necessarily congruent.

51. Strategy   Draw a sketch of the two triangles and determine if one of the rules of congruence is satisfied.

Solution

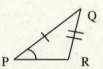

$\angle B = \angle P$, $BC = PQ$, and $AC = QR$.
The triangles do not satisfy the SSS rule, the SAS rule, or the ASA rule. The triangles are not necessarily congruent.

53. Yes

Critical Thinking 9.3

55. To determine if a 25-foot ladder is long enough to reach 24 ft up the side of the home when the bottom of the ladder is 6 ft from the base of the side of the house, use the Pythagorean Theorem to find the hypotenuse of a right triangle with legs that measure 24 ft and 6 ft.
$$c^2 = a^2 + b^2$$
$$c^2 = 24^2 + 6^2$$
$$c^2 = 576 + 36$$
$$c^2 = 612$$
$$c \approx 24.74$$
Compare the leg of the hypotenuse with 25. If the hypotenuse is shorter than 25 ft, the ladder will reach the gutter. If the hypotenuse is longer than 25 feet, the ladder will not reach the gutter.
$$24.74 < 25$$
The hypotenuse is shorter than 25 ft. The ladder will reach the gutters.

Projects and Group Activities 9.3

57.

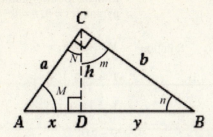

Triangles ACD and DCB are right triangles.
Angle M = angle m and angle N = angle n.
Because corresponding angles are equal,
the triangles are similar triangles.

Triangles ABC and ADC are right triangles.
Angle M = angle M and angle N = angle n.
Because corresponding angles are equal,
the triangles are similar triangles.

Triangles ABC and DCB are right triangles.
Angle n = angle n and angle M = angle m.
Because corresponding angles are equal,
the triangles are similar triangles.

Section 9.4

Concept Check

1. **a.** cone

   b. cube

   c. sphere

   d. cylinder

3. $s^2 + 2sl$; l; s

Objective A Exercises

5. Strategy   To find the volume, use the
              formula for the volume of a
              rectangular solid.
              $L = 14$, $W = 10$, $H = 6$.

   Solution $V = LWH$
              $V = 14(10)(6)$
              $V = 840$
              The volume is 840 in^3.

7. Strategy   To find the volume, use the
              formula for the volume of a
              pyramid. $s = 3$, $h = 5$.

   Solution $V = \frac{1}{3}s^2h$

              $V = \frac{1}{3}(3^2)(5)$

              $V = \frac{1}{3}(9)(5)$

              $V = 15$
              The volume is 15 ft^3.

9. Strategy   To find the volume:
              →Find the radius of the
                sphere. $d = 3$.
              →Use the formula for the
                volume of sphere.

   Solution $r = \frac{1}{2}d = \frac{1}{2}(3) = 1.5$

              $V = \frac{4}{3}\pi r^3$

              $V = \frac{4}{3}\pi(1.5)^3$

              $V = \frac{4}{3}\pi(3.375)$

              $V = 4.5\pi$

              $V \approx 14.14$
              The volume is 4.5π cm^3.
              The volume is approximately
              14.14 cm^3.

11. Strategy To find the volume, use the
              formula for the volume of a
              rectangle solid.
              $L = 6.8$, $W = 2.5$, $H = 2$.

    Solution $V = LWH$
              $V = 6.8(2.5)(2)$
              $V = 34$
              The volume of the rectangular
              solid is 34 m^3.

13. Strategy To find the volume, use the
              formula for the volume of a
              cube. $s = 2.5$.

    Solution $V = s^3$

              $V = (2.5)^3$

              $V = 15.625$
              The volume of the cube is
              15.625 in^3.

15. Strategy   To find the volume:
→Find the radius of the sphere. $d = 6$.
→Use the formula for the volume of a sphere.

Solution $r = \frac{1}{2}d = \frac{1}{2}(6) = 3$

$V = \frac{4}{3}\pi r^3$

$V = \frac{4}{3}\pi(3)^3$

$V = \frac{4}{3}\pi(27)$

$V = 36\pi$

The volume of the sphere is 36π ft^3.

17. Strategy   To find the volume:
→Find the radius of the cylinder. $d = 24$.
→Use the formula for the volume of a cylinder. $h = 18$.

Solution $r = \frac{1}{2}d = \frac{1}{2}(24) = 12$

$V = \pi r^2 h$

$V = \pi(12^2)(18)$

$V = \pi(144)(18)$

$V = 2,592\pi$

$V \approx 8,143.01$

The volume of the cylinder is approximately 8,143.01 cm^3.

19. Strategy   To find the volume:
→Find the radius of the base of the cone. $d = 10$.
→Use the formula for the volume of a cone. $h = 15$.

Solution $r = \frac{1}{2}d = \frac{1}{2}(10) = 5$

$V = \frac{1}{3}\pi r^2 h$

$V = \frac{1}{3}\pi(5)^2(15)$

$V = \frac{1}{3}\pi(25)(15)$

$V = 125\pi$

$V \approx 392.70$

The volume of the cone is approximately 392.70 cm^3.

21. Strategy   To find the volume, use the formula for the volume of a pyramid. $s = 9, h = 8$.

Solution $V = \frac{1}{3}s^2 h$

$V = \frac{1}{3}(9^2)(8)$

$V = \frac{1}{3}(81)(8)$

$V = 216$

The volume of the pyramid is 216 m^3.

23. Strategy   To find the height, use the formula for the volume of rectangular solid.
$V = 1,836, L = 18, W = 12$.

Solution $V = LWH$

$1,836 = 18(12)(H)$

$1,836 = 216H$

$8.5 = H$

The height of the aquarium is 8.5 in.

25. Strategy   To find the height:
→Find the radius of the base of the cylinder.
$d = 14$.
→Use the formula for the volume of a cylinder.
$V = 2,310$.

Solution $r = \frac{1}{2}d = \frac{1}{2}(14) = 7$

$V = \pi r^2 h$

$2,310 = \pi(7)^2 h$

$2,310 = \pi(49)h$

$15.01 \approx h$

The height of the cone is approximately 15.01 cm.

27. Strategy   To find the amount not being used for storage:

→Find the radius of the base of the cylinder. $d = 16$.

→Use the formula for the volume of a cylinder. $h = 30$.

→Multiply the volume of the cylinder by $\frac{1}{4}$, which is the portion not being used.

$$\left(1 - \frac{3}{4}\right)$$

Solution $r = \frac{1}{2}d = \frac{1}{2}(16) = 8$

$V = \pi r^2 h$

$V = \pi(8)^2(30)$

$V = \pi(64)(30)$

$V = 1{,}920\pi$

$\frac{1}{4}V = \frac{1}{4}(1{,}920\pi)$

$= 480\pi = 1{,}507.96$

The portion of the silo that is not used is approximately $1{,}507.96 \text{ ft}^3$.

29. Strategy   To find the volume, use the formula for the volume of rectangular solid. $L = 1000, W = 110, H = 60$.

Solution $V = LWH$

$V = 1000(110)(60)$

$V = 6{,}600{,}000$

The volume of the Panama Canal is $6{,}600{,}000 \text{ ft}^3$.

31. Yes

33. No

35. Strategy   To find the volume, use the formula for the volume of rectangular solid. $L = 360, W = 160, H = 3$.

Solution $V = LWH$

$V = 360(160)(3)$

$V = 172{,}800$

The volume of the guacamole is $172{,}800 \text{ ft}^3$.

Objective B Exercises

37. Strategy   To find the surface area, use the formula for the surface area of a rectangular solid. $L = 5, W = 4, H = 3$.

Solution $SA = 2LW + 2LH + 2WH$

$SA = 2(5)(4) + 2(5)(3) + 2(4)(3)$

$SA = 40 + 30 + 24$

$SA = 94$

The surface area of the rectangular solid is 94 m^2.

39. Strategy   To find the surface area, use the formula for the surface area of a pyramid. $s = 4$, $l = 5$.

Solution $SA = s^2 + 2sl$

$SA = 4^2 + 2(4)(5)$

$SA = 16 + 40$

$SA = 56$

The surface area of the pyramid is 56 m^2.

41. Strategy   To find the surface area, use the formula for the surface area of a cylinder. $r = 6$, $h = 2$.

Solution $SA = 2\pi r^2 + 2\pi rh$

$SA = 2\pi(6^2) + 2\pi(6)(2)$

$SA = 2\pi(36) + 24\pi$

$SA = 72\pi + 24\pi$

$SA = 96\pi$

$SA \approx 301.59$

The surface area of the cylinder is $96\pi \text{ in}^2$. The surface area of the cylinder is approximately 301.59 in^2.

43. Strategy To find the surface area, use the formula for the surface area of a rectangular solid. $H = 5, L = 8, W = 4$.

Solution
$SA = 2LW + 2LH + 2WH$
$SA = 2(8)(4) + 2(8)(5)$
$\qquad + 2(4)(5)$
$SA = 64 + 80 + 40$
$SA = 184$
The surface area of the rectangular solid is 184 ft^2.

45. Strategy To find the surface area, use the formula for the surface area of a cube. $s = 3.4$.

Solution
$SA = 6s^2$
$SA = 6(3.4)^2$
$SA = 6(11.56)$
The surface area of the cube is 69.36 m^2.

47. Strategy To find the surface area:
→Find the radius of the sphere. $d = 15$.
→Use the formula for the surface area of a sphere.

Solution
$r = \frac{1}{2}d = \frac{1}{2}(15) = 7.5$
$SA = 4\pi r^2$
$SA = 4\pi(7.5)^2$
$SA = 4\pi(56.25)$
$SA = 225\pi$
The surface area of the sphere is 225π cm^2.

49. Strategy To find the surface area, use the formula for the surface area of a cylinder. $r = 4$, $h = 12$.

Solution
$SA = 2\pi r^2 + 2\pi rh$
$SA = 2\pi(4^2) + 2\pi(4)(12)$
$SA = 2\pi(16) + 96\pi$
$SA = 32\pi + 96\pi$
$SA = 128\pi$
$SA \approx 402.12$
The surface area of the cylinder is approximately 402.12 in^2.

51. Strategy To find the surface area, use the formula for the surface area of a cone. $r = 1.5, l = 2.5$.

Solution
$SA = \pi r^2 + \pi rl$
$SA = \pi(1.5^2) + \pi(1.5)(2.5)$
$SA = \pi(2.25) + 3.75\pi$
$SA = 6\pi$
The surface area of the cone is 6π ft^2.

53. Strategy To find the surface area, use the formula for the surface area of a pyramid. $s = 9$, $l = 12$.

Solution
$SA = s^2 + 2sl$
$SA = 9^2 + 2(9)(12)$
$SA = 81 + 216$
$SA = 297$
The surface area of the pyramid is 297 in^2.

55. Strategy To find the width, use the formula for the surface area of a rectangular solid. $SA = 108, L = 6$, and $H = 4$.

Solution
$SA = 2LW + 2LH + 2WH$
$108 = 2(6)W + 2(6)(4) + 2W(4)$
$108 = 12W + 48 + 8W$
$108 = 20W + 48$
$60 = 20W$
$3 = W$
The width of the rectangular solid is 3 cm.

57. Strategy
To find the amount of fabric:
→Find the radius of the sphere. $d = 32$.
→Use the formula for the surface area of a sphere.

Solution
$r = \frac{1}{2}d = \frac{1}{2}(32) = 16$
$SA = 4\pi r^2$
$SA = 4\pi(16^2)$
$SA = 4\pi(256)$
$SA = 1024\pi$
$SA \approx 3,217$
Approximately 3,217 ft² of fabric was used to construct the balloon.

59. Strategy
To find the amount of glass, use the formula for the surface area of a rectangular solid. Omit the top of the fish tank. The formula becomes $SA = LW + 2LH + 2WH$. $L = 12, W = 8, H = 9$.

Solution
$SA = LW + 2LH + 2WH$
$SA = 12(8) + 2(12)(9)$
$+ 2(8)(9)$
$SA = 96 + 216 + 144$
$SA = 456$
The fish tank requires 456 in² of glass.

Critical Thinking 9.4

61. a. The distance from the edge of the base to the vertex of a regular pyramid is longer than the distance, perpendicular to the base, from the base to the vertex. The statement is always true.

b. The distance from the edge of the base of a cone to the vertex is longer than the distance, perpendicular to the base, from the base to the vertex. The statement is never true.

c. The four triangular faces of a regular pyramid could be equilateral triangles, but they could be isosceles triangles that are not equilateral. The statement is sometimes true.

63. three

Projects and Group Activities 9.4

65. Many of your students may already be familiar with M.C. Escher's work, since it now appears on everything from postcards to T-shirts. You might have each student select a different work to report on. Some examples include Relativity, House of Stairs, Whirlpools, Stars, Rind, Belvedere, Another World II, Ascending and Descending, and Waterfall.

Chapter Review Exercises

1. Strategy
→To find the measure of $\angle c$, use the fact that the sum of an interior and exterior angle is 180°. $V = LWH - \pi r^2 h$
→To find the measure of $\angle x$, use the fact that the sum of the measurements of the interior angles of a triangle is 180°.
→To find the measure of $\angle y$, use the fact that the sum of an interior and exterior angle is 180°.

Solution
$\angle a + \angle c = 180°$
$74° + \angle c = 180°$
$\angle c = 106°$
$\angle b + \angle c + \angle x = 180°$
$52° + 106° + \angle x = 180°$
$158° + \angle x = 180°$
$\angle x = 22°$
$\angle x + \angle y = 180°$
$22° + \angle y = 180°$
$\angle y = 158°$

2. Strategy — To find the perimeter:
→Find AC by writing a proportion, using the fact that in similar triangles, the ratios of corresponding sides are equal.
→Use the formula for finding the perimeter of a triangle.

Solution

$$\frac{AC}{DF} = \frac{BC}{EF}$$
$$\frac{AC}{12} = \frac{6}{9}$$
$9(AC) = 12(6)$
$9(AC) = 72$
$AC = 8$
$P = AB + BC + AC$
$P = 10 + 6 + 8$
$P = 24$
The perimeter of the triangle is 24 in.

3. Strategy — The volume is equal to the volume of the rectangular solid with dimensions 7 by 8 by 3.

Solution

$V = LWH$
$V = 7(8)(3)$
$V = 168$
The volume of the solid is 168 in^3.

4. Strategy — To find the measure of $\angle x$, use the fact that adjacent angles of intersecting lines are supplementary.

Solution

$112° + \angle x = 180°$
$\angle x = 68°$
The measure of $\angle x$ is 68°.

5. Strategy — To determine if the triangles are congruent, determine if one of the rules for congruence is satisfied.

Solution

$BC = DE, AC = DF,$
$\angle C = \angle D$
Two sides and the included angle of one triangle equal two sides and the included angle of the other triangle.
The triangles are congruent by the SAS rule.

6. Strategy — The total surface area equals the surface area of a cylinder. The radius of the sphere is one half the diameter of the cylinder.

Solution

$r = \frac{1}{2}d = \frac{1}{2}(4) = 2$
$SA = 2\pi r^2 + 2\pi rh$
$SA = 2\pi(2^2) + 2\pi(2)(8)$
$SA = 2\pi(4) + 32\pi$
$SA = 8\pi + 32\pi$
$SA = 40\pi$
$SA \approx 125.66$
The surface area of the solid is approximately 125.66 m^2.

7. $AC = AB + BC$
$AC = 3(BC) + BC$
$AC = 4(BC)$
$AC = 4(11)$
$AC = 44$
The length of AC is 44 cm.

8. Strategy — The sum of the measures of the three angles shown is 180°. To find x, write an equation and solve for x.

Solution

$4x + 3x + (x + 28°) = 180°$
$8x + 28° = 180°$
$8x = 152°$
$x = 19°$
The measure of x is 19°.

9. Strategy — The area is equal to the area of the rectangle.

Solution

$A = LW$
$A = 8(4)$
$A = 32$
The area is 32 in^2.

10. Strategy — To find the volume, use the formula for the volume of a pyramid. $s = 6, h = 8$.

Solution

$V = \frac{1}{3}s^2h$
$V = \frac{1}{3}(6^2)(8)$
$V = \frac{1}{3}(36)(8)$
$V = 96$ cm^3.
The volume of the pyramid is 96 cm^3.

11. Strategy   The perimeter is equal to three sides of a triangle.

Solution $P = a + b + c$
$P = 16 + 16 + 10$
$P = 42$
The perimeter is 42 in.

12. Strategy $\angle a = 138°$ because alternate interior angles of parallel lines are equal.
$\angle a + \angle b = 180°$ because adjacent angles of intersecting lines are supplementary.

Solution $\angle a = 138°$
$\angle a + \angle b = 180°$
$138° + \angle b = 180°$
$\angle b = 42°$
The measure of $\angle b$ is 42°.

13. Strategy   To find the surface area, use the formula for the surface area of a rectangular solid.
$L = 10, W = 5, H = 4.$

Solution $SA = 2LW + 2LH + 2WH$
$SA = 2(10)(5) + 2(10)(4) + 2(5)(4)$
$SA = 100 + 80 + 40$
$SA = 220$
The surface area of the solid is 220 ft^2.

14. Strategy   To find the measure of the other leg, use the Pythagorean Theorem.
$c = 12, a = 7.$

Solution $a^2 + b^2 = c^2$
$7^2 + b^2 = 12^2$
$49 + b^2 = 144$
$b^2 = 95$
$b = \sqrt{95}$
$b \approx 9.75$
The other leg is approximately 9.75 ft.

15. Strategy   To find the volume, use the formula for the volume of a cube. $s = 3.5.$

Solution $V = s^3$
$V = (3.5)^3$
$V = 42.875$
The volume of the cube is 42.875 in^3.

16. Strategy   Supplementary angles are two angles whose sum is 180°. To find the supplement, let x represent the supplement of a 32° angle. Write an equation and solve for x.

Solution $32° + x = 180°$
$x = 148°$
The supplement of a 32° angle is a 148° angle.

17. Strategy   To find the volume, use the formula for the volume of a rectangular solid.
$L = 6.5, W = 2, H = 3.$

Solution $V = LWH$
$V = (6.5)(2)(3)$
$V = 39$
The volume of the solid is 39 ft^3.

18. Strategy   To find the third angle, use the fact that the sum of the measures of the interior angles of a triangle is 180°. Let $x =$ the third angle.

Solution $37° + 48° + x = 180°$
$85° + x = 180°$
$x = 95°$
The third angle is 95°.

19. Strategy   To find the base, use the formula for the area of a triangle. Substitute 7 for h and 28 for A and solve for b.

Solution $A = \frac{1}{2}bh$
$28 = \frac{1}{2}b(7)$
$56 = 7b$
$8 = b$
The base of the triangle is 8 cm.

20. Strategy   To find the volume, use the formula for the volume of a sphere. The radius of the sphere is one half the diameter.

Solution

$$r = \frac{1}{2}d = \frac{1}{2}(12) = 6$$

$$V = \frac{4}{3}\pi r^3$$

$$V = \frac{4}{3}\pi(6^3)$$

$$V = \frac{4}{3}\pi(216)$$

$$V = 288\pi$$

The volume of the sphere is 288π mm^3.

21. Strategy   To find the length of each side, use the formula for the perimeter of a square. $P = 86$.

Solution

$P = 4s$

$86 = 4s$

$21.5 = s$

A side of the square is 21.5 cm.

22. Strategy   To find the number of cans of paint:
→Find the surface area by using the formula for the surface area of a cylinder.
→Divide the surface area by 200.

Solution

$$SA = 2\pi r^2 + 2\pi rh$$

$$SA = 2\pi(6^2) + 2\pi(6)(15)$$

$$SA = 2\pi(36) + 180\pi$$

$$SA = 72\pi + 180\pi$$

$$SA = 252\pi$$

$$252\pi \div 200 \approx 3.96$$

Because a portion of a fourth can is needed, 4 cans of paint should purchased.

23. Strategy   To find the amount of fencing, use the formula for the perimeter of a rectangle.

Solution

$P = 2L + 2W$

$P = 2(56) + 2(48)$

$P = 112 + 96$

$P = 208$

208 yd of fencing are needed to fence the park.

24. Strategy   To find the area, use the formula for the area of a square. $s = 9.5$.

Solution

$A = s^2$

$A = (9.5)^2$

$A = 90.25$

The area of the patio is 90.25 m^2.

25. Strategy   To find the area of the walkway:
→Find the length and width of the total area.
→Subtract the area of the plot of grass from the total area.

Solution

$L_1 = 40 + 2 + 2 = 44$

$W_1 = 25 + 2 + 2 = 29$

$A = L_1 W_1 - L_2 W_2$

$A = 44(29) - 40(25)$

$A = 1{,}276 - 1{,}000$

$A = 276$

The area of the walkway is 276 m^2.

Chapter Test

1. Strategy   To find the measure of the other leg, use the Pythagorean Theorem. $c = 11$, $a = 8$.

Solution

$a^2 + b^2 = c^2$

$8^2 + b^2 = 11^2$

$64 + b^2 = 121$

$b^2 = 57$

$b = \sqrt{57}$

$b \approx 7.55$

The other leg is approximately 7.55 cm.

2. Strategy   To determine if the triangles are congruent, determine if one of the rules for congruence is satisfied.

Solution $AC = AC, AB = AB,$
$\angle A = \angle A$
Two sides and the included angle of one triangle equal two sides and the included angle of the other triangle.
The triangles are congruent by the SAS rule.

3. Strategy   To find the area, use the formula for the area of a rectangle. Substitute 15 for L and 7.4 for W. Solve for A.

Solution $A = LW$
$A = 15(7.4)$
$A = 111$
The area is 111 m^2.

4. Strategy   To find the area, use the formula for the area of a triangle. Substitute 7 for b and 12 for h. Solve for A.

Solution $A = \frac{1}{2}bh$
$A = \frac{1}{2}(7)(12)$
$A = 42$
The area is 42 ft^2.

5. Strategy   To find the volume, use the formula for the volume of a cone. $r = 7, h = 16$.

Solution $V = \frac{1}{3}\pi r^2 h$
$V = \frac{1}{3}\pi (7)^2 (16)$
$V = \frac{1}{3}\pi (49)(16)$
$V = \frac{784\pi}{3}$
The volume is $\frac{784\pi}{3}$ cm^3.

6. Strategy   To find the surface area, use the formula for the surface area of a pyramid. $s = 3$, $l = 11$.

Solution $SA = s^2 + 2sl$
$SA = 3^2 + 2(3)(11)$
$SA = 9 + 66$
$SA = 75$
The surface area of the pyramid is 75 m^2.

7. Strategy   The volume is equal to the volume of the cylinder. $h = 30, r = 7$

Solution $V = \pi r^2 h$
$V = \pi (7^2)(30)$
$V = \pi (49)(30)$
$V = 1,470\pi$
$V \approx 4,618.14$
The volume of the solid is approximately 4 618.14 cm^3.

8. Strategy   To find the area, use the formula for the area of a trapezoid. Substitute 20 for b_1, 33 for b_2, and 6 for h. Solve for A.

Solution $A = \frac{1}{2}h(b_1 + b_2)$
$A = \frac{1}{2} \cdot 6(20 + 33)$
$A = 159$
The area is 159 in^2.

9. Strategy   The angles labeled are alternate interior angles of intersecting lines and are, therefore, supplementary angles. To find x, write an equation and solve for x.

Solution $3x + 6x = 180°$
$9x = 180°$
$x = 20°$
The measure of x is 20°.

10. Strategy   The total surface area equals the surface area of the pyramid.

Solution $SA = s^2 + 2sl$
$SA = 5^2 + 2(5)(5)$
$SA = 25 + 50$
$SA = 75$
The surface area of the pyramid is 75 m^2.

11. Strategy   The angles labeled are adjacent angles of intersecting lines and are, therefore, supplementary angles. To find x, write an equation and solve for x.

Solution $4x + 10° + x = 180°$
$5x = 170°$
$x = 34°$
The measure of x is 34°.

12. The polygon has 8 sides.
The polygon is an octagon.

13. Strategy   To determine whether the triangles are congruent, determine whether one of the rules for congruence is satisfied.

Solution   The triangles do not satisfy the SSS Rule, the SAS Rule, or the ASA Rule. The triangles are not necessarily congruent.

14. Strategy   To find the volume, use the formula for the volume of a rectangular solid. $L = 6, W = 7, H = 4$.

Solution $V = LWH$
$V = 6(7)(4) = 168$
The volume of the rectangular solid is 168 ft^3.

15. Strategy   To find the hypotenuse, use the Pythagorean Theorem.
$a = 4, b = 7$

Solution $c^2 = a^2 + b^2$
$c^2 = 4^2 + 7^2$
$c^2 = 16 + 49 = 65$
$c = \sqrt{65}$
$c \approx 8.06$
The length of the hypotenuse if approximately 8.06 m.

16. Strategy

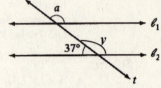

$y = a$ because corresponding angles have the same measure.
$y + 37° = 180°$ because adjacent angles of intersecting lines are supplementary angles. Substitute a for y and solve for a.

Solution $a + 37° = 180°$
$a = 143°$
The measure of a is 143°.

17. Strategy   To find the surface area, use the formula for the surface area of a cylinder. $r = 10$, $h = 15$.

Solution $SA = 2\pi r^2 + 2\pi rh$
$SA = 2\pi\left(10^2\right) + 2\pi(10)(15)$
$SA = 2\pi(100) + 300\pi$
$SA = 200\pi + 300\pi = 500\pi$
The surface area of the cylinder is 500π cm^2.

18. Strategy

→To find the measure of
∠*y*, use the fact that
∠*y* and 159° are
supplemental angles.
→To find the measure of
∠*a* by using the fact that the
sum of the interior angles
of a triangle is 180°.

Solution

$\angle y + 159° = 180°$
$\angle y = 21°$
$21° + 98° + \angle a = 180°$
$119° + \angle a = 180°$
$\angle a = 61°$
The measure of ∠*a* is 61°.

19. Strategy

To find the length of side *FG*,
write a proportion using the
fact that, in similar triangles,
the ratio of corresponding sides
equals the ratio of
corresponding heights. Solve
the proportion for the height *h*.

Solution

$\dfrac{EF}{BC} = \dfrac{h}{5}$

$\dfrac{12}{9} = \dfrac{h}{5}$

$5(12) = 9h$
$60 = 9h$
$6.67 \approx h$
The length of line segment *FG*
is approximately 6.67 ft.

20. Strategy

To find *BC*, write a proportion
using the fact that in similar
triangles, the ratios of
corresponding sides are equal.
Solve the proportion for *BC*.

Solution

$\dfrac{BC}{EF} = \dfrac{AB}{DE}$

$\dfrac{BC}{8} = \dfrac{8}{15}$

$15BC = (8)8$
$15BC = 64$
$BC \approx 4.27$
The length of *BC* is
approximately 4.27 ft.

21. Strategy

To find the perimeter, use the
formula for the perimeter of a
square. Substitute 5 for *s*. Solve
for *P*.

Solution

$P = 4s$
$P = 4(5)$
$P = 20$
The perimeter is 20 m.

22. Strategy

To find the perimeter, use the
formula for the perimeter of a
rectangle. Substitute 8 for *L*
and 5 for *W*. Solve for *P*.

Solution

$P = 2L + 2W$
$P = 2(8) + 2(5)$
$P = 16 + 10$
$P = 26$
The perimeter is 26 cm.

23. Strategy   To find the perimeter:
→Use the Pythagorean Theorem to find the hypotenuse of the triangle. $a = 12, b = 18$
→Use the formula for the perimeter of a triangle to find the perimeter.

Solution $c^2 = a^2 + b^2$
$c^2 = 12^2 + 18^2$
$c^2 = 144 + 324$
$c^2 = 468$
$c = \sqrt{468}$
$c \approx 21.6$
$P = a + b + c$
$P = 12 + 18 + 21.6$
$P = 51.6$
The perimeter is approximately 51.6 ft.

24. Strategy   To find the third angle, use the fact that the sum of the measures of the interior angles of a triangle is 180°. Let $x =$ the third angle.

Solution $41° + 37° + x = 180°$
$78° + x = 180°$
$x = 102°$
The third angle is 102°.

25. Strategy   Supplementary angles are two angles whose sum is 180°. To find the supplement, let x represent the supplement of a 41° angle. Write an equation and solve for x.

Solution $41° + x = 180°$
$x = 139°$
The supplement of a 41° angle is a 139° angle.

Cumulative Review Exercises

1. Strategy   To find the amount, use the basic percent equation.
Percent = 8.5% = 0.085,
base = 2,400, amount = n.

Solution   Percent · base = amount
$0.085(2,400) = n$
$204 = n$
8.5% of 2,400 is 204.

2. $78 \div 1 = 78$
$78 \div 2 = 39$
$78 \div 3 = 26$
$78 \div 6 = 13$
$78 \div 13 = 6$
The factors of 78 are 1, 2, 3, 6, 13, 26, 39, and 78.

3. $4\frac{2}{3} \div 5\frac{3}{4} = \frac{14}{3} \div \frac{28}{5}$

$= \frac{14}{3} \cdot \frac{5}{28}$

$= \frac{14 \cdot 5}{3 \cdot 28}$

$= \frac{2 \cdot 7 \cdot 5}{3 \cdot 2 \cdot 2 \cdot 7}$

$= \frac{5}{6}$

4. $(3x^2 + 5x - 2) + (4x^2 - x + 7)$

$= (3x^2 + 4x^2) + (5x - x) + (-2 + 7)$

$= 7x^2 + 4x + 5$

5. $82.93 \div 6.5 \approx 12.8$

6. $0.000029 = 2.9 \times 10^{-5}$

7. Strategy   To find the measure of $\angle x$, use the fact that adjacent angles of intersecting lines are supplementary.

Solution $\angle x + 49° = 180°$
$\angle x = 131°$
The measure of $\angle x$ is 131°.

8. Strategy   To find the hypotenuse, use the Pythagorean Theorem.
$a = 10, b = 24$.

Solution $c^2 = a^2 + b^2$
$c^2 = 10^2 + 24^2$
$c^2 = 100 + 576$
$c^2 = 676$
$c = \sqrt{676}$
→C is the square root of 676.
$c = 26$
The length of the hypotenuse is 26 cm.

9. Strategy   To find the area, use the formula for the area of a triangle. $h = 7$, $b = 16$.

Solution
$$A = \frac{1}{2}bh$$
$$A = \frac{1}{2}(16)(7)$$
$$A = (8)(7) = 56$$
The area is 56 in^2.

10. Strategy   Corresponding angles have the same measure. $x + 20° + 3x = 180°$ because adjacent angles of intersecting lines are supplementary angles.

Solution
$$x + 20° + 3x = 180°$$
$$4x + 20° = 180°$$
$$4x = 160°$$
$$x = 40°$$
The measure of x is 40°.

11. $(4x^2y^2)(-3x^3y)$
$$= [4(-3)](x^2 \cdot x^3)(y^2 \cdot y)$$
$$= -12x^5y^3$$

12. $3(2x + 5) = 18$
$$6x + 15 = 18$$
$$6x = 3$$
$$x = \frac{1}{2}$$
The solution is $\frac{1}{2}$.

13. Strategy   To find the third angle, use the fact that the sum of the measures of the interior angles of a triangle is 180°. Let x = the third angle.

Solution
$$32° + 90° + x = 180°$$
$$122° + x = 180°$$
$$x = 58°$$
The third angle is 58°.

14. The real numbers greater than –3 are to the right of –3 on the number line. Draw a parenthesis at –3. Draw a heavy line to the right of –3. Draw an arrow at the right of the line.

$$\begin{array}{c} \longleftarrow +\ +\ +\ (\!\!+\ +\ +\ +\ +\ +\ +\ + \longrightarrow \\ -6\ -5\ -4\ -3\ -2\ -1\ \ 0\ \ 1\ \ 2\ \ 3\ \ 4\ \ 5\ \ 6 \end{array}$$

15. $5(2x + 4) - (3x + 2) = 10x + 20 - 3x - 2$
$$= 7x + 18$$

16. $2x + 3y^2z$
$$2(5) + 3(-1)^2(-4) = 2(5) + 3(1)(-4)$$
$$= 10 + (-12)$$
$$= -2$$

17. $x^2y - 2z$
$$\left(\frac{1}{2}\right)^2\left(\frac{4}{5}\right) - 2\left(-\frac{3}{10}\right) = \frac{1}{4}\left(\frac{4}{5}\right) - 2\left(\frac{-3}{10}\right)$$
$$= \frac{1}{5} - \frac{-3}{5}$$
$$= \frac{1}{5} + \frac{3}{5}$$
$$= \frac{4}{5}$$

18. $60\text{mph} = \dfrac{60\,\text{mi}}{\text{h}} \cdot \dfrac{1.61\,\text{km}}{1\,\text{mi}} = 96.6\,\text{km/h}$

19. $4x + 2 = 6x - 8$
$$4x - 6x + 2 = 6x - 6x - 8$$
$$-2x + 2 = -8$$
$$-2x + 2 - 2 = -8 - 2$$
$$-2x = -10$$
$$\frac{-2x}{-2} = \frac{-10}{-2}$$
$$x = 5$$
The solution is 5.

20.

x	y
2	0
0	3
–2	6

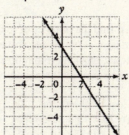

21. 3,482 m = 3.482 km

22. $\dfrac{3}{8} = \dfrac{3}{8}(100\%) = 37.5\%$

23. Strategy   To find the simple interest, solve the simple interest formula $I = Prt$ for I.

$P = 20,000$, $t = \dfrac{270}{365}$,

$r = 0.08875$

Solution $I = Prt$

$I = (20,000)(0.08875)\left(\dfrac{270}{365}\right)$

$I = 1,313.01$

The interest on the loan is $1,313.01.

24. Strategy   To find the amount of coffee:
→Multiply the number of people (250) by the amount of coffee consumed by each person (12).
→Use the following conversion factors: $\dfrac{1c}{8\ oz}$,

$\dfrac{1\ pt}{2\ c}$, $\dfrac{1\ qt}{2\ pt}$ and $\dfrac{1\ gal}{4\ qt}$.

Solution $250 \cdot 12 = 3,000$ oz

$3,000$ oz $=$

$\dfrac{3,000\ oz}{1} \cdot \dfrac{1c}{8\ oz} \cdot \dfrac{1\ pt}{2\ c} \cdot \dfrac{1\ qt}{2\ pt} \cdot \dfrac{1\ gal}{4\ qt} =$

23.4375 gal

23 gal of coffee should be prepared for the reception.

25. Strategy   To find the number of text messages, write and solve an equation using n to represent the number of text messages.

Solution   Cost $= 79 + 0.35n$

$86.70 = 79 + 0.35n$

$7.70 = 0.35n$

$22 = n$

22 text messages were sent or received.

26. Strategy   To find the sales tax, write and solve a proportion using n to represent the sales tax.

Solution $\dfrac{0.75}{12.50} = \dfrac{n}{75}$

$0.75(75) = 12.50 \cdot n$

$56.25 = 12.50n$

$\dfrac{56.25}{12.50} = \dfrac{12.50n}{12.50}$

$4.50 = n$

The sales tax is $4.50.

27. Strategy   To find the percent increase:
→Subtract 1.48 from 1.56 to find the amount of increase.
→Use the basic percent equation. Percent $= n$, base $= 1.48$, amount $=$ amount of increase.

Solution $1.56 - 1.48 = 0.08$

Percent $\cdot$ base $=$ amount

$n \cdot 1.48 = 0.08$

$n = \dfrac{0.08}{1.48}$

$n \approx 0.054 = 5.4\%$

The percent increase is 5.4%.

28. Strategy   To find the height of the box, use the formula for the volume of a rectangular solid.

$V = 144$, $L = 12$, $W = 4$.

Solution $V = LWH$

$144 = 12(4)H$

$144 = 48H$

$3 = H$

The height of the box is 3 ft.

29. Strategy   To find the pressure, substitute 35 for P in the given equation and solve for D.

Solution
$$P = 15 + \frac{1}{2}D$$
$$35 = 15 + \frac{1}{2}D$$
$$20 = \frac{1}{2}D$$
$$40 = D$$
The depth is 40 ft.

30. Strategy   To find the distance, substitute 196 for E and 49 for S and solve for d.

Solution
$$d = 4{,}000\sqrt{\frac{E}{S}} = 4{,}000$$
$$d = 4{,}000\sqrt{\frac{196}{49}} - 4{,}000$$
$$d = 4{,}000\sqrt{4} - 4{,}000$$
$$d = 4{,}000(2) - 4{,}000$$
$$d = 8{,}000 - 4{,}000$$
$$d = 4{,}000$$
The explorer is 4,000 mi above the surface.

Chapter 10: Statistics and Probability

Prep Test

1. $\dfrac{3}{2+7} = \dfrac{3}{9} = \dfrac{1}{3}$

2. $\sqrt{13} \approx 3.606$

3. $\dfrac{49}{102} \approx .0480 = 48.0\%$

4. $\dfrac{45}{27} = \dfrac{5}{3}$

5. a. $1{,}500{,}000 \cdot 0.20 = 300{,}000$ women

   b. $\dfrac{300{,}000}{1{,}500{,}000} = \dfrac{1}{5}$

Section 10.1

Concept Check

1. In their descriptions of a frequency table, students should explain that it is a method of organizing data and that the data is organized into classes so that the frequency of each class is readily apparent.

3. a. 98; 55; 43

   b. 43; 5; 9

   c. 55; 55; 9; 64

   d. 65; 65; 9; 74

   e. 1

5. class midpoint

Objective A Exercises

7. **Strategy** To prepare the frequency distribution table, find the range. Then divide the range by 8, the number of classes. If necessary, round the quotient to a whole number. This is the class width.

   Solution range $= 96 - 32 = 64$

   class width $= \dfrac{64}{8} = 8$

   Annual Tuition of Universities

Classes	Tally	Frequency
32–40	////	4
41–49	/////	5
50–58	/////	5
59–67	/////	5
68–76	////	4
77–85	////	4
86–94	/////////	9
95–103	////	4

9. **Strategy** To find how many universities charge a tuition between $7,700 and $8,500, refer to the frequency table in Exercise 7.

   Solution 4 universities charge a tuition between $7,700 and $8,500.

11. **Strategy** To find how many universities charge a tuition less than or equal to $6,700, refer to the frequency table in Exercise 7.

    Solution Number of universities
    $= 4 + 5 + 5 + 5 = 19$
    19 universities charge a tuition less than $6,700.

13. Strategy   To find the percent:
→Use Exercise 7 to find the number of universities that charge a tuition between $3,200 and $4,000.
→Use the basic percent equation.

Solution   Number of universities = 4
Percent · base = amount
$p(40) = 4$
$p = \dfrac{4}{40}$
$p = 0.10$
10% of the universities charge between $3,200 and $4,000.

15. Strategy   To find the number of hotels with room rates between $179 and $189, refer to the frequency table in Exercise 14.

Solution   7 hotels charge a corporate room rate between $179 and $189.

17. Strategy   To find the number of hotels:
→Refer to the frequency table in Exercise 14 to find the number of hotels with room rates between $212 and $222 and between $223 and $233.
→Add the numbers.

Solution   Number of hotels with rates between:
$212 and $222: 6
$223 and $233: 1
$6 + 1 = 7$
7 hotels charge a corporate room rate between $212 and $233.

19. Strategy   To find the percent:
→Refer to the frequency table in Exercise 14 to find the number of hotels with room rates between $201 and $211
→Use the basic percent equation.

Solution   Number of hotels: 10
Percent · base = amount
$p(50) = 10$
$p = \dfrac{10}{50} = 0.20$
20% of the room rates are between $201 to $211.

21. Strategy   To find the percent:
→Refer to the frequency table in Exercise 14 to find the number of hotels with room rates greater than or equal to $201.
→Use the basic percent equation.

Solution   Number of hotels with rates between:
$201–$211: 10
$212–$222: 6
$223–$233: 1
$10 + 6 + 1 = 17$
Percent · base = amount
$p(50) = 17$
$p = \dfrac{17}{50} = 0.34$
34% of the hotels have corporate room rates equal to or greater than $201.

23. Strategy   To find the ratio:
→Refer to the frequency table in Exercise 14 to find the number of hotels with room rates between $179 and $189 and between $190 and $200.
→Write the ratio:

  Solution   Number of hotels with rates between:
$179–$189: 7
$190–$200: 11

$$\frac{\text{Number of hotels with rates between \$79–\$89}}{\text{Number of hotels with rates between \$90–\$100}} = \frac{7}{11}$$

The ratio of the number of hotels with room rates between $179 and $189 to the number with room rates between $190 and $200 is $\frac{7}{11}$ or 7 to 11.

Objective B Exercises

25. Strategy   Read the histogram to find the number of account balances between $2500 and $3000.

  Solution   There are 7 account balances between $2500 and $3000.

27. Strategy   To find the number of account balances:
→Read the histogram to find the number of account balances between $0–$500, $500–$1000, $1000–$1500, and $1500–$2000.
→Add the numbers.

  Solution   Number of account balances between:
$0–$500: 2
$500–$1000: 7
$1000–$1500: 10
$1500–$2000: 13
$13 + 10 + 7 + 2 = 32$
There were 32 account balances less than $2000.

29. Strategy   To find the percent:
→Read the histogram to find the number of account balances $2000 to $2500.
→Use the basic percent equation.

  Solution   Number of accounts: 11
Percent · base = amount
$p(50) = 11$
$p = \frac{11}{50} = 0.22$
22% of the account balances were between $2000 and $2500.

31. Strategy   To find the ratio:
→Read the histogram to find the number of runners with times between 150 min–155 min and 175 min–180 min.
→Write the ratio in simplest form.

  Solution   Number of runners with times between:
150 min–155 min: 5
175 min–180 min: 10

$$\frac{\text{Number of runners with times between 150–155 min}}{\text{Number of runners with times between 175 min–180 min}} = \frac{5}{10} = \frac{1}{2}$$

The ratio is 1 to 2.

33. Strategy   To find the percent:
            →Read the histogram to find the number of runners with times between
               165 min–170 min, 170 min–175 min, and 175 min–180 min.
            →Add the numbers.
            →Use the basic percent equation.

     Solution   Number of runners with times between:
            165 min–170 min: 30
            170 min–175 min: 20
            175 min–180 min: 10
            $30 + 20 + 10 = 60$
            Percent · base = amount
            $p(100) = 60$
            $p = \dfrac{60}{100} = 0.60$
            60% of the runners had times greater than 165 min.

35. 155–160 and 160–165

37. Strategy   To find the percent:
            →Read the histogram to find number of apartments with rents between $1250 and $1500.
            →Use the basic percent equation.

     Solution   Number of apartments with rents between $1250 and $1500: 5
            Percent · base = amount
            $p(40) = 5$
            $p = \dfrac{5}{40} = 0.125$
            12.5% of the apartments have rents between $1250 and $1500.

39. Strategy   To find the percent:
            →Read the histogram to find the number of apartments with rents between
              $1000–$1250, $1250–$1500, and $1500–$1750.
            →Add the numbers.
            →Use the basic percent equation.

     Solution   Number of apartments with rents between:
            $1000–$1250: 15
            $1250–$1500: 5
            $1500–$1750: 2
            $15 + 5 + 2 = 22$
            Percent · base = amount
            $p(40) = 22$
            $p = \dfrac{22}{40} = 0.55$
            55% of the apartments have rents over $1000.

Objective C Exercises

41. 80; 90

43. Strategy   To find the number of nurses:
→Read the frequency polygon to find the number of nurses whose score was between 80–90 and between 90–100.
→Add the numbers.

Solution   Number of nurses whose score was between:
80–90: 18
90–100: 4
$18 + 4 = 22$
22 of the nurses had scores greater than 80.

45. Strategy   To find the percent:
→Read the frequency polygon to find the number of nurses whose score was between 70–80 and between 80–90.
→Add the numbers.
→Use the basic percent equation.

Solution   Number of nurses whose score was between:
70–80: 15
80–90: 18
$15 + 18 = 33$
Percent · base = amount
$p(50) = 33$
$p = \dfrac{33}{50} = 0.66$
66% of the nurses had scores between 70 and 90.

47. Frequency Distribution for Nursing Board Test Scores

Classes (test score)	Frequency
50–60	5
60–70	8
70–80	15
80–90	18
90–100	4

49. Strategy   To find the ratio:
→Read the frequency polygon to find the response times between 6 min and 9 min and between 15 min and 18 min.
→Write the ratio in simplest form.

Solution   Number of response times between:
6 min–9 min: 18
15 min–18 min: 3
$\dfrac{\text{response times between } 6-9 \text{ min}}{\text{response times between } 15-18 \text{ min}} = \dfrac{18}{3} = \dfrac{6}{1}$
The ratio is 6 to 1.

51. Strategy   To find the percent:
→Read the frequency polygon to find the response time between 9 min and 12 min, between 12 min and 15 min, and between 15 min and 18 min.
→Add the numbers.
→Use the basic percent equation.

Solution   Number of response times between:
9 min–12 min: 20
12 min–15 min: 17
15 min–18 min: 3
20 + 17 + 3 = 40
Percent · base = amount
$p(75) = 40$
$p = \dfrac{40}{75}$
$p \approx 0.533$
Approximately 53.3% of the response times are greater than 9 min.

Critical Thinking 10.1

53. A frequency distribution of nonnumerical observations, such as the type of vehicle a prospective buyer would consider, can be presented in the form of a bar graph, but not a histogram. Students might note that the classes given in the frequency table cannot be changed; different ranges cannot be formed. There is no midpoint for a class. It does not represent continuous data. A histogram is a graph of the frequencies of all outcomes for a single variable; we cannot represent this data as the outcomes for a single variable.

Section 10.2

Concept Check

1. a. Median

   b. Mean

   c. Mode

   d. Median

   e. Mode

   f. Mean

3. a. 62; 75; 87; 54; 70; 36; 6

   b. 384; 64

   c. 64

5. a. 6; 7; 8; 9; 10; 12; 13; 13; 15; 18; 21; the median is 12

   b. 8

   c. 15

7. a. $\dfrac{9+13+15+8+10}{5} = \dfrac{55}{5} = 11$

   b. $9 - 11 = -2;\ (-2)^2 = 4$

   $13 - 11 = 2;\ (2)^2 = 4$

   $15 - 11 = 4;\ (4)^2 = 16$

   $8 - 11 = -3;\ (-3)^2 = 9$

   $10 - 11 = -1;\ (-1)^2 = 1$

   c.   34

   d. $\dfrac{34}{5} = 6.8$

   e.   6.8; 2.608

Objective A Exercises

9. Strategy   To find the mean number of occupied seats:
→Determine the sum of the occupied seats
→Divide the sum by 16.
To find the median number of seats occupied:
→Arrange the numbers from smallest to largest.
→Because there is an even number of values, the median is the sum of the two middle numbers divided by 2.

Solution   The sum of the numbers is 6,105.
$$\bar{x} = \frac{6,105}{16} = 381.5625$$
The mean number of occupied seats is 381.5625.
309 319 330 352 367 387 389 391 398 399 401 408 410 411 412 422
$$\text{median} = \frac{391+398}{2} = 394.5$$
The median is 394.5 occupied seats.

11. Strategy   To find the mean of the cost of the items:
→Determine the sum of the costs.
→Divide the sum by 8.
To find the median cost of the items:
→Arrange the numbers from smallest to largest.
→Because there is an even number of values, the median is the sum of the two middle numbers divided by 2.

Solution   The sum of the numbers is 364.92.
$$\bar{x} = \frac{364.92}{8} = 45.615$$
The mean cost of the items is $45.615.
40.67 41.43 42.45 45.82 45.89 47.81 48.73 51.12
$$\text{median} = \frac{45.82+45.89}{2} = 45.855$$
The median cost of the items is $45.855.

13. Strategy   To find the number of yards to be gained:
→Let n represent the number of yards to be gained.
→Use the formula for the mean to solve for n.

Solution $100 = \dfrac{98+105+120+90+111+104+n}{7}$

$700 = 628 + n$
$72 = n$
The running back must gain 72 yd.

15. Strategy   To find the score:
→Let n represent the score on the sixth round.
→Use the formula for the mean to solve for n.

Solution $78 = \dfrac{78+82+75+77+79+n}{6}$

$468 = 391 + n$
$77 = n$
The score must be 77 on the sixth round.

17. Strategy   To find the modal response, write down the category that received the most responses.

Solution   Because a response of satisfactory was recorded most frequently, the modal response was satisfactory.

19. the mode

21. True

Objective B Exercises

23. Strategy   To draw the box-and-whiskers plot:
→Arrange the data from smallest to largest, and then find the median.
→Find Q_1, the median of the lower half of the data.
→Find Q_3, the median of the upper half of the data.
→Determine the smallest and largest data values.
→Draw the box-and-whiskers plot.

Solution   172 183 185 198 208 211 214 221 233 251 254 258 292 375
median $= \dfrac{214+221}{2} = 217.5$
$Q_1 = 198, Q_3 = 254$
Smallest value: 172
Largest value: 375

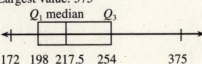

Q_1 median Q_3

172 198 217.5 254   375

25. Strategy   To draw the box-and-whiskers plot:

→Arrange the data from smallest to largest, and then find the median.

→Find Q_1, the median of the lower half of the data.

→Find Q_3, the median of the upper half of the data.

→Determine the smallest and largest data values.

→Draw the box-and-whiskers plot.

Solution   24 24 25 26 26 27 28 28 29 30 34 34 35 37 42 43 46

median = 29

$Q_1 = \dfrac{26+26}{2} = 26$

$Q_3 = \dfrac{35+37}{2} = 36$

Smallest value: 24

Largest value: 46

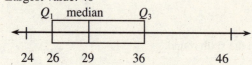

24   26   29   36   46

27. Strategy   To draw the box-and-whiskers plot:

→Arrange the data from smallest to largest, and then find the median.

→Find Q_1, the median of the lower half of the data.

→Find Q_3, the median of the upper half of the data.

→Determine the smallest and largest data values.

→Draw the box-and-whiskers plot.

Solution   789   890   905   986   992   998   1,010   1,020   1,050   1,100
1,106   1,180   1,200   1,235   1,268   1,298   1,309   1,381   1,390   1,400

median $= \dfrac{1,100+1,106}{2} = 1,103$

$Q_1 = \dfrac{992+998}{2} = 995$

$Q_3 = \dfrac{1,268+1,298}{2} = 1,283$

Smallest value: 789

Largest value: 1,400

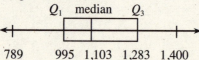

789   995   1,103   1,283   1,400

Objective C Exercises

29. Strategy To calculate the standard deviation:
→Find the mean of the weights.
→Use the procedure for calculating standard deviation.

Solution $\bar{x} = \dfrac{96+105+84+90+102+99}{6} = 96$

x	$(x-\bar{x})^2$	
96	$(96-96)^2$	0
105	$(105-96)^2$	81
84	$(84-96)^2$	144
90	$(90-96)^2$	36
102	$(102-96)^2$	36
99	$(99-96)^2$	9

Total = 306

$\dfrac{306}{6} = 51$

$\sigma = \sqrt{51} \approx 7.141$

The standard deviation of the weights is approximately 7.141 oz.

31. Strategy To calculate the standard deviation:
→Find the mean number of heads.
→Use the procedure for calculating standard deviation.

Solution $\bar{x} = \dfrac{56+63+49+50+48+53+52}{7} = \dfrac{371}{7} = 53$

x	$(x-\bar{x})^2$	
56	$(56-53)^2$	9
63	$(63-53)^2$	100
49	$(49-53)^2$	16
50	$(50-53)^2$	9
48	$(48-53)^2$	25
53	$(53-53)^2$	0
52	$(52-53)^2$	1

Total = 160

$\dfrac{160}{7} \approx 22.857$

$\sigma = \sqrt{22.857} \approx 4.781$

The standard deviation of the number of heads is approximately 4.781 heads.

33. **Strategy** To determine which basketball team has the greater standard deviation of scores:
→Find the means of the two sets of scores.
→Use the procedure for calculating standard deviation.
→Compare the answers and write the greater standard deviation.

Solution $\bar{x} = \dfrac{56+68+60+72+64}{5} = \dfrac{320}{5} = 64$

x	$(x-\bar{x})^2$	
56	$(56-64)^2$	64
68	$(68-64)^2$	16
60	$(60-64)^2$	16
72	$(72-64)^2$	64
64	$(64-64)^2$	0
	Total	= 160

$\dfrac{160}{5} = 32$

$\sigma = \sqrt{32} = 5.657$

$\bar{x} = \dfrac{106+118+110+122+114}{5} = \dfrac{570}{5} = 114$

x	$(x-\bar{x})^2$	
106	$(106-114)^2$	64
118	$(118-114)^2$	16
110	$(110-114)^2$	16
122	$(122-114)^2$	64
114	$(114-114)^2$	0
	Total	= 160

$\dfrac{160}{5} = 32$

$\sigma = \sqrt{32} = 5.657$

The standard deviations are the same.

Critical Thinking 10.2

35. a. Always true. If the data contain an odd number of values, once arranged from smallest to largest, the median is the middle number of the data set.

b. Sometimes true. If the data contain an even number of values, the median is the sum of the two middle numbers divided by 2. If the two middle numbers are the same, then the median *is* one of the numbers in the set. If the two middle numbers are *not* the same, then the median is not one of the numbers in the set.

37. Answers will vary. For example, 20, 21, 22, 24, 26, 27, 29, 31, 31, 32, 32, 33, 33, 36, 37, 37, 39, 40, 41, 43, 45, 46, 50, 54, 57

Check Your Progress: Chapter 10

1. Strategy   To find the mean salary:
   →Determine the sum of the salaries.
   →Divide the sum by 5.
   To find the median salary:
   →Arrange the numbers from smallest to largest.
   →Because there is an odd number of values, the median is the middle value.

   Solution   The sum of the numbers is 206.500.

   $$\bar{x} = \frac{206,500}{5} = 41,300$$

   The mean salary is $41,300.
   38,000   39,500   41,250
   43,750   44,000
   median = 41,250
   The median salary is $41,250.

2. Strategy   To find the modal response, write down the number that received the most responses.

   Solution   Because 23 bags recorded most frequently, the modal response was 23 bags of flour.

3. **a.** $8-$10

   b. Number of employees with an hourly wage between $12 and $14: 6
   Percent · base = amount
   $p(30) = 6$
   $$p = \frac{6}{30} = 0.20$$
   20% of the employees earned between $12 and $14 per hour.

4. Strategy   To find the number of subscribers:
   →Read the histogram to find the number of subscribers requiring between 25-30 and 30-35 seconds,
   →Add the numbers.

   Solution   Number of subscribers between:
   25-30: 190
   30-35: 180
   190 + 180 = 370
   There were 370 subscribers.

5. Strategy To calculate the standard deviation:
   →Find the mean number of rooms.
   →Use the procedure for calculating standard deviation.

   Solution $\bar{x} = \dfrac{6.2+6.4+7.1+5.9+8.3+5.3+7.5+9.3}{8} = \dfrac{56}{8} = 7$

x	$(x-\bar{x})^2$	
6.2	$(6.2-7)^2$	0.64
6.4	$(6.4-7)^2$	0.36
7.1	$(7.1-7)^2$	0.01
5.9	$(5.9-7)^2$	1.21
8.3	$(8.3-7)^2$	1.69
5.3	$(5.3-7)^2$	2.89
7.5	$(7.5-7)^2$	0.25
9.3	$(9.3-7)^2$	5.29

Total $= 12.34$

$\dfrac{12.34}{8} = 1.5425$

$\alpha = \sqrt{1.5425} \approx 1.24197$

The standard deviation of the number of hours is approximately 1.242 hours.

6. **a.** Number of responses between:
   $0-$15: 4
   $15-$30: 6
   $4 + 6 = 10$
   10 students spend less than $30 per week dining out.

   b. Number of response times between $45-$60: 12.
   Number of response times between $75-$90: 3.
   $\dfrac{\text{number of reponses between \$45-\$60}}{\text{number of reponses between \$75-\$90}} = \dfrac{12}{3} = \dfrac{4}{1}$
   The ratio is 4 to 1.

   c. Number of response times between:
   $60-$75: 7
   $75-$90: 3
   $7 + 3 = 10$
   Percent · base = amount
   $p(40) = 10$
   $p = \dfrac{10}{40} = 0.25$
   25% of students spend more than $60 a week dining out.

7. a. Range = $73 - 26 = 47$

   b. 26 32 36 36 37 39 39 40 40 41 42 42 43
        45 45 48 48 49 50 53 53 56 58 62 73
        mean = 43

$$Q_1 = \frac{39+39}{2} = 39 \,,\; Q_3 = \frac{50+53}{2} = 51.5$$

Interquartile range: $51.5 - 39 = 21.5$

   c.

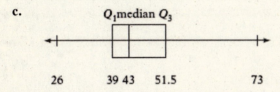

     26   39 43   51.5   73

8. Strategy   To find the score:
                      →Let n represent the score on the fifth exam.
                      →Use the formula for the mean to solve for n.

   Solution
$$80 = \frac{89+65+76+70+n}{5}$$
$$400 = 300 + n$$
$$100 = n$$
The score must be 100 on the fifth exam.

Section 10.3

Concept Check

1. The probability of an event cannot be greater than 1. The probability of an event cannot be $\frac{5}{3}$ because

   $\frac{5}{3} = 1\frac{2}{3}$ and $1\frac{2}{3} > 1$.

3. a. favorable; unfavorable

   b. unfavorable; favorable

Objective A Exercises

5. Possible outcomes of having three colors:

Q_1	Q_2	Q_3
R	G	B
R	B	G
G	R	B
G	B	R
B	R	G
B	G	R

7. Possible outcomes of tossing 4 coins:

Q_1	Q_2	Q_3	Q_4
H	H	H	H
H	H	H	T
H	H	T	H
H	T	H	H
T	H	H	H
H	H	T	T
H	T	T	H
T	T	H	H
H	T	H	T
T	H	T	H
T	H	H	T
H	T	T	T
T	T	T	H
T	T	H	T
T	H	T	T
T	T	T	T

9. Strategy   To find the probability:
→Refer to Exercise 5 to count the number of possible outcomes of the experiment.
→Count the outcomes of the experiment that are favorable to the event HHTT.
→Use the probability formula.

Solution   There are 16 possible outcomes.
There is 1 outcome favorable to E.

$$P(E) = \frac{\text{number of favorable outcome}}{\text{number of possible outcomes}} \quad P(E) = \frac{1}{16}$$

The probability of HHTT is $\frac{1}{16}$.

11. Strategy   To find the probability:
→Refer to Exercise 5 to count the number of possible outcomes of the experiment.
→Count the outcomes of the experiment that are favorable to the event.
→Use the probability formula.

Solution   There are 16 possible outcomes.
There are 6 outcomes favorable to E: HHTT, HTTH, TTHH, HTHT, THTH, THHT

$$P(E) = \frac{\text{number of favorable outcome}}{\text{number of possible outcomes}}$$

$$P(E) = \frac{1}{16} = \frac{3}{8}$$

The probability of the event is $\frac{3}{8}$.

13. Strategy   To find the probability:
→Refer to the table on page 628 to count the number of possible outcomes of the experiment.
→Count the outcomes of the experiment that are favorable to the event the sum is 5.
→Use the probability formula.

Solution   There are 36 possible outcomes.
There are 4 outcomes favorable to E: $(1, 4), (2, 3), (3, 2), (4, 1)$

$$P(E) = \frac{\text{number of favorable outcome}}{\text{number of possible outcomes}}$$

$$P(E) = \frac{4}{36} = \frac{1}{9}$$

The probability that the sum is 5 is $\frac{1}{9}$.

15. Strategy   To find the probability:
→Refer to the table on page 628 to count the number of possible outcomes of the experiment.
→Count the outcomes of the experiment that are favorable to the event the sum is 15.
→Use the probability formula.

Solution   There are 36 possible outcomes.

$$P(E) = \frac{\text{number of favorable outcome}}{\text{number of possible outcomes}}$$

$$P(E) = \frac{0}{36} = 0$$

The probability that the sum is 15 is 0.

17. Strategy   To find the probability:

→Refer to the table on page 628 to count the number of possible outcomes of the experiment.

→Count the outcomes of the experiment that are favorable to the event the sum is 2.

→Use the probability formula.

Solution   There are 36 possible outcomes.

There is 1 outcome favorable to E: (1, 1)

$$P(E) = \frac{\text{number of favorable outcomes}}{\text{number of possible outcomes}}$$

$$P(E) = \frac{1}{36}$$

The probability that the sum is 2 is $\frac{1}{36}$.

19. Strategy   To find the probability:

→Count the number of possible outcomes of the experiment.

→Count the outcomes of the experiment that produce an 11.

→Use the probability formula.

Solution   A dodecahedral die has 12 sides.

There is 1 outcome favorable to E: 11

$$P(E) = \frac{\text{number of favorable outcomes}}{\text{number of possible outcomes}}$$

$$P(E) = \frac{1}{12}$$

The probability that the number is 11 is $\frac{1}{12}$.

21. Strategy   To find the probability:

→Refer to Exercise 7 to count the number of possible outcomes of the experiment.

→Count the outcomes of the experiment that are favorable to the event the sum is 4.

→Use the probability formula.

Solution   There are 16 possible outcomes.

There are 3 outcomes favorable to E: (1, 3), (2, 2), (3, 1)

$$P(E) = \frac{\text{number of favorable outcomes}}{\text{number of possible outcomes}}$$

$$P(E) = \frac{3}{16}$$

The probability that the sum is 4 is $\frac{3}{16}$.

23. Strategy   To find the probability:

→Count the number of possible outcomes of the experiment.

→Count the outcomes of the experiment that produce a number divisible by 4.

→Use the probability formula.

Solution   A dodecahedral die has 12 sides.

There are 3 outcomes favorable to E: 4, 8, 12

$$P(E) = \frac{\text{number of favorable outcomes}}{\text{number of possible outcomes}}$$

$$P(E) = \frac{3}{12} = \frac{1}{4}$$

The probability that the side is divisible by 4 is $\frac{1}{4}$.

25. Strategy To find the empirical probability, use the probability formula and divide the number of observations of E (37) by the total number of observations (95).

Solution

$$P(E) = \frac{\text{number of observations of } E}{\text{total number of observations}}$$

$$P(E) = \frac{37}{95} \approx 38.9\%$$

The probability is 38.9% that a person prefers a cash discount.

27. Strategy To find the probability:
→Count the number of possible outcomes of the experiment.
→Count the number of outcomes of the experiment favorable to the event E, the light is green.
→Use the probability formula.

Solution There are $5\frac{1}{4}\left(3 + \frac{1}{4} + 2\right)$ min for the lights to proceed through a complete cycle.

The green light lasts for a duration of 3 min.

$$P(E) = \frac{3}{5\frac{1}{4}} = \frac{3}{\frac{21}{4}} = 3 \div \frac{21}{4} = \frac{3}{1} \cdot \frac{4}{21} = \frac{4}{7}$$

The probability of having a green light is $\frac{4}{7}$.

29. Strategy To find the probability:
→Count the number of possible outcomes of the experiment.
→Count the number of outcomes of the experiment favorable to the event E, the service is satisfactory or excellent.
→Use the probability formula.

Solution There are 377(98 + 87 + 129 + 42 + 21) outcomes of the survey.
There are 185(98 + 87) outcomes favorable to E.

$$P(E) = \frac{185}{377} \approx 49.1\%$$

The probability is 49.1% that the cable service is rated satisfactory or excellent.

31. $\dfrac{1}{1000} = 0.001$

As 0.00092 < 0.001, the decimal probably is less than one chance in a thousand.

33. No, because some areas are larger than others, some numbers will occur more often than other numbers.

35. It represents the probability that a single paper chosen from this class received a D or an F.

Objective B Exercises

37. iii

39. Strategy   To find the odds:
→Count the favorable outcomes.
→Count the unfavorable outcomes.
→Use the odds in favor of an event formula.

Solution   Number of favorable outcomes: 1
Number of unfavorable outcomes: 1

$$\text{Odds in favor} = \frac{\text{number of favorable outcomes}}{\text{number of unfavorable outcome}}$$

$$\text{Odds in favor} = \frac{1}{1}$$

The odds of showing heads is 1 to 1.

41. Strategy   To calculate the probability of winning, use the odds in favor fraction. The probability of winning is the ratio of the numerator to the sum of the numerator and denominator.

Solution $\text{Probability of winning} = \frac{3}{3+2} = \frac{3}{5}$

The probability of winning the election is $\frac{3}{5}$.

43. Strategy   To find the odds:
→Count the favorable outcomes of the experiment.
→Count the unfavorable outcomes of the experiment.
→Use the formula for the odds in favor of an event.

Solution   Use the table on page 629 to:
Count the favorable outcomes: 6
Count the unfavorable outcomes: 30

$\text{Odds in favor} = \frac{6}{30} = \frac{1}{5}$

The odds in favor of rolling a 7 are $\frac{1}{5}$.

45. Strategy   To find the odds:
→Determine the unfavorable outcomes.
→Determine the favorable outcomes.
→Use the formula for the odds against an event.

Solution   There are 48 ways of not picking an ace.
There are 4 way of picking an ace.

$\text{Odds against} = \frac{48}{4} = \frac{12}{1}$

The odds against picking an ace are $\frac{12}{1}$.

47. Strategy To calculate the probability of winning:
→Restate the odds against as odds in favor.
→Using the odds in favor fraction, the probability of winning is the ratio of the numerator to the sum of the numerator and denominator.

Solution The odds against winning are 40 to 1. Therefore, the odds in favor of winning are 1 to 40.
Probability of winning:
$$\frac{1}{1+40} = \frac{1}{41}$$
The probability of winning the Super Bowl is $\frac{1}{41}$.

49. Strategy To calculate the probability of the stock going down:
→Restate the odds of the stock going up as the odds of the stock going down.
→Using the odds in favor of the stock going down, the probability of going down is the ratio of the numerator to the sum of the numerator and denominator.

Solution The odds in favor of the stock going up: 2 to 1
The odds in favor of the stock going down: 1 to 2
Probability of going down:
$$= \frac{1}{1+2} = \frac{1}{3}$$
The probability of the stock going down is $\frac{1}{3}$.

Critical Thinking 10.3

51. $1 - p$

53. $0 \le p \le 1$

Projects and Group Activities 10.3

55. a.

Digit	Frequency
0	12
1	12
2	19
3	10
4	13
5	16
6	16
7	16
8	16
9	20

b. Answers will vary depending on students' heights.

c. In the random number table shown, there are only 7 numbers whose first two digits are even: 40784, 42228, 002218, 42658, 69249, 08506, and 64795. Because we defined "heads" as being an even number and because we are only looking at the first two digits of the numbers in the table, these are the only numbers that represent the outcome of two heads. Thus out of the 30 simulated coin tosses, we would say that there were 7 outcomes of two heads. This shows that the probability of two heads in this simulation is $\frac{7}{30} = 0.2\overline{3}$. However, when two coins are tossed, the possible outcomes are: HH, HT, TH, and TT. Because there are four possible outcomes and only one way to get two heads the actual probability is $\frac{1}{4} = 0.25$.

d. Answers will vary.

Chapter Review Exercises

1. Strategy   To prepare the frequency distribution table, use 7 as the class width and 12 as the lower class boundary.

   Solution   Number of Students in Math Classes

Classes	Tally	Frequency
12–19	/////	5
20–27	/////////	9
28–35	///////////	11
36–43	/////////	9
44–51	////	4
52–59	//	2

2. Strategy   To find the class with the greatest frequency, refer to the frequency table in Exercise 1.

   Solution   The class with the greatest frequency is 28–35.

3. Strategy   To find the number of classes:
   → Use Exercise 1 to find the number of classes with 12–19 students, 20–27 students, and 28–35 students.
   → Add the numbers.

   Solution   Number of classes with 12–19 students: 5
   Number of classes with 20–27 students: 9
   Number of classes with 28–35 students: 11
   $5 + 9 + 11 = 25$
   25 of the math classes have 35 or fewer students.

4. Strategy   To find the percent:
   → Use Exercise 1 to find the number of classes with 44–51 students and with 52–59 students.
   → Add the numbers.
   → Use the basic percent equation.

   Solution   Number of classes with 44–51 students: 4
   Number of classes with 52–59 students: 2
   $4 + 2 = 6$
   $pB = A$
   $P(40) = 6$
   $p = \dfrac{6}{40} = 0.15$
   15% of the math classes have 44 or more students.

5. Strategy   To find the percent:
   → Use Exercise 1 to find the number of classes with 12–19 students and with 20–27 students.
   → Add the numbers.
   → Use the basic percent equation.

   Solution   Number of classes with 12–19 students: 5
   Number of classes with 20–27 students: 9
   $5 + 9 = 14$
   $pB = A$
   $p(40) = 14$
   $p = \dfrac{14}{40} = 0.35$
   35% of the math classes have 27 or fewer students.

6. Strategy To find the number of days the temperature was 45° or above:
→Read the histogram to find the number of days the temperature was 45°–50° and the number of days the temperature was 50°–55°.
→Add the numbers.

Solution Number of days the temperature was 45°–50°: 15
Number of days the temperature was 50°–55°: 10
15 + 10 = 25
The temperature was 45° or above on 25 days.

7. Strategy To find the number of days the temperature was 25° or below:
→Read the histogram to find the number of days the temperature was 10°–15°, the number of days the temperature was 15°–20°, and the number of days the temperature was 20°–25°.
→Add the numbers.

Solution Number of days the temperature was 10°–15°: 2
Number of days the temperature was 15°–20°: 15
Number of days the temperature was 20°–25°: 12
2 + 15 + 12 = 29
The temperature was 25° or below on 29 days.

8. Strategy To calculate the mean:
→Calculate the sum of the cholesterol levels.
→Divide by 11.
To calculate the median:
→Arrange the numbers from smallest to largest. The median is the middle number.

Solution The sum of the numbers is 2,360.
$$\bar{x} = \frac{2,360}{11} = 214.\overline{54}$$
The mean of the cholesterol levels is $215.\overline{54}$.
160 180 190 200 210 210
220 230 230 250 280

The median cholesterol level is 210.

9. Strategy To calculate the mean:
→Calculate the sum of the weights.
→Divide by 10.
To calculate the median:
→Arrange the numbers from smallest to largest.
→Because there is an even number of values, the median is the sum of the two middle numbers divided by 2.

Solution The sum of the numbers is 71.7.
$$\bar{x} = \frac{71.7}{10} = 7.17$$
The mean weight of the babies is 7.17 lb.
5.6 5.9 6.3 6.5 6.9
7.2 7.2 8.1 8.9 9.1
$$\text{median} = \frac{6.9 + 7.2}{2} = 7.05$$
The median weight of the babies is 7.05 lb.

10. Strategy   To find the modal response:
→Find the response that was recorded most frequently.

Solution   The response "good" was mentioned most frequently and thus is the modal response.

11. Strategy   To find the number of shares:
→Read the frequency polygon to find the number of shares sold between 7 A.M.–8 A.M., 8 A.M.–9 A.M., and 9 A.M.–10 A.M.
→Add the numbers.

Solution   Number of shares sold 7 A.M.–8 A.M.: 25
Number of shares sold 8 A.M.–9 A.M.: 13
Number of shares sold 9 A.M.–10 A.M.: 17
$25 + 13 + 17 = 55$
55 million shares of stock were sold between 7 A.M. and 10 A.M.

12. Strategy   To find the number of shares:
→Read the frequency polygon to determine when less than 15 million shares sold.

Solution   Less than 15 million shares sold between 8 A.M. and 9 A.M.

13. Strategy   To find the ratio:
→Read the frequency polygon to find the number of shares sold between 10 A.M.–11 A.M. and between 11 A.M.–12 P.M.
→Write the ratio in lowest terms.

Solution   Number of shares sold 10 A.M.–11 A.M.: 15
Number of shares sold 11 A.M.–12 P.M.: 25

$$\frac{15 \text{ million}}{25 \text{ million}} = \frac{15}{25} = \frac{3}{5}$$

The ratio is $\frac{3}{5}$.

14. Strategy   To prepare the box-and-whiskers plot:
→Arrange the data from smallest to largest. Then find the median.
→Find Q_1, the median of the lower half of the data.
→Find Q_3, the median of the upper half of the data.
→Determine the smallest and largest data values.
→Draw the box-and-whiskers plot.

Solution   89 99 102 105 109 110 110 111 116 120 121 124 124 131 134
median = 111
$Q_1 = 105$
$Q_3 = 124$
Smallest value: 89
Largest value: 134

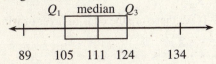

89   105   111 124   134

15. Strategy   To calculate the standard deviation:
→Find the mean of the average miles per gallon.
→Use the procedure for calculating the standard deviation.

Solution $\bar{x} = \dfrac{177}{6} = 29.5$

x	$(x - \bar{x})^2$	
24	$(24 - 29.5)^2$	30.25
28	$(28 - 29.5)^2$	2.25
22	$(22 - 29.5)^2$	56.25
35	$(35 - 29.5)^2$	30.25
41	$(41 - 29.5)^2$	132.25
27	$(27 - 29.5)^2$	6.25
	Total	= 257.5

$\dfrac{257.5}{6} \approx 42.9167$

$\sigma = \sqrt{42.9167} \approx 6.55$
The standard deviation is 6.551 mpg.

16. Strategy   Use the probability formula.

      Solution   There are 2500 possible outcomes.
There are 5 outcomes favorable to E.

$$P(E) = \frac{\text{number of favorable outcome}}{\text{total number of outcomes}}$$

$$P(E) = \frac{5}{2500} = \frac{1}{500}$$

The probability of winning the television is $\frac{1}{500}$.

17. Strategy   Use the formula for finding the odds in favor.

      Solution   Number of favorable outcomes: 15
Number of unfavorable outcomes: $50 - 15 = 35$

$$\text{Odds in Favor} = \frac{\text{number of favorable outcomes}}{\text{number of unfavorable outcome}}$$

$$= \frac{15}{35} = \frac{3}{7}$$

The odds of the ball being red are $\frac{3}{7}$.

18. Strategy   To calculate the probability:
→Restate the odds against as odds in favor.
→Using the odds in favor fraction, the probability of winning is the ratio of the numerator to the sum of the numerator and denominator.

      Solution   The odds against winning are 5 TO 2, therefore the odds in favor of winning are 2 TO 5.

$$\text{Probability of winning} = \frac{2}{2+5} = \frac{2}{7}$$

The probability of winning is $\frac{2}{7}$.

19. Strategy   To calculate the probability:
→Count the number of possible outcomes.
→Count the number of favorable outcomes of the experiment E.
→Use the probability formula.

      Solution   There are 12 possible outcomes of the experiment.
There are 2 outcomes favorable to E: 6, 12

$$P(E) = \frac{2}{12} = \frac{1}{6}$$

The probability is $\frac{1}{6}$ that the number will be divisible by 6.

20. Strategy   To calculate the probability:
→Find the number of possible outcomes.
→Find the number of favorable outcomes.
→Use the probability formula.

      Solution   There are $14(3 + 4 + 5 + 2)$ possible outcomes.
There are 5 favorable outcomes.

$$P(E) = \frac{5}{14}$$

The probability is $\frac{5}{14}$ that the student is a junior.

Chapter Test

1. Strategy   To find the number of residences where the cost was $60 or above:
→Read the histogram to find the number of customers who had monthly cost of over $60 was $60–$80 and the number of customers when the cost was $80–$100 and the number of customers when the cost was $100–$120.
→Add the numbers.

Solution   Number of customers when the cost was $60–$80: 40
Number of customers when the cost was $80–$100: 15
Number of customers when the cost was $100–$120: 10
$40 + 15 + 10 = 65$
The number of residences having costs for telephone service of $60 or more is 65.

2. Strategy   To find the total gross sales:
→Read the frequency polygon to find the total gross sales in January and February.
→Add the numbers.

Solution   Total gross sales in January: 25,000
Total gross sales in February: 30,000
$25,000 + 30,000 = 55,000$
The total gross sales for Jan. and Feb. was $55,000.

3. Strategy   To find the percent:
→Use the frequency table to find the number of restaurants that had annual sales between $750,000 and $1,000,000.
→Use the basic percent equation.

Solution   Number of restaurants = 8
Percent · base = amount
$p(50) = 8$
$p = \dfrac{8}{50} = 0.16$
16% of the restaurants had annual sales between $750,000 and $1,000,000.

4. Strategy   To find the mean bowling score:
→Determine the sum of the bowling scores.
→Divide the sum by 8.

Solution   The sum of the scores is 1223.
$\bar{x} = \dfrac{1223}{8} = 152.875$
The mean bowling score is 152.875.

5. Strategy   To find the median response time:
→Arrange the numbers from smallest to largest.
→The median is the middle term.

Solution   8   8   11   11   14   15   17   21   22
median = 14
The median response time is 14 minutes.

6. Strategy   To find the modal response, write down the category that received the most responses.

Solution   Because a response of very good was recorded most frequently, the modal response is very good.

7. Strategy   To find the number to be sold:
→Let n represent the number to be sold.
→Use the formula for the mean to solve for n.

Solution $35 = \dfrac{34 + 28 + 31 + 36 + 38 + n}{6}$

$210 = 167 + n$
$43 = n$
In the sixth month, 43 eReaders must be sold.

8. Strategy   To find the first quartile:
→Arrange the data from smallest to largest, and then find the median.
→Find Q_1, the median of the lower half of the data.

Solution   6.5   8.6   9.3   9.8   9.8   10.5   10.5   11.2   11.9   17.3   18.5   19.6   20.3   2.10

median $= \dfrac{10.5 + 11.2}{2} = 10.85$

$Q_1 = 9.8$
The first quartile is 9.8.

9. Strategy
a. To find the range, subtract the smallest value from the largest value.

b. To find the median, look at the median of the box-and-whisker plot.

Solution
a. $26 - 4 = 22$
The range is 22.

b. The median is 14 vacation days.

10. Strategy   To draw the box-and-whiskers plot:
→Arrange the data from smallest to largest, and then find the median.
→Find Q_1, the median of the lower half of the data.
→Find Q_3, the median of the upper half of the data.
→Determine the smallest and largest data values.
→Draw the box-and-whiskers plot.

Solution   68   69   70   70   70   71   72   73   73   74   74   75   76   80

median $= \dfrac{72 + 73}{2} = 72.5$

$Q_1 = 70$
$Q_3 = 74$
Smallest value: 68
Largest value: 80

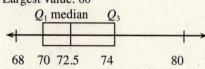

11. Strategy   To calculate the standard deviation:

→Find the mean number of incorrect answers.

→Use the procedure for calculating standard deviation.

Solution $\bar{x} = \dfrac{2+0+3+1+0+4+5+1+3+1}{10} = \dfrac{20}{10} = 2$

x	$(x-\bar{x})^2$	
2	$(2-2)^2$	0
0	$(0-2)^2$	4
3	$(3-2)^2$	1
1	$(1-2)^2$	1
0	$(0-2)^2$	4
4	$(4-2)^2$	4
5	$(5-2)^2$	9
1	$(1-2)^2$	1
3	$(3-2)^2$	1
1	$(1-2)^2$	1
	Total	= 26

$\dfrac{26}{10} = 2.6$

$\sigma = \sqrt{2.6} \approx 1.612$

The standard deviation of the number of incorrect answers is approximately 1.612.

12. The possible outcomes of tossing a coin and then a regular die:

(H, 1), (H, 2), (H, 3), (H, 4), (H, 5), (H, 6), (T, 1), (T, 2), (T, 3), (T, 4), (T, 5), (T, 6).

There are 12 elements in the sample space.

13. The possible outcomes of stacking a nickel, dime and quarter:

(N, D, Q), (N, Q, D), (D, N, Q), (D, Q, N), (Q, N, D), (Q, D, N)

14. Strategy   To calculate the probability:

→Find the number of possible outcomes.

→Find the number of favorable outcomes.

→Use the probability formula.

Solution   There are 248(14 + 32 + 202) possible outcomes.

There are 32 favorable outcomes.

$P(E) = \dfrac{32}{248} = \dfrac{4}{31}$

The probability is $\dfrac{4}{31}$ that the person is in business class.

15. Strategy To calculate the probability:
    →Find the number of possible outcomes.
    →Find the number of favorable outcomes.
    →Use the probability formula.

    Solution There are 6 possible outcomes:
    $(A, K, Q), (A, Q, K),$
    $(K, A, Q), (K, Q, A),$
    $(Q, A, K), (Q, K, A).$
    There are 2 favorable outcomes.

    $$P(E) = \frac{2}{6} = \frac{1}{3}$$

    The probability is $\frac{1}{3}$ that the ace is on top of the stack.

16. Strategy To calculate the probability:
    →Find the number of possible outcomes.
    →Find the number of favorable outcomes.
    →Use the probability formula.

    Solution There are 8 possible outcomes.
    There is 1 favorable outcomes.

    $$P(E) = \frac{1}{8}$$

    The probability is $\frac{1}{8}$ that the student will answer all three questions correctly.

17. Strategy To calculate the probability:
    →Find the number of possible outcomes.
    →Find the number of favorable outcomes.
    →Use the probability formula.

    Solution There are $45(15 + 20 + 10)$ possible outcomes.
    There are $30(20 + 10)$ favorable outcomes.

    $$P(E) = \frac{30}{45} = \frac{2}{3}$$

    The probability is $\frac{2}{3}$ that the seed is not for a red flower.

18. Strategy To calculate the probability of winning by using the odds in favor fraction, the probability of winning is the ratio of the numerator to the sum of the numerator and denominator.

    Solution Probability of winning:

    $$\frac{1}{1+12} = \frac{1}{13}$$

    The probability of winning the lottery is $\frac{1}{13}$.

19. Strategy To find the odds:
    →Find the unfavorable outcomes.
    →Use the odds in favor of an event formula.

    Solution There is 1 favorable outcome. There are $8(9-1)$ unfavorable outcomes:

    Odds in favor $= \frac{1}{8}$

    The odds of in favor of rolling a nine is 1 to 8.

20. Strategy To calculate the probability:
    →Find the number of possible outcomes.
    →Find the number of favorable outcomes.
    →Use the probability formula.

    Solution There are 12 possible outcomes
    There are 5 favorable outcomes

    $$P(E) = \frac{5}{12}$$

    The probability is $\frac{5}{12}$ that the number on the upward face is less than six.

Cumulative Review Exercises

1. $\sqrt{200} = \sqrt{100 \cdot 2} = \sqrt{100} \cdot \sqrt{2} = 10\sqrt{2}$

2. $7p - 2(3p - 1) = 5p + 6$
$7p - 6p + 2 = 5p + 6$
$p + 2 = 5p + 6$
$p - 5p + 2 = 5p - 5p + 6$
$-4p + 2 = 6$
$-4p + 2 - 2 = 6 - 2$
$-4p = 4$
$\dfrac{-4p}{-4} = \dfrac{4}{-4}$
$p = -1$
The solution is -1.

3. $3a^2b - 4ab^2$
$3(-1)^2(2) - 4(-1)(2^2)$
$= 3(1)(2) - 4(-1)(4)$
$= 3(2) - 4(-1)(4)$
$= 6 - 4(-1)(4)$
$= 6 - (-4)(4)$
$= 6 - (-16)$
$= 6 + 16$
$= 22$

4. $-2[2 - 4(3x - 1) + 2(3x - 1)]$
$= -2[2 - 12x + 4 + 6x - 2]$
$= -2[-6x + 4]$
$= 12x - 8$

5. $-\dfrac{2}{3}y - 5 = 7$
$-\dfrac{2}{3}y - 5 + 5 = 7 + 5$
$-\dfrac{2}{3}y = 12$
$-\dfrac{3}{2}\left(-\dfrac{2}{3}\right)y = -\dfrac{3}{2}(12)$
$y = -18$
The solution is -18.

6. $-\dfrac{4}{5}\left(\dfrac{3}{4} - \dfrac{7}{8} - \left(\dfrac{2}{3}\right)^2\right) = -\dfrac{4}{5}\left(\dfrac{3}{4} - \dfrac{7}{8} - \dfrac{4}{9}\right)$
$= -\dfrac{4}{5}\left(\dfrac{54}{72} - \dfrac{63}{72} + \dfrac{-32}{72}\right)$
$= -\dfrac{4}{5}\left(\dfrac{54 - 63 - 32}{72}\right)$
$= -\dfrac{4}{5}\left(\dfrac{-41}{72}\right)$
$= \dfrac{41}{90}$

7.

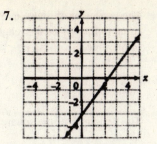

8.

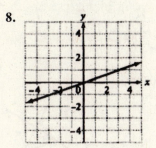

9. $\left(7y^2 + 5y - 8\right) - \left(4y^2 - 3y + 1\right)$
$= \left(7y^2 + 5y - 8\right) + \left(-4y^2 + 3y - 1\right)$
$= 3y^2 + 8y - 9$

10. $\left(4a^2b\right)^3 = 4^{1 \cdot 3}a^{2 \cdot 3}b^{1 \cdot 3} = 4^3 a^6 b^3$
$= 64a^6 b^3$

11. Strategy   To find the base, solve the basic percent equation.
Percent $= 16\dfrac{2}{3}\% = \dfrac{50}{300} = \dfrac{1}{6}$,
base $= n$, amount $= 24$.

     Solution   Percent $\cdot$ base $=$ amount
$\dfrac{1}{6}n = 24$
$\left(\dfrac{6}{1}\right)\left(\dfrac{1}{6}\right)n = \left(\dfrac{6}{1}\right)(24)$
$n = 144$
$16\dfrac{2}{3}\%$ of 144 is 24.

12. $\dfrac{9}{8} = \dfrac{3}{n}$
$9n = 8(3)$
$9n = 24$
$n = \dfrac{24}{9}$
$n = \dfrac{8}{3}$
The solution is $\dfrac{8}{3}$.

13. $87{,}600{,}000{,}000 = 8.76 \times 10^{10}$

14. Strategy   The perimeter of the square is equal to four times the length of one side.

Solution $P = 4s$
$P = 4(12)$
$P = 48$
The perimeter of the square is 48 in.

15. $(5c^2d^4)(-3cd^6)$
$= [5(-3)](c^{2+1})(d^{4+6})$
$= -15c^3d^{10}$

16. 40 km = 40,000 m

17. Strategy

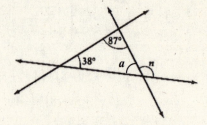

→To find the measure of $\angle a$, use the fact that the sum of the interior angles of a triangle is 180°.
→To find the measure of $\angle n$, use the fact that the sum of an interior and an exterior angle is 180°.

Solution $38° + 87° + \angle a = 180°$
$125° + \angle a = 180°$
$\angle a = 55°$
$\angle a + \angle n = 180°$
$55° + \angle n = 180°$
$\angle n = 125°$
The measure of $\angle n$ is 125°.

18. Strategy   To find the area, use the formula for the area of a parallelogram. Substitute 8 for b and 4 for h. Solve for A.

Solution $A = bh$
$A = 8(4)$
$A = 32$
The area of the parallelogram is 32 m^2.

19. Strategy   To find the simple interest, solve the formula $I = Prt$ for I.
$P = 25,000, r = 0.075, t = \dfrac{3}{12}$

Solution $I = Prt$
$I = 25,000(0.075)\left(\dfrac{3}{12}\right)$
$I = 468.75$
The simple interest on the loan is $468.75.

20. Strategy   To calculate the probability:
→Count the number of possible outcomes of the experiment.
→Count the number of outcomes that are favorable to the event of the ball is not white.
→Use the probability formula.

Solution   There are 36(12 + 15 + 9) possible outcomes.
There are 24(15 + 9) outcomes favorable to E.
$P(E) = \dfrac{24}{36} = \dfrac{2}{3}$
The probability is $\dfrac{2}{3}$ that the ball is not white.

21. Strategy   To calculate the mean score:
→Calculate the sum of the scores.
→Divide by the number of scores.
To find the median score:
→Arrange the scores from smallest to largest.
→Because there is an even number of values, the median is the sum of the two middle numbers divided by 2.

Solution   The sum of the numbers is 186.

$$\bar{x} = \frac{186}{6} = 31$$

The mean score on the six tests is 31.

22   24   31
34   37   38

$$\text{median} = \frac{31+34}{2} = 32.5$$

The median score on the six tests is 32.5.

22. Strategy   To find the percent:
→Find the number of voters registered who did not vote.
→Use the basic percent equation.
Percent = n, base = 230,000, amount = number not voting.

Solution $230{,}000 - 55{,}000 - 175{,}000$
percent · base = amount
$n \cdot 230{,}000 = 175{,}000$

$$n = \frac{175{,}000}{230{,}000} \approx 0.761$$

76.1% of the registered voters did not vote.

23. Strategy   To find the circumference of the earth, write and solve a proportion using n to represent the circumference.

Solution

$$\frac{7.5}{360} = \frac{1600}{n}$$

$7.5n = 360(1600)$
$7.5n = 576{,}000$

$$n = \frac{576{,}000}{7.5} = 76{,}800$$

The circumference of the earth is approximately 76,800 km.

24. Strategy   To find the standard deviation:
→Find the mean of the rainfall totals.
→Use the procedure for calculating the standard deviation.

Solution

$$\bar{x} = \frac{12+16+20+18+14}{5} = 16$$

x	$(x-\bar{x})^2$	
12	$(12-16)^2$	16
16	$(16-16)^2$	0
20	$(20-16)^2$	16
18	$(18-16)^2$	4
14	$(14-16)^2$	4
	Total	= 40

$$\frac{40}{5} = 8$$

$$\sigma = \sqrt{8} = 2.828$$

The standard deviation of the rainfall totals is 2.828 in.

25. Strategy   To find the wage before the increase, write and solve an equation, using n to represent the wage before the increase.

Solution $0.10n + n = 19.80$
$1.10n = 19.80$

$$n = \frac{19.80}{1.10}$$

$n \approx 18.00$
The hourly wage before the increase was $18.00 per hour.

FINAL EXAMINATION

1.
$$
\begin{array}{rcl}
672 & \to & 700 \\
843 & \to & 800 \\
509 & \to & 500 \\
417 & \to & +400 \\
\hline
 & & 2400
\end{array}
$$

2. $18 + 3(6-4)^2 \div 2 = 18 + 3(2)^2 \div 2$
$= 18 + 3(4) \div 2$
$= 18 + 12 \div 2$
$= 18 + 6$
$= 24$

3. $-8 - (-13) - 10 + 7 = -8 + 13 + (-10) + 7$
$= 5 + (-10) + 7$
$= -5 + 7$
$= 2$

4. $|a - b| - 3bc^3$

$|-2 - 4| - 3(4)(-1)^3 = |-6| - 3(4)(-1)^3$
$= 6 - 3(4)(-1)^3$
$= 6 - 3(4)(-1)$
$= 6 - (12)(-1)$
$= 6 - (-12)$
$= 6 + 12$
$= 18$

5. $5\frac{3}{8} - 2\frac{11}{16} = 5\frac{6}{16} - 2\frac{11}{16}$
$= 4\frac{22}{16} - 2\frac{11}{16}$
$= 2\frac{11}{16}$

6. $\dfrac{7}{9} \div \dfrac{5}{6} = \dfrac{7}{9} \cdot \dfrac{6}{5}$
$= \dfrac{7 \cdot 6}{9 \cdot 5}$
$= \dfrac{7 \cdot 2 \cdot 3}{3 \cdot 3 \cdot 5}$
$= \dfrac{14}{15}$

7. $\dfrac{\frac{3}{4} - \frac{1}{2}}{\frac{5}{8} + \frac{1}{2}} = \dfrac{\frac{3}{4} - \frac{2}{4}}{\frac{5}{8} + \frac{4}{8}}$

$= \dfrac{\frac{1}{4}}{\frac{9}{8}} = \dfrac{1}{4} \div \dfrac{9}{8}$

$= \dfrac{1}{4} \cdot \dfrac{8}{9}$

$= \dfrac{1 \cdot 8}{4 \cdot 9}$

$= \dfrac{1 \cdot 2 \cdot 2 \cdot 2}{2 \cdot 2 \cdot 3 \cdot 3}$

$= \dfrac{2}{9}$

8. $\dfrac{5}{16} = 0.3125$

$0.3125 < 0.313$

$\dfrac{5}{6} < 0.313$

9. $-10qr$
$-10(-8.1)(-9.5) = 81(-9.5)$
$= -769.5$

10. $-15.32 \div 4.67 \approx -3.28$

11. $-90y = 45$
$-90(-0.5) \,|\, 45$
$45 = 45$
Yes, -0.5 is a solution of the equation.

12. $\sqrt{162} = \sqrt{81 \cdot 2}$
$= \sqrt{81} \cdot \sqrt{2}$
$= 9\sqrt{2}$

13. Draw a bracket at -4.
Draw a heavy line to the right of -4.
Draw an arrow at the right of the line.

14. $-\dfrac{5}{6}(-12t) = 10t$

15. $2(x - 3y) - 4(x + 2y) = 2x - 6y - 4x - 8y$
$= (2x - 4x) + (-6y - 8y)$
$= -2x - 14y$

16. $(5z^3 + 2z^2 - 1) - (4z^3 + 6z - 8)$
$= (5z^3 + 2z^2 - 1) + (-4z^3 - 6z + 8)$
$= z^3 + 2z^2 - 6z + 7$

17. $\left(4x^2\right)\left(2x^5 y\right) = \left(4\cdot 2\right)\left(x^{2+5}\right)(y) = 8x^7 y$

18. $2a^2 b^2\left(5a^2 - 3ab + 4b^2\right)$
$= 2a^2 b^2\left(5a^2\right) - \left(2a^2 b^2\right)(3ab)$
$\quad + \left(2a^2 b^2\right)\left(4b^2\right)$
$= 10a^4 b^2 - 6a^3 b^3 + 8a^2 b^4$

19. $(3x - 2)(5x + 3)$
$= 3x(5x) + 3x(3) - 2(5x) - 2(3)$
$= 15x^2 + 9x - 10x - 6$
$= 15x^2 - x - 6$

20. $\left(3x^2 y\right)^4 = 3^{1\cdot 4} x^{2\cdot 4} y^{1\cdot 4} = 3^4 x^8 y^4$
$= 81x^8 y^4$

21. $4^{-3} = \dfrac{1}{4^3} = \dfrac{1}{64}$

22. $\dfrac{m^5 n^8}{m^3 n^4} = m^{5-3} n^{8-4} = m^2 n^4$

23. $2 - \dfrac{4}{3} y = 10$

$2 - 2 - \dfrac{4}{3} y = 10 - 2$

$-\dfrac{4}{3} y = 8$

$\left(-\dfrac{3}{4}\right)\left(-\dfrac{4}{3}\right) y = -\dfrac{3}{4}(8)$

$y = -6$
The solution is -6.

24. $6z + 8 = 5 - 3z$
$6z + 3z + 8 = 5 - 3z + 3z$
$9z + 8 = 5$
$9z + 8 - 8 = 5 - 8$
$9z = -3$
$\dfrac{9z}{9} = \dfrac{-3}{9}$

$z = -\dfrac{1}{3}$
The solution is $-\dfrac{1}{3}$.

25. $8 + 2(6c - 7) = 4$
$8 + 12c - 14 = 4$
$12c - 6 = 4$
$12c - 6 + 6 = 4 + 6$
$12c = 10$
$\dfrac{12c}{12} = \dfrac{10}{12}$

$c = \dfrac{5}{6}$

The solution is $\dfrac{5}{6}$.

26. $2.48\text{m} = 248\text{ cm}$

27. $2.6\text{ mi} = \dfrac{2.6\text{ mi}}{1} \cdot \dfrac{5{,}280\text{ ft}}{1\text{ mi}} = 13{,}728\text{ ft}$

28. $\dfrac{n+2}{8} = \dfrac{5}{12}$
$(n + 2) \cdot 12 = 8 \cdot 5$
$12n + 24 = 40$
$12n = 16$
$n = \dfrac{16}{12}$
$n = \dfrac{4}{3}$

The solution is $\dfrac{4}{3}$.

29. Strategy

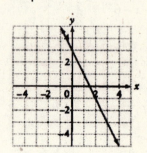

→To find the measure of $\angle a$, use the fact that $\angle a$ and the 74° angle are vertical angles.

→Find the measure of $\angle c$ by using the fact that adjacent angles of intersecting lines are supplementary.

→To find the measure of $\angle b$, use the fact that $\angle b$ and $\angle c$ are alternate interior angles of parallel lines.

Solution

$\angle a = 74°$
$\angle a + \angle c = 180°$
$74° + \angle c + 180°$
$\angle c = 106°$
$\angle b = 106°$
The measure of $\angle a$ is 74°.
The measure of $\angle b$ is 106°.

30. Strategy

To find the hypotenuse, use the Pythagorean Theorem.
$a = 7, b = 8$.

Solution

$c^2 = a^2 + b^2$
$c^2 = 7^2 + 8^2$
$c^2 = 49 + 64$
$c^2 = 113$
$c = \sqrt{113}$
$c \approx 10.6$
The length of the hypotenuse is approximately 10.6 ft.

31. Strategy

The perimeter is equal to twice the width plus twice the length.

Solution

$P = 2W + 2L$
$P = 2(7) + 2(6)$
$P = 14 + 12$
$P = 26$
The perimeter is 26 cm.

32. Strategy

Find the volume of the rectangular solid.

Solution

$V = LWH$
$V = 8(4)(3)$
$V = 96$
The volume of the solid is 96 in^3.

33.

x	y
−1	5
1	1
3	−3

34.

x	y
0	−4
5	−1
2	$-2\frac{4}{5}$

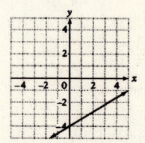

35. Strategy

To find the ground speed, substitute 22 for h and 386 for a in the given formula and solve for g.

Solution

$g = a - h$
$g = 386 - 22$
$g = 364$
The ground speed of the airplane is 364 mph.

36. Strategy   To find the number products:
→Convert 8 h to minutes using the conversion factor $\dfrac{60 \text{ min}}{1 \text{ h}}$.
→Divide the total number of minutes by $1\frac{1}{2}$.

Solution $8 \text{ h} = \dfrac{8 \text{ h}}{1} \cdot \dfrac{60 \text{ min}}{1 \text{ h}} = 480 \text{ min}$

$480 \div 1\frac{1}{2} = 480 \div \frac{3}{2}$

$= 480 \cdot \frac{2}{3}$

$= 320$

The worker can inspect 320 products in one day.

37. Strategy   To find the difference, subtract the melting point of bromine ($-7.2°$) from the boiling point of bromine ($58.78°$).

Solution $58.78° - (-7.2°)$
$= 58.78° + 7.2°$
$= 65.98°$
The difference between the melting point and the boiling point is 65.98°C.

38. $5,880,000,000,000 = 5.88 \times 10^{12}$

39. Strategy   The distance of the fulcrum from the 50-pound child: x
The distance of the fulcrum from the 75-pound child: $10 - x$
To find the placement of the fulcrum, replace the variables F_1, F_2, and d by the given variables and solve for x.

Solution $F_1 x = F_2(d - x)$
$50 \cdot x = 75(10 - x)$
$50x = 750 - 75x$
$125x = 750$
$x = \dfrac{750}{125}$
$x = 6$
The fulcrum is 6 ft from the 50-pound child.

40. Strategy   To find the number of tickets, write and solve an equation, using x to represent the number of tickets.

Solution $10.50 + 52.50x = 325.50$
$52.50x = 315$
$x = \dfrac{315}{52.50}$
$x = 6$
You purchased 6 tickets.

41. Strategy   To find the property tax, write and solve a proportion using n to represent the amount of tax.

Solution $\dfrac{3750}{250,000} = \dfrac{n}{314,000}$
$3750(314,000) = 250,000(n)$
$117,750,000 = 250,000n$
$\dfrac{117,750,000}{250,000} = n$
$4710 = n$
The property tax is $4710.

42. Strategy   To find the percent of the states that have a land area of 75,000 mi^2 or more:
→Add the number of states that have a land area of 75,000–100,000 mi^2 (8) and the number that have a land area of 100,000 mi^2 (8).
→Use the basic percent equation.
Percent = n, base = 50, amount = the sum found in Step 1

Solution $8 + 8 = 16$
Percent · base = amount
$n \cdot 50 = 16$
$n = \dfrac{16}{50}$
$n = 0.32 = 32\%$
32% of the states have a land area of 75,000 mi^2 or more.

43. Strategy   To find the revolutions per minute:
→Write the basic inverse variation equation, replace the variables by the given values, and solve for k.
→Write the inverse variation equation, replacing k by its value. Substitute 24 for the number of teeth and solve for the number of revolutions per minute.

Solution $s = \dfrac{k}{t}$

$12 = \dfrac{k}{32}$

$12(32) = k$

$384 = k$

$s = \dfrac{384}{t}$

$s = \dfrac{384}{24}$

$s = 16$

The gear will make 16 revolutions per minute.

44. Strategy   To find the total cost of the car:
→Find the sales tax by using the basic percent equation.
Percent = 5.5% = 0.055, base = 32,500, amount = n
→Add the tax to 32,500.

Solution   Percent · base = amount
$0.055(32,500) = n$
$1787.50 = n$
$1787.50 + 32,500 = 34,287.50$
The total cost is $34,287.50.

45. Strategy   To find the percent decrease:
→Subtract 96 from 124 to find the decrease.
→Use the basic percent equation.
Percent = n, base = 124, amount = the decrease.

Solution $124 - 96 = 28$
Percent · base = amount
$n \cdot 124 = 28$
$n = \dfrac{28}{124}$

$n \approx 0.226$
The housing starts decreased approximately 22.6%.

46. Strategy   To find the sale price, solve the formula
$S = (1 - r)R$ for S.
$r = 35\%, R = 245$.

Solution $S = (1 - r)R$
$S = (1 - 0.35)(245)$
$S = 0.65(245)$
$S = 159.25$
The sale price of the necklace is $159.25.

47. Strategy   To find the simple interest, solve the formula $I = Prt$ for I.
$P = 25,000$ $r = 0.086$,
$t = \dfrac{9}{12}$.

Solution $I = Prt$

$I = 25,000(0.086)\left(\dfrac{9}{12}\right)$

$I = 1612.50$
The interest on the loan is $1612.50.

48. Strategy   To find the percent:
→Use the line graph to find the number of students that work more than 15 hours per week.
→Solve the basic percent equation for percent.

Solution   Number of students who work
15–20 h: 15
Number of students who work
20–25 h: 10
Number of students who work
25–30 h: 5
Total $15 + 10 + 5 = 30$
Percent · base = amount
$p(80) = 30$
$p = \dfrac{30}{80}$

$p = 0.375$
37.5% of the students work more than 15 h per week.

49. Strategy   To calculate the mean rate:
→Calculate the sum of the rates for the insurance.
→Divide by the number of quotes.
To calculate the median rate:
→Arrange the numbers from smallest to largest.
→The median is the middle number.

Solution   The sum of the rates is 1674.

$$\bar{x} = \frac{1674}{5} = 334.8$$

The mean rate for the insurance is $334.80.

281 297 309
362 425

median = 309
The median rate for the insurance is $309.

50. Strategy   To calculate the probability:
→Refer to the table on page 629 to count the number of possible outcomes.
→Count the outcomes of the experiment that are favorable to the event that the sum is divisible by 3.
→Use the probability formula

Solution   There are 36 possible outcomes of the experiment.
There are 12 outcomes favorable to E:
$(1, 2), (2, 1), (1, 5), (2, 4), (3, 3), (4, 2), (5, 1), (3, 6), (4, 5), (5, 4), (6, 3), (6, 6)$

$$P(E) = \frac{12}{36} = \frac{1}{3}$$

The probability is $\frac{1}{3}$ that the sum will be divisible by 3.